PETS AROUND THE WORLD

Also Available From

Bloomsbury Academic

Extraordinary Expeditions: The People and Journeys That Changed History
Anne Della Guardia

The Geography of Disease: How Location and Culture Influence Health in Today's World
Jing Luo

The World's Oceans: Geography, History, and Environment, 2nd edition
Rainer F. Buschmann and Lance Nolde, editors

PETS AROUND THE WORLD

A CULTURAL ENCYCLOPEDIA OF THE HUMAN-ANIMAL BOND

Margo DeMello

BLOOMSBURY ACADEMIC
NEW YORK • LONDON • OXFORD • NEW DELHI • SYDNEY

BLOOMSBURY ACADEMIC
Bloomsbury Publishing Inc, 1359 Broadway, New York, NY 10018, USA
Bloomsbury Publishing Plc, 50 Bedford Square, London, WC1B 3DP, UK
Bloomsbury Publishing Ireland, 29 Earlsfort Terrace, Dublin 2, D02 AY28, Ireland

First published in the United States of America 2026

Cover design by Akihiro Nakayama
Cover images © clockwise: Jacob Lund/ajr_images/Karori Production/
Deborah Kolb/Veera/all via adobe Stock, AleksandarNakic via Getty Images.

Library of Congress Cataloging-in-Publication Data

Names: DeMello, Margo author
Title: Pets around the world : a cultural encyclopedia of the human-animal bond / Margo DeMello.
Description: New York : Bloomsbury Academic, 2026. | Includes index.
Identifiers: LCCN 2025032871 (print) | LCCN 2025032872 (ebook) | ISBN 9798765114940 hardback | ISBN 9798765114957 adobe pdf | ISBN 9798765114964 epub
Subjects: LCSH: Pets--Encyclopedias
Classification: LCC SF411.5 .D46 2026 (print) | LCC SF411.5 (ebook)
LC record available at https://lccn.loc.gov/2025032871
LC ebook record available at https://lccn.loc.gov/2025032872

ISBN: HB: 979-8-7651-1494-0
ePDF: 979-8-7651-1495-7
eBook: 979-8-7651-1496-4

Typeset by Deanta Global Publishing Services, Chennai, India
Printed and bound in the United States of America

For product safety related questions contact productsafety@bloomsbury.com.

For the hundreds of animals who have brought me comfort and enriched my life beyond measure

CONTENTS

PREFACE

Families around the world are changing and have been for some time. The change I am referring to is the expansion of the family to include members of other species. This has been a gradual process, happening at different rates in different parts of the world, but it is clear that families are becoming more formally inter- and multispecies. As such, pets have moved from something that was once seen as trivial, a waste of money, and associated with the frivolity of women and elites, to a firmly middle-class and mainstream reality.

This encyclopedia looks at the rise of pets around the world, focusing on the cultural, social, economic, and political causes of, and implications of, this development. Entries are organized alphabetically, and are crossed-referenced with other similar topics. Each entry is heavily referenced, and reflects the most up-to-date information on companion animals available today, and all in one place.

This encyclopedia is a useful reference tool for anyone interested in one of the most monumental changes to happen to human beings in our history on this earth—the inclusion of nonhuman animals into virtually every aspect of the human experience.

Introduction

Pets are becoming ever more popular around the world. In the United States, where this author resides, approximately 66 percent of households host at least one pet (American Pet Products (APPA) National Pet Owners Survey of 2023 and 2024). While acquiring reliable statistics on global pet ownership is challenging, a comparatively recent study (GFK 2016) indicates that around 57 percent of people in twenty-two countries across Europe, Asia, Australia, and the Americas may have a pet.

However, these numbers need to be contextualized in order to make sense, especially given the variability in how different cultures define what constitutes a pet. For instance, dogs, the most prevalent pets in the United States, inhabit every continent; nevertheless, not all of them meet the American standard of a "pet." Many serve primarily as working dogs or exist as free-roaming street or village dogs, receiving communal care without being tied to a specific household.

This raises the critical question: What qualifies as a pet?

A pet, much like every other classification of animals, is a social construction. Specifically, pets—just like livestock, wildlife, or service animals—do not exist in isolation within nature. Certainly, animals exist in the environment—they are physical beings, just as humans are. However, once integrated into human societies, these animals are attributed specific roles, which guide their categorization. It is these classifications (e.g., lab animals, pets, livestock) that influence perceptions, treatment, and usage of the animals within society.

Moreover, one might further inquire: What is an animal independent of cultural context? As sociologist Keith Tester articulates, "A fish is only a fish if you classify it as one" (1991, 46). Additionally, such classifications are not impartial—they carry political implications that tend to benefit certain groups (humans, select animals) at the detriment of others (various other animals).

To illustrate, a rabbit may be classified in multiple ways: as a source of meat, a laboratory animal, a wild creature, an attraction at a petting zoo, a classroom companion, a magician's assistant, or finally, as a pet. In the United States, rabbits complicate matters further, as they are legally addressed as "multi-purpose animals," with their treatment contingent upon their designated function and the context surrounding their use. The classification and utilization of animals in contemporary society are interconnected. However, the reverse is also true; how we categorize these beings influences how they are treated. For instance, should a rabbit reside with a family in the United States, it is classified as a pet, consequently shielding it from severe forms of cruelty under state anti-cruelty legislation. Transitioning that same rabbit to a laboratory context means that it loses this protection and instead falls under the federal Animal Welfare Act framework. While killing a pet rabbit is illegal in most states due to anti-cruelty laws, in a laboratory setting, such actions may be legally permissible.

So how does an animal earn the designation of a pet, given that pet ownership is a socially constructed phenomenon? The term "pet," or more recently "companion animal," refers to an animal characterized by its intimate rapport with human beings. Similar to other classifications of animals—such as livestock, working animals, and laboratory animals—the primary distinction of pets lies in their selection by humans for companionship. Consider the array of animals identified as pets. While many are domesticated and have historically cohabitated with humans, numerous others remain undomesticated, essentially wild. Birds, fish, and various species recognized as exotic pets are typically wild animals relocated into domestic places.

Historian Keith Thomas (1983), tracing pet ownership in England from the sixteenth to the nineteenth centuries, suggested that a pet was an animal that received a name, was allowed within the home, and was never consumed. Although these criteria might aptly describe many pets, they do not universally apply worldwide. The genesis of most pets can be attributed to their breeding and birth circumstances. Domestic animals, especially companion animals, result from regulated breeding processes aligned with human purpose—practices that date back several millennia, influencing their physical traits as well as behavioral characteristics and interactions with humans.

While livestock are "produced" by industrial agriculture, today, companion animals originate either through intentional breeding by pet breeders aiming for sale or via breeding conducted (intentionally or not) by pet owners. When intentionally bred, pets may originate from small-scale facilities or homes, or they may come from larger operations colloquially referred to as "puppy mills" or "kitten mills" in the United States (Fumarola 1998; Towsey 2010).

These facilities are typically large, often unlicensed, and frequently in poor conditions, housing numerous puppies, kittens, rabbits, and even birds bred en masse, usually destined for pet stores through brokers. The sheer volume of animals bred annually and the significant number abandoned by human caregivers lead many pets to animal shelters or private rescue organizations, where fortunate animals may be "recycled" back into homes as pets (Morris and Gies 2014).

If we consider a pet to be an animal inhabiting or near a human domicile, a critical factor in distinguishing it as a pet is its naming. In the Global North, an animal cannot be deemed a pet without having a name. Naming fosters the animal's integration into the human social structure, allowing for both direct address and reference (ArluKe and Sanders 1996). Consequently, humans can communicate with their pets as they do with family and friends, and engage in discussions about the animal, clarifying its history and personality. Thus, pets can be characterized as animals primarily bred to assume the role of companions, residing in or near human domiciles, and generally displaying manageable behavior while being cared for by humans. Furthermore, they may be either domesticated or possess a degree of tameness. The most popular pet types generally exhibit temperaments conducive to human interaction; many, such as dogs, demonstrate a genuine affinity for human companionship. Ideally, pets are envisioned by animal lovers to enjoy lives filled with affection and attention.

The term "pet" traces back to the fifteenth century and originally referred to a "spoiled child," likely derived from the French "petit," meaning "little," and evolving to signify anything or anyone graced with indulgence. Nevertheless, notable exceptions exist: exotic pets, typically captured from the wild and sold through often illegal means within the exotic pet industry, lack domestication, specific breeding intentions, and often even tameness. Additionally, while we frequently classify pets by their primary function as companions, numerous pets remain outside—restricted to cages or chains—and thus offer no companionship to human family members. Others are maintained primarily for distinct purposes, such as security or economic gain (as observed in specialized breeding or animal shows). Moreover, some species, including snakes and fish, lack any semblance of companionability.

For urbanized individuals in the Western world, relationships with pets frequently represent the only interaction they have with nonhuman animals. Data from the American Pet Products Association (APPA) reveals that as of 2019, a notable 67 percent of Americans live with a companion animal—an increase of 5 percent over the past decade—compared to just 41 percent who cohabitate with children. The APPA further indicates that American expenditures on pets exceeded 75 billion dollars in 2019, with two-thirds to three-fourths of pet owners purchasing gifts for their animals (notably, dogs receive gifts more frequently than other species). Furthermore, 30–40 percent of pet owners (and half of horse owners) include their animals in their wills. That year, the American household pet population soared to approximately 393 million across 85 million homes.

Historically in the West, pet ownership was considered an extravagant or irrational indulgence. Scholars produced various theories to rationalize the phenomenon of pets. For instance, Konrad Lorenz (1970) and other animal behaviorists contended that pets evolved to possess appealing physical characteristics designed to elicit parental instincts from humans. A parallel hypothesis is the notion of anthropomorphism, in which humans project their own desires and thoughts onto their pets, thereby forging substitute familial bonds. This concept speculated that individuals forming attachments to animals were deficient in their human relationships, leading them to create artificial connections with pets. Two studies conducted in the 1960s and 1970s even sought to demonstrate that pet ownership stemmed from psychological deficits. However, modern research has disproven the validity of these studies.

Contemporary scholars broadly acknowledge that humans cohabitate with animals for a considerably more straightforward reason: mutual benefits. As previously stated, the predominant motivation for keeping companion animals remains companionship. For instance, the American Pet Products Association discovered that half of United States dogs share sleeping quarters at night with their human caretakers—either in or on the bed—while more American cats, which were traditionally outdoor animals, are now primarily housed indoors and regarded as integral family members. Additionally, rabbits, once commonly appreciated solely as food sources, are now embraced by thousands of households as companion animals living indoors, often sleeping alongside their caregivers. The close bonds formed between humans and their animals testify to their significant role as companions.

Empirical studies examining the human–animal bond reveal that interactions with animals yield substantial emotional, psychological, and even physical rewards. Moreover, the Victorian rationale for pet ownership—fostering positive traits such as kindness and empathy in children—still holds significance today. Research suggests that forming attachments to companion animals can stimulate nurturing behaviors and empathy in children, with those possessing pets exhibiting greater levels of empathy than their peers (see Taylor and Signal 2005; Ascione and Weber 1996; Ascione 1992).

Companion animals occupy a "social position" within our families, households, and daily rituals. By including them in our breakfast gatherings, television sessions, and vacations, these creatures become true members of our families. The human–pet dynamic diverges from most other human–animal associations in that it is not primarily utility-driven. Instead, it constitutes a reciprocal relationship in which both parties—human and animal—play crucial roles. In engaging with a companion animal, we interact with a being perceived as an individual, positioned in our lives as a friend, companion, and family. Ideally, this relationship is structured not solely around human desires but also considers the needs and well-being of the animal. However, the manner in which individuals relate to companion animals is not universally consistent.

Psychologist Michael Fox (1979) identified four typologies of pet-owner relationships: the object-oriented relationship, in which the pet is perceived merely as a novelty or decorative item; the utilitarian relationship, where the animal provides a specific benefit, such as serving as a watchdog; the need-dependency relationship, where the animal satisfies a human's companionship needs; and the actualizing relationship, characterized by equality and mutual respect between the owner and pet. Nevertheless, the primary distinction in attitudes toward pets is likely cultural rather than individual, facilitating rapid changes in pet ownership as societal views on nonhuman animals and our connections to them evolve.

References and Further Reading

ArlUKe, A., and C. Sanders. (1996). *Regarding Animals*. Philadelphia: Temple University Press.

Ascione, F. R. (1992). "Enhancing Children's Attitudes About the Humane Treatment of Animals: Generalization to Human-Directed Empathy." *Anthrozoös* 5, no. 3: 176–91.

Ascione, F. R., and C. V. Weber. 1996. "Children's Attitudes About the Humane Treatment of Animals and Empathy: One-Year Follow Up of a School-Based Intervention." *Anthrozoös* 9, no. 4: 188–95.

Fox, M. W. 1979. "The Values and Uses of Pets." In *The Handbook of Animal Welfare: Biomedical, Psychological, and Ecological Aspects of Pet Problems and Control*, ed. W. H. Westbrook and R. D. Allen. New York: Garland STPM.

Fumarola, A.J. 1998. "With Best Friends Like Us Who Needs Enemies—The Phenomenon of the Puppy Mill, the Failure of Legal Regimes to Manage It, and the Positive Prospects of Animal Rights." *Buffalo Environmental Law Journal* 6, no. 2: 253–89.

Lorenz, K. 1970. *Studies in Animal and Human Behavior*. London: Methuen.

Morris, K. N., and D. L. Gies. 2014. "Trends in Intake and Outcome Data for Animal Shelters in a Large US Metropolitan Area, 1989 to 2010." *Journal of Applied Animal Welfare Science* 17, no. 1: 59–72.

Taylor, N., and T. D. Signal. 2005. "Empathy and Attitudes to Animals." *Anthrozoös* 18, no. 1: 18–27.

Tester, K. 1991. *Animals and Society: The Humanity of Animal Rights*. London: Routledge.

Thomas, K. 1983. *Man and the Natural World: A History of the Modern Sensibility*. New York: Pantheon

Towsey, M. 2010. "Something Stinks: The Need for Environmental Regulation of Puppy Mills." *Villanova Environmental Law Journal* 21, no. 1: 159–88.

A

ADULTS AND PETS

While pets have long been associated with children, recent research on pet ownership among adults has highlighted the significant psychological, social, and cognitive benefits associated with maintaining pet companionship, especially for older adults and those living alone.

A substantial body of evidence suggests that pet ownership correlates positively with enhanced mental health. For instance, a study by McDonough et al. identified that pet ownership is linked to greater cognitive and brain health throughout the adult lifespan, suggesting that community-based pet programs can support cognitive health in older adults (McDonough et al., 2022). Similarly, research by Li et al. found that pet ownership may mitigate cognitive decline in older individuals living alone, potentially offsetting the negative cognitive impacts associated with solitary living conditions (Li et al., 2023).

The emotional benefits of pet ownership also emerge prominently in the literature. The care responsibilities associated with pet keeping can foster a sense of purpose and structure, encouraging physical activity and social interaction among older adults. For example, Obradović et al. discuss how pet ownership can enhance daily engagement and facilitate independence for community-dwelling older adults, reinforcing their emotional and physical well-being (Obradović et al., 2021). Additionally, Meier and Maurer emphasize that community strategies promoting pet ownership may significantly alleviate feelings of loneliness and social isolation prevalent among older adults, thereby enhancing their mental health (Meier and Maurer, 2022).

However, while the benefits of pet ownership are extensively documented, some challenges and complexities have also been noted. Hawkins et al. point out that for certain individuals, the obligations of pet ownership can evoke feelings of guilt and exacerbate mental health challenges, indicating a nuanced relationship that warrants careful consideration within the scope of pet ownership studies (Hawkins et al., 2021). Furthermore, the financial and logistical burdens of pet care can pose challenges, particularly among low-income populations or those with preexisting health issues. Brooks et al. highlight that although pet ownership generally has a positive impact on mental health, there can be negative aspects that arise from the financial and emotional burdens involved in caring for a pet, particularly during stressful periods, such as the Covid-19 pandemic (Brooks et al., 2018).

In addition, it is clear that living arrangements significantly influence pet ownership dynamics, with households comprising multiple adults showing higher rates of pet ownership compared to individuals living alone (Rathish et al., 2022).

While pet ownership among adults is primarily associated with significant cognitive and emotional benefits, numerous factors, including living conditions, economic status, and individual health profiles, interplay within this dynamic. Future research should continue to address these multifaceted relationships, particularly focusing on interventions that can optimize the benefits of pet ownership while mitigating potential drawbacks.

See also: Bond, Human-Animal; Children and Pets; Human Health, Pets And; Mental Health, Pets And

References and Further Reading

Brooks, H., K. Rushton, K. Lovell, P. Bee, L. Walker, L. Grant, and A. Rogers (2018). "The Power of Support from Companion Animals for People Living with Mental Health Problems: A Systematic Review and Narrative Synthesis of the Evidence." *BMC Psychiatry*, 18 (31).

Hawkins, R. D., Hawkins, E. L., & Tip, L. (2021). "I Can't Give Up When I Have Them to Care for": People's Experiences of Pets and Their Mental Health. *Anthrozoös*, 34 (4): 543–62.

Li, Y., W. Wang, L. Zhu, L. Yang, H. Wu, X. Zhang, L. Guo, and C. Lu (2023). "Pet Ownership, Living Alone, and Cognitive Decline Among Adults 50 Years and Older." *Jama Network Open*, 6 (12): e2349241.

McDonough, I., H. Erwin, N. Sin, and R. Allen (2022). "Pet Ownership is Associated with Greater Cognitive and Brain Health in a Cross-Sectional Sample Across the Adult Lifespan." *Frontiers in Aging Neuroscience*, 14: 953889.

Meier, C., and J. Maurer (2022). "Buddy or Burden? Patterns, Perceptions, and Experiences of Pet Ownership Among Older Adults in Switzerland." *European Journal of Ageing*, 19 (4): 1201–12.

Obradović, N., É. Lagueux, K. Latulippe, and V. Provencher (2021). "Understanding the Benefits, Challenges, and the Role of Pet Ownership in the Daily Lives of Community-Dwelling Older Adults: A Case Study." *Animals*, 11 (9): 2628.

Rathish, D., J. Rajapakse, and K. Weerakoon (2022). "Household Preferences for Pet Keeping: Findings from a Rural District of Sri Lanka." *Plos One*, 17 (11): e0277108.

ADVERTISING, PETS IN

Animals, and pets in particular, are extremely effective tools to use in advertising a variety of products. Recent research on pet ownership has increasingly examined its implications for consumer behavior, particularly within advertising and marketing frameworks. A notable finding by Liang et al. indicates that pet ownership enhances consumers' willingness to pay a premium for products labeled with animal welfare certifications, thereby influencing their consumption intentions (Liang et al., 2024). This aligns with other studies suggesting that people who own pets are more inclined to support animal welfare initiatives, indicating that marketers might better appeal to this demographic by emphasizing ethical sourcing and animal welfare in their advertising campaigns.

The psychological benefits associated with pet ownership also appear to translate into marketing dynamics. Martin et al. suggest that owning a dog during stressful periods, such as the Covid-19 pandemic, may positively influence overall happiness and reduce feelings of anxiety and depression (Martin et al., 2021). This suggests that marketing strategies that promote companion animals could capitalize on the emotional support that pets provide, appealing to potential pet owners by highlighting the mental health benefits associated with owning pets.

Furthermore, the significance of pets in affecting consumer spending goes beyond mere marketing strategies; it has broader implications for societal welfare. White et al. emphasize that spending on pets can enhance pet owners' happiness, which cascades into beneficial outcomes for society, such as increased social interaction or community engagement (White et al., 2021). This suggests that pet-related products could be marketed not just as consumer goods, but also as avenues for enhancing personal and societal well-being.

Moreover, changes in societal attitudes toward pets, influenced by trends such as pet humanization, suggest that consumers increasingly view pets as family members rather than just animals. This shift emphasizes the importance of incorporating pet care practices into branding and advertising strategies. Amiot et al. report that the relevance of pets in households has seen a resurgence in visibility, especially during periods characterized by societal stress, prompting brands to adapt their offerings to cater to this evolving pet-centric landscape (Amiot, Gagné, and Bastian, 2022).

See also: Healthcare Spending on Pets; Television, Pets On

References and Further Reading

Amiot, C. E., C. Gagné, and B. Bastian (2022). "Pet Ownership and Psychological Well-Being During the COVID-19 Pandemic." *Scientific Reports*, 12 (1): 6091.

Liang, Y., C. Meng, R. Chen, Y. Yang, and Y. Zeng (2024). "Pet Ownership and Its Influence on Animal Welfare Attitudes and Consumption Intentions Among Chinese University Students." *Animals*, 14 (22): 3242.

Martin, F., K. E. Bachert, L. Snow, H. W. Tu, J. Belahbib, and S. A. Lyn (2021). "Depression, Anxiety, and Happiness in Dog Owners and Potential Dog Owners during the COVID-19 Pandemic in the United States." *PLoS One*, 16 (12): e0260676.

White, M., N. Khan, J. Deren, J. Sim, and E. Majka (2021). "Give a Dog a Bone: Spending Money on Pets Promotes Happiness." *The Journal of Positive Psychology*, 17 (4): 589–95.

AFFECTION AND CONTROL

We love our pets and shower them with affection. But we also control our pets—including where they live, what they eat, who they associate with, whether or not they breed, and much more. This contradiction between affection and control is central to the notion of the pet (Tuan, 1984).

The English word "pet" dates to the fifteenth century and meant "spoiled child." At that time, and even today, people would sometimes use the term when referring to a child or an adult woman. And while it may be used as a term of endearment, it also serves to trivialize and minimize the person to whom it is directed. This is an example of paternalism, which is a system in which those with more authority can control the lives of those with less authority by asserting that the authority figure knows what is best for those that he controls. What historians call the patron-client relationship, found in feudal Europe as well as more modern systems of servitude, is an example of paternalism: the landlord or boss controls the lives of those who serve him, but at the same time offers small benefits to keep those workers somewhat appeased.

The modern pet perfectly embodies that mixture of affection and control that we see in these ancient paternalistic relationships. Our dogs, cats, and other animals are both loved by us, but also are controlled entirely by their human owners—even the term owner, which is most commonly used to refer to the human part of the human–animal relationship, reinforces a much older idea that the nonhuman animals with whom we live are not free but are owned and controlled entirely by a human.

Historian Harriet Ritvo (1987) writes about how the rise of organized and expanded pet keeping in the nineteenth century was linked with a new set of attitudes about animals. Prior to this time period, animals were largely seen in functional terms, and the idea of keeping an animal in your home when that animal does not provide a clear service simply did not make sense for most people, most of whom were living hand to mouth. Because of the economics of feeding a non-working animal, it was, for the most part, only elites who kept pets.

But in the nineteenth century, starting in England and the United States, pet keeping started to spread from the upper class into the rapidly expanding middle classes. Pets were still trivialized by many, in part because they quickly began to be associated with women and children, both of whom were still controlled entirely by men.

For the first time in this period, we see the linkage of affection toward pets on the one hand with the notion of control, on the one hand, and domination, on the other. But like the patron-client relationships of old, the control of and domination over pets is obscured by the many justifications for these systems of control, justifications that suggest that all of these methods are in the pet's best interest. Whether we are talking about spaying and neutering our pets to control overpopulation and modify behavior, or hitting our dogs in order to "train" them, we are able to justify almost any kind of (mis)treatment. It should not surprise us that the dog training methods developed during this period were based on domination and control.

The early pet fanciers were also pet breeders, and the breeding of pets is one of the most concrete, corporeal ways in which humans exercise control over animals. This is one reason that for so many years, and still today in fancy pet circles, mixed-breed animals were viewed with such distaste—animals that are allowed to have sex on their own, with their own partners, and create their own "mongrel" offspring are seen as vulgar and uncontrolled.

Today we see this level of control not only in the controlled breeding and genetic manipulation of pets, but also in the forms of surgery that they undergo as well. The reliance on cosmetic surgery for dogs is one result of the breeder's focus on breed perfection. For example, certain breeds of dogs require, for example, in order to conform to breed standards, docked tails and/or cropped ears in order to conform to breed standards. Some dog behaviorists worry that because dogs use their tails to communicate with other dogs, tail docking puts them at a disadvantage when socializing and might impair physical functions such as stability as well (Bennett and Perini, 2003; Mellor, 2018). Ritvo (1987) suggests that historically, cats have historically been less popular than dogs as pets in part because they could not be as easily controlled.

Philosopher Yi-Fu Tuan, in his book *Dominance and Affection: The Making of Pets* (1984), deals with the question of dominance with respect to our relationship with pets. Many modern breeds of companion animals are bred for qualities that we find attractive, but that are harmful to the health of the animal. Is it cruelty or playfulness to breed a variety of goldfish with dysfunctional bulging eyes? Even without specific genetic defects associated with certain dog or cat breeds, many modern breeds of dog or cat are unable to survive without close human attention. While dependency has been bred into domestic animals since the earliest days of domestication, it has accelerated in recent years with the production of animals such as Chihuahuas, who are physically and temperamentally unsuited to survival outside of the most sheltered of environments.

More disturbing are cats that go by the name of Twisty Cats or Kangaroo Cats, all of whom have a genetic abnormality that results in dramatically shortened forelegs or sometimes a flipper-like paw rather than a normal front leg, and who are being selectively bred by a handful of breeders. How much do we really love our pets, Tuan asks, if we continue to breed them in a way that makes them less healthy and that shortens their life expectancy?

We can list a whole host of other ways in which humans have, throughout history, sought to control animals, including castration, tail docking, de-clawing, tattooing, microchipping, vocal cord cutting, and ear cropping. But one might argue that more important than all of these secondary control measures is selective breeding. By creating animals that are smaller, and more docile, more colorful, and more dependent, humans exercise the ultimate form of control because we control the very nature of the animal.

The reality is that most people don't know, and many do not really care that we have bred health problems and dependency into our pets, because we're not looking to create independent animals anyway. We train our dogs to "sit" and to "lie down," demonstrating that their role is to do what we ask them to do and to not cause any trouble. We like our pets to be quiet and unobtrusive, like a piece of furniture. For Tuan, a dog's submission to command is their most important attribute.

See also: Attachment; Cruelty; Ethics of Pet Ownership; Selective Breeding

References and Further Reading

Bennett, P. C., and E. Perini (2003). "Tail Docking in Dogs: A Review of the Issues." *Australian Veterinary Journal*, 81 (4): 208–18.
Mellor, D. J. (2018). "Tail Docking of Canine Puppies: Reassessment of the Tail's Role in Communication, the Acute Pain Caused by Docking and Interpretation of Behavioural Responses." *Animals*, 8 (6): 82.
Ritvo, H. (1987). *The Animal Estate: The English and Other Creatures in the Victorian Age*. Cambridge, MA: Harvard University Press.
Tuan, Y. (1984). *Dominance and Affection: The Making of Pets*. New Haven, CT: Yale University Press.

AFRICA, PETS IN

Pets are found throughout Africa, in a variety of cultural contexts. Recent research on pet keeping in Africa reveals diverse cultural practices, attitudes, and socioeconomic factors affecting the perception and management of pets across various regions. While pet ownership is common, the specific species kept, as well as the implications for social dynamics and conservation, vary significantly across different cultures and environments.

One key area of focus is the cultural significance of pet ownership. In Cameroon, for instance, keeping birds such as parrots as pets is relatively popular in urban centers; however, this practice is less prevalent in rural areas and when compared to developed countries like those in Europe and the United States. One recent study indicates that the motivations for keeping birds can often be tied to social status and cultural practices, highlighting how pet ownership can reflect broader societal values and individual identity Tamungang (2016).

Moreover, the advent of urbanization in Africa appears to influence pet keeping patterns. As urban populations grow, there is an increasing interest in various pets, including exotic animals. However, this trend raises concerns about biodiversity conservation and the risks associated with the wildlife trade. The pressures of urban living can lead to challenges such as inadequate resources for proper pet care, affecting the welfare of animals and potentially leading to negative environmental impacts, particularly as people opt for easily obtainable exotic species (Measey, Hui, and Somers, 2020).

In addition to animal welfare and conservation issues, pet keeping has been linked to various social and health benefits. While specific studies directly focusing on African populations are limited, general trends suggest that pets contribute positively to psychological well-being, especially among vulnerable populations. Animals are often recognized for their role in providing companionship, reducing loneliness, and positively impacting mental health (Applebaum et al., 2021). For instance, during crises such as the Covid-19 pandemic, pets provided emotional support and played a crucial role in alleviating stress for individuals, including older adults (Applebaum et al., 2021).

Particularly among young adults and children, pet ownership has shown potential therapeutic effects, fostering responsibility, empathy, and social skills. The emotional intelligence cultivated by caring for pets can have lasting developmental benefits

Budgies and other small birds for sale at a market in Bloemfontein, South Africa. Keeping birds can be tied to social status in some African countries. (Vladimira Pufflerova/Dreamstime.com)

(Purewal et al., 2019). In Africa, where children often grow up in less structured environments, providing them with pets can contribute positively to their development.

However, challenges persist regarding policies and practices around pet ownership that can sometimes discourage responsible pet keeping. Issues related to housing regulations, particularly in urban settings where renting policies may not be favorable to pet owners, suggest a need for greater awareness and improved policies to accommodate pet ownership as a normal aspect of family life (Power, 2016).

In conclusion, pet keeping in Africa encompasses a rich tapestry of cultural significance, social impact, and environmental challenges. Continued research is essential to understand the complex interplay between these factors, offering insights into how pet ownership can be better integrated into societal values while promoting animal welfare and conservation efforts.

See also: Indigenous Pet Keeping; Trends in Pet Ownership

References and Further Reading

Applebaum, J. W., C. Ellison, L. Struckmeyer, B. A. Zsembik, and S. E. McDonald (2021). "The Impact of Pets on Everyday Life for Older Adults During the COVID-19 Pandemic." *Frontiers in Public Health*, 9: 652610.

Measey, J., C. Hui, and M. J. Somers (2020). "Terrestrial Vertebrate Invasions in South Africa." *Biological Invasions in South Africa*, 14: 787–830.
Power, E. R. (2017). "Renting with Pets: A Pathway to Housing Insecurity?" *Housing Studies*, 32 (3): 336–60.
Purewal, R., R. Christley, K. Kordas, C. Joinson, K. Meints, N. Gee, and C. Westgarth (2019). "Socio-demographic Factors Associated with Pet Ownership Amongst Adolescents from a UK Birth Cohort." *BMC Veterinary Research*, 15 (1): 334.
Tamungang, S. (2016). "Challenges and 'Conservation Implications of the Parrot Trade in Cameroon.'" *International Journal of Biological and Chemical Sciences*, 10 (3): 1210.

ALLERGIES

Allergies to nonhuman animals, or in particular, to the fur or dander that those animals produce, are very common in human societies and contributes to the disposal of many companion animals.

The body of research addressing the relationship between pet exposure and allergic disease encompasses a variety of studies spanning diverse populations and age groups. Many studies have focused on quantifying allergenic proteins such as Fel d 1 from cats and Can f 1 from dogs, demonstrating that both direct contact and environmental dispersion of pet dander contribute to sensitization among children and adults (Luo et al., 2018).

Several studies have explored the paradox that early life exposure to pet allergens may, under particular conditions, induce immunological tolerance rather than sensitization. In this context, research has suggested that infancy or early childhood exposure to cats and dogs might be inversely associated with the development of asthma in later life (Indolfi et al., 2023; Hesselmar et al., 1999; Carlsen et al., 2012).

Epidemiological investigations have yielded a mixed set of results. In some European and Asian cohorts, pet exposure in early life was associated with lower risks of developing asthma. However, other studies have reported that families with an allergic predisposition may intentionally avoid pet ownership, suggesting that reverse causation (avoidance behavior) might confound the observed associations (Carlsen et al., 2012; Wegienka et al., 2010; Collin et al., 2014). Further complicating the picture, certain studies in rural settings or among farming communities have pointed toward protective effects linked to a broader environmental exposure profile, which includes pets (Levin et al., 2020; Leynaert et al., 2001).

In addition to childhood studies, research has addressed the persistence and evolution of pet allergen sensitization into adulthood (Park et al., 2013). Longitudinal analyses have shown that early pet exposure may influence the trajectory of allergic responses over time, with some adults remaining sensitized despite a history of tolerance-inducing exposures in childhood (Park et al., 2013; Wegienka et al., 2010). Moreover, emerging evidence indicates that pet ownership may interact with other environmental and lifestyle factors—such as tobacco smoke—further highlighting the need for an integrated approach to understanding allergy.

Research examining the role of allergies as a driver for pet relinquishment to animal shelters indicates that allergic reactions—particularly to cats—are significant factors leading some owners to surrender their companion animals. Recent studies show that severe allergic responses, including dangerous asthma attacks attributed to feline exposure, can trigger critical medical events and prompt owners to part with their pets (Dance, 2020). In this context, allergies are not only regarded as a health hazard but are also linked to the broader decision-making process of pet relinquishment, where medical advice to remove the allergen source, that is, the pet, becomes a pivotal factor.

Shelter intake records, and in particular, owner surrender forms, have identified allergies as one component among multiple owner-related factors influencing the decision to surrender animals (Kisley et al., 2024). Furthermore, research has highlighted that in some studies, up to 40 percent of pet owners experiencing allergic symptoms are advised to remove the pet from their home as part of their management strategy, thus increasing the likelihood that allergies contribute directly to relinquishment decisions (Kudryavtseva et al., 2025).

In summary, the research literature underscores that allergies, notably those triggered by cats, are a significant determinant in decisions to surrender pets to shelters. Additionally, to complicate matters further, recent research on the intersection of allergies with race, ethnicity, and class indicates that allergic diseases, including food allergies, asthma, and hay fever, do not affect all populations uniformly. Studies have shown that nonwhite populations often experience a higher prevalence and severity of allergic conditions compared to white populations, even after adjusting for socioeconomic and clinical covariates. For instance, Gupta et al. (2019) reported that food allergy rates are significantly elevated among adults of color relative to their white counterparts.

In addition to prevalence differences, socioeconomic factors play a crucial role in shaping allergy outcomes. Research indicates that differences in environmental exposures, access to diagnostic services, and quality of care may mediate the relationship between race/ethnicity and allergic disease severity. These findings reinforce the notion that socially mediated experiences—including class, discrimination, and environmental injustice—intertwine with biological predispositions to impact both allergy prevalence and health-related quality of life.

See also: Breeds, Popularity Of; Children and Pets; Health Risks of Pet Ownership; Human Health, Pets And

References and Further Reading

Ahlbom, A., A. Bäckman, J. Bakke, T. Foucard, S. Halken, N. Kjellman, L. Malm, S. Skerfving, J. Sundell, and O. Zetterström (1998). "Pets Indoors – A Risk Factor for or Protection Against Sensitisation/Allergy. A Nordic Interdisciplinary Review of the Scientific Literature Concerning the Relationship Between the Exposure to Pets at Home, Sensitisation and the Development of Allergy." *Indoor Air*, 8 (4): 219–35.

Bottema, R., N. Reijmerink, M. Kerkhof, G. Koppelman, F. Stelma, J. Gerritsen, C. Thijs, B. Brunekreef, C. P. van Schayck, and D. Postma (2008). "InterleUKin 13, CD14, Pet and Tobacco Smoke Influence Atopy in Three Dutch Cohorts: The Allergenic Study." *European Respiratory Journal*, 32 (3): 593–602.

Carlsen, K., S. Roll, K. Carlsen, P. Mowinckel, A. Wijga, B. Brunekreef, and T. Keil (2012). "Does Pet Ownership in Infancy Lead to Asthma or Allergy at School Age? Pooled Analysis of Individual Participant Data from 11 European Birth Cohorts." *Plos One*, 7 (8): e43214.

Chen, J., N. Krieger, S. Eeden, and C. Quesenberry (2002). "Different Slopes for Different Folks: Socioeconomic and Racial/Ethnic Disparities in Asthma and Hay Fever Among 173, 859 US Men and Women." *Environmental Health Perspectives*, 110 (suppl 2): 211–16.
Collin, S, R. Granell, C. Westgarth, J. Murray, E. Paul, J. Sterne, and A. Henderson (2014). "Pet Ownership is Associated with Increased Risk of Non-Atopic Asthma and Reduced Risk of Atopy in Childhood: Findings from a UK Birth Cohort." *Clinical & Experimental Allergy*, 45 (1): 200–10.
Dance, A. (2020). "The Race to Deliver the Hypoallergenic Cat." *Nature*, 588 (7836): S7–S9.
Gupta, N., P. Agarwal, A. Sachdev, and D. Gupta (2019). "Allergy Testing—an Overview." *Indian Pediatrics*, 56: 951–7.
Hesselmar, B., N. Åberg, B. Åberg, B. Eriksson, and B. Björkstén (1999). "Does Early Exposure to Cat or Dog Protect Against Later Allergy Development?" *Clinical & Experimental Allergy*, 29 (5): 611–17.
Indolfi, C., E. D'Addio, C. L. Bencivenga, G. Rivetti, I. Bettini, A. Licari, ... and A. Klain (2023). "The Primary Prevention of Atopy: Does Early Exposure to Cats and Dogs Prevent the Development of Allergy and Asthma in Children? A Comprehensive Analysis of the Literature." *Life*, 13 (9): 1859.
Kisley, M. A., E. J. Chung, and H. Levitt (2024). "Investigating the Reasons Behind Companion Animal Relinquishment: A Systematic Content Analysis of Shelter Records for Cats and Dogs, 2018–2023." *Animals*, 14 (17): 2606.
Korppi, M., M. Hyvärinen, A. Kotaniemi-Syrjänen, E. Piippo-Savolainen, and T. Reijonen (2008). "Early Exposure and Sensitization to Cat and Dog: Different Effects on Asthma Risk after Wheezing in Infancy." *Pediatric Allergy and Immunology*, 19 (8): 696–701.
Kudryavtseva, A., F. Fluer, L. Khachatryan, S. Makarova, O. Osipenko, E. Ryzhii, ... and K. Gelezhe (2025). "Staphylococcus Strains in Atopic Dermatitis in Children: Toxins Production and Resistance Properties." *Life*, 15 (7): 1120.
Lodge, C., K. Allen, S. Luo, Y. Sun, J. Hou, X. Kong, P. Wang, Q. Zhang, ... J. Sundell (2018). "Pet Keeping in Childhood and Asthma and Allergy Among Children in Tianjin Area, China." *Plos One*, 13 (5): e0197274.
Luo, S., Y. Sun, J. Hou, X. Kong, P. Wang, Q. Zhang, and J. Sundell (2018). "Pet Keeping in Childhood and Asthma and Allergy Among Children in Tianjin Area, China." PLOS One, 13 (5): e0197274.
Medjo, B., M. Atanasković-Marković, D. Nikolić, B. Spasojević, I. Ivanovski, and S. DjUKić (2013). "Association Between Pet-Keeping and Asthma in School Children." *Pediatrics International*, 55 (2): 133–7.
Park, Y.-B., E.-K. Mo, J.-Y. Lee, J.-H. Kim, C.-H. Kim, I.-G. Hyun, and J.-H. Choi (2013)."Association Between Pet Ownership and the Sensitization to Pet Allergens in Adults with Various Allergic Diseases." Allergy, Asthma & Immunology Research, 5 (5): 295–300.
Wegienka, G., C. C. Johnson, S. Havstad, D. R. Ownby, and E. M. Zoratti (2010). "Indoor Pet Exposure and the Outcomes of Total IgE and Sensitization at Age 18 years." *Journal of Allergy and Clinical Immunology*, 126 (2): 274–9.

ANIMAL CAFES

Animal cafes are cafes that patrons can visit, can order food and drink, and can, at the same time, visit with animals such as cats, rabbits, or even wild animals like owls.

The origin of animal cafes can be traced back to Japan in the early 2000s, with the first facilities enabling customers to view and interact with exotic species such as otters and owls around 2004. These cafes have proliferated in countries including Cambodia, Hong Kong, Indonesia, China, South Korea, Taiwan, Thailand, Vietnam, and Japan, with Japan having the highest number of animal cafés. Recent research highlights that this development is part of a broader trend in exotic animal cafés (EACs), which combine aspects of public entertainment and the exotic pet trade (Sigaud et al., 2023).

There may be as many as four hundred animal cafes across Asia, of which 27 percent house wild species—mammals, birds, and reptiles—often of conservation concern. Nearly half of these exotic species are classified as threatened according to IUCN criteria (McMillan et al., 2020). This trend raises concerns regarding biodiversity, as the demand for such species may contribute to unsustainable practices within the exotic pet trade (Sigaud, Kitade, and Sarabian, 2023; McMillan et al., 2020).

The current regulatory landscape governing animal cafés depends entirely on the country in which the café exists. In many Asian regions, regulatory oversight is either just starting or poorly enforced (Sigaud, Kitade, and Sarabian, 2023). Although some countries have begun to implement or consider regulations and welfare guidelines, these are often not standardized across the region or between countries (Sigaud, Kitade, and Sarabian, 2023). Calls for clearer accountability and improved welfare strategies are growing, urging authorities to consider both local cultural practices and international standards in animal welfare.

Welfare concerns in animal cafés have been a particular focus of concern. Studies assessing the captive environments of oriental small-clawed otters in Japanese cafés reveal discrepancies between current conditions and established welfare guidelines, indicating that animals may be housed under conditions that do not fully cater to their species-specific needs (Ushine et al., 2024). Similarly, investigations into owl cafés have documented elevated stress levels in owls, attributed to factors such as inappropriate lighting, tethering, and frequent human interactions (Urita, Kusuda, and Rooney, 2022). These findings underscore the importance of rigorous animal welfare assessments and highlight the potential for long-term negative impacts on animal health when welfare standards are not met.

See also: China, Pets In; Japan, Pets In

References and Further Reading

McMillan, S. E., C. Dingle, J. A. Allcock, and T. C. Bonebrake (2021). "Exotic Animal Cafes are Increasingly Home to Threatened Biodiversity." *Conservation Letters*, 14 (1): e12760.

Sigaud, M., T. Kitade, and C. Sarabian (2023). "Exotic Animal Cafés in Japan: A New Fashion with Potential Implications for Biodiversity, Global Health, and Animal Welfare." *Conservation Science and Practice*, 5 (2): e12867.

Urita, C., S. Kusuda, and N. Rooney (2022). "Physiological and Behavioural Assessments of Stress Levels in Owls Housed at Owl Cafes." *Animal Welfare*, 31 (3): 283–92.

Ushine, N., A. Kamitaki, A. SuzUKi, and S. Hayama (2024). "Assessment of Captive Environment for Oriental Small-Clawed Otters (aonyx cinereus) in Otter Cafés in Japan." *Animals*, 14 (16): 2412.

ANIMAL PROTECTION MOVEMENT

The animal protection movement began in the mid-nineteenth century with two separate movements: the prevention of cruelty movement and the anti-vivisectionist movement. These two distinct but overlapping movements initiatives can be together referred to as the first wave of the modern animal protection movement. This movement also overlapped extensively with the first wave of the women's rights movement—the suffrage movement—as well as the final stages of the abolitionist movement prior to the American Civil War. Many of the same people, in fact, were working to overthrow slavery, to give women the vote, and to end the suffering of animals. The individuals and organizations active during this period, in both the United Kingdom and the United States, would pass the first (very narrow) laws against animal cruelty in the modern world.

British activists' first attempt at getting an anti-cruelty act passed in Parliament was in 1800, but, after months of debate, it did not pass; another attempt was made in 1802, which was also unsuccessful. These bills were focused on bullbaiting, a form of entertainment in which dogs would be set on a bull to harass them. A more general bill, for "preventing wanton and malicious cruelty to animals," was proposed by Lord Erskine in 1809, but that too, though passing the House of Lords, it, too, failed to pass the House of Commons, both in 1809 and in 1810 when he proposed it again.

It would not be until 1822 that Richard Martin introduced, and saw passed, the nation's first anti-cruelty law in two hundred years. This act prevented the "cruel and improper Treatment of Horses, Mares, Geldings, Mules, Asses, Cows, Heifers, Steers, Oxen, Sheep, and other Cattle," subject to a minimum ten shilling fine or imprisonment up to three months in jail. The fact that the animals listed in this and the other early bills were what we would consider today to be farm animals is not an aberration. Instead, it reflects the fact that by the early nineteenth century, pets were still rarely found outside of aristocratic households, and thus their treatment was not a concern to those worried about animal welfare. Instead, it was the working and food animals whose lives—and deaths—were most visible who were the focus of this early legislation.

Richard Martin, the Irish member of Parliament who ushered that 1822 legislation (commonly known as "Martin's Act") through the House of Commons, realized that the police would not have the manpower or will to enforce the bill. The Reverend Arthur Broome stepped up and in 1824 founded the Royal Society for the Prevention of Cruelty to Animals (RSPCA), which took as its mission the protection of animals and the enforcement of what was commonly called Martin's Act. Only one other animal protection organization was ever founded in England prior to that time (the Society for the Suppression and Prevention of Wanton Cruelty to Animals, founded in 1809), but that organization did not last long. The RSPCA became the world's longest-lived animal protection organization, first focusing primarily on large working animals, and only in the twentieth century switching their focus to pets.

The RSPCA also tried to ban dog fighting (which, along with cockfighting, was finally prohibited in 1835), bull fighting, and bullbaiting, and worked with police officers to educate them on the importance of Martin's Act. (See Beers 2006 for more on this history.)

The second part of the early animal protection movement was anti-vivisection. Starting in the seventeenth century, experimentation on animals became a standard part of medical research, with wild animals, farm animals, and domestic cats and dogs being used as subjects. By the nineteenth century, knowledge about these experiments had made many members of the public uneasy. Even before this time, a number of important European thinkers such as Alexander Pope and Samuel Johnson had been critical of animal experimentation. But it wasn't until the nineteenth century that anti-vivisectionists organized themselves, forming the first anti-vivisection societies in England and leading to the first investigation of animal experimentation in England by a Royal Commission in 1875.

From the beginning, leading figures from other social justice movements of the period were involved in the fight to end animal experimentation. The work of these organizations (which included The Society for the Protection of Animals Liable to Vivisection, the Society for the Abolition of Vivisection, the International Association, and the London Anti-Vivisection Society) led to the passage of the world's first anti-vivisection legislation, the Cruelty to Animals Act, in 1876. The law did not abolish animal experimentation, but instead, it mandated that physicians be licensed by the Home Secretary and that facilities conducting experimentation be inspected. (For a full discussion of the anti-vivisection movement in Victorian England, see French 2019.) The movement picked up steam toward the end of the century, and it is not coincidental that by that time, pets were becoming much more popular. In fact, one of the biggest anti-vivisection protests in this period was what is now called the Brown Dog Riots. These events were inspired by the death of a dog whom two female medical students claimed had been experimented on multiple times, contrary to the conditions of the Cruelty to Animals Act. The dog had not only been subjected to multiple procedures, but no pain medication nor sedation was provided to the dog, whose suffering shocked the women into action. They later installed a memorial to the dog in a park in Battersea, England, which became the focus of the battle between pro-vivisectionists—mostly medical students—and anti-vivisectionists—made up of feminists, trade unionists, and socialists.

While the anti-vivisection movement began to lose steam in the twentieth century, the anti-cruelty movement would continue to expand.

In America, the anti-cruelty movement was strongly influenced by what was happening in England at the time and can be largely traced to the work of one man, Henry Bergh. Bergh was a New York City gentleman who traveled extensively and witnessed the abuse of animals around the world. He visited London in 1865 and met with the then-president of the RSPCA, a meeting that led to his founding, in 1866, of the American Society for the Prevention of Cruelty to Animals (ASPCA) upon his return. Bergh found, as did his contemporaries in England, that the animals who suffered the most at that time were the large working animals—especially horses. They were being beaten and worked almost to death and starved when they could work no more. Also, as in England, there were a number of popular pastimes that involved cruelty to animals, such as bear beating, dog fighting, and cockfighting.

Within months of the founding of the ASPCA, New York passed a major anti-cruelty law, which banned the cruel treatment of animals and included over-driving and over-loading horses, cockfighting, bullbaiting, dog fighting, and depriving animals of food or water, and animal abandonment. In addition, the state gave the ASPCA (and Bergh himself) the power to enforce the new law. Bergh was known as "The Great Meddler" for involving himself in what had once been the private practices of Americans. When cart drivers now beat their horses, Bergh would intervene, letting the drivers know that these practices were now illegal. (See Beers, 2006.)

New York's laws became models for laws that were passed around the country. Unlike England, which has a strong centralized government, the United States has a relatively weak federal government, with most of the powers of the nation taken over by the states. That is why the ASPCA only had—and still has today—jurisdiction within the state of New York, and why every other state had to develop their own SPCAs and humane societies in order to protect animals in those regions, with groups in Pennsylvania, Massachusetts, and San Francisco among the first. By 1890, thirty-one states had such organizations. Originally concerned with enforcing anti-cruelty laws, many groups soon began running animal shelters following a model developed in Philadelphia.

During this time, the federal government passed its very first (and still one of only a handful of federal laws) animal protection law, the Twenty-Eight Hour Law of 1873. This law mandated that animals being shipped by rail be provided with rest, food, and water after twenty-eight hours of transit. (Interestingly, it was only in 2005 that, thanks to the work of animal rights organizations, the USDA recognized that the dominant mode of transit for farm animals today is truck, not train, and began enforcing the law in that context.)

Ironically, it took one of the animal anti-cruelty laws that Bergh helped to pass to aid children suffering from abuse in the United States. Prior to 1873, children were like animals—they were simply the property of their parents and had no rights of their own nor protection under the law. Bergh was introduced to a foster child named Mary Ellen Wilson, who was being both starved and beaten by her foster mother. Because no American law could intervene in the "private" activities between parents and children, the ASPCA brought the case before a court and convinced it that the child was an animal, and thus deserved protection under the law of 1867. Bergh then received so many pleas on behalf of children that in 1875 he founded the American Society for the Prevention of Cruelty to Children to advocate on their behalf. Two years later, the American Humane Association was founded in order to protect both children and animals. (See Herman, 2011 for more of this history.)

As in England, the other major arm of the growing animal protection movement in America was found in the anti-vivisection movement. The first US organization in America dedicated to the abolition of vivisection was the American Anti-Vivisection Society (AAVS), founded in 1883 by Caroline Earle White. Others quickly followed: the New England Anti-Vivisection Society in 1895, the Vivisection Reform Society in 1903, and the New York Anti-Vivisection Society in 1908. But unlike in England, America's anti-vivisection movement resulted in no laws at all—the American medical establishment, which opposed any regulations or restrictions on vivisection, was too powerful. In fact, the first law regulating animal experimentation in America would not appear until 1966 with the passage of the Laboratory Animal Welfare Act (later shortened to the Animal Welfare Act, and famously passed because of the theft and death of a pet Dalmatian named

Pepper, demonstrating again how important pets are to us). Still, this movement played an important role in promoting the view that animals were God's creatures who should not be tortured.

As in England, many of the leading American anti-vivisectionists were drawn from other social justice movements. Caroline Earle White, for example, was a Quaker whose attorney father represented free Blacks in his legal practice, and she was actively involved in the abolitionist movement. White was also engaged in an example of a type of nineteenth-century pursuit known as "social feminism," in which women like White worked to expand the role of women in public institutions through the creation of separate institutions. For example, twenty years before the founding of the AAVS, she was the leading force in organizing the Pennsylvania Society for the Prevention of Cruelty to Animals, but because of her gender, she was not allowed to serve on the board. (For more on women in the American anti-vivisection movement, see Buettinger, 1997.)

By the 1920s, with industrialization and the invention of the automobile, the use of horses as transport declined, so the movement's focus shifted from large animals such as horses and cattle to small animals, like cats and dogs. At the same time, pet keeping was increasing in importance in both countries, and issues relating to dogs and cats became ever more important to the public.

If the nineteenth-century animal protection movement was centered around opposition to vivisection, on the one hand, and protection of large working animals, on the other hand, the movement of the twentieth century began with concerns about companion animals. America, in particular, was the focus of vast economic and demographic changes, with fewer and fewer people working on farms or living in rural communities, and more people living in cities and, increasingly, suburbs, far away from animal agriculture.

As Americans grew more distant from both wild and farmed animals, companion animals became more important, a trend which really picked up after the end of the Second World War and has continued, unabated, ever since. The expansion of pet keeping led in part to the expansion of humane attitudes toward animals and greater concern toward companion animals in particular.

The mid-twentieth century also saw the founding of a handful of national humane organizations, including The American Welfare Institute in 1950, The Humane Society of the United States, which split in 1954, from the American Humane Association, and The Society for Animal Protective Legislation in 1955. These organizations, and others like them, were responsible for every major animal protection law in the United States, including the Humane Methods of Slaughter Act (1958), the Laboratory Animal Welfare Act (1966), the Endangered Species Act (1969), the Wild Horse and Burro Protection Act (1970), and the Marine Mammal Protection Act (1972). It was in this era that Americans began to view certain kinds of treatments against certain kinds of animals as a social problem, similar to the ways in which the feminist and civil rights movements constructed sexism and racism as social problems.

Although activists had been working to protect animals in England and America since the early nineteenth century, it really wasn't until the 1970s, though, that what most people think of as the animal rights movement was born. This movement, or the second wave of animal rights, arose with the publication of philosopher Peter Singer's groundbreaking book, *Animal Liberation*, in 1975. Singer's work, and in the 1980s, the work of Tom Regan, provided the philosophical underpinnings for the modern animal rights movement and gave rise to the current generation of animal activists.

Today, the animal rights movement is an international movement that can be roughly divided into three areas: philosophical debate regarding the status of animals, legislative work, and direct action. Whereas the nineteenth century had been about humane treatment of working animals plus the abolition of (or restrictions on) vivisection, and the early twentieth century was about the humane treatment of companion animals, the modern animal rights movement encompasses a variety of issues and concerns, including the end of the use of animals in circuses, the promotion of a non-animal diet, opposition to the wearing of fur or leather, and more.

Whereas those who define themselves primarily as animal protection advocates, or animal welfare advocates, seek to protect animals from abuse and harm, the animal rights movement seeks an end to the status of animals as property and an end to the use of animals in research and for food, clothing, and entertainment. This conflict within the movement—between animal rights activists or abolitionists and animal welfare advocates, as well as between those, sometimes called pragmatists, who argue for the special treatment of certain species such as apes and dolphins—has been present for decades.

Except for the Humane Society of the United States (HSUS), founded in 1954 and now a leader in the modern animal rights movement, all of the other American organizations at the center of this movement were founded after the 1970s. This includes the Animal Legal Defense Fund, established in 1979 and at the forefront of working to gain animals legal victories, People for the Ethical Treatment of Animals (PETA) in 1980, the Farm Animal Reform Movement in 1981, Farm Sanctuary in 1986, and Compassion over Killing in 1995. These organizations work on companion animal issues but tend to focus more on the issues affecting non-pets, such as meat production, the fur industry, or the biomedical industry and its use of "animal models."

The 1970s, 1980s, and 1990s were the years of protests, demonstrations, public outreach, and direct action. Animal rights activists, either alone or with the support of the major organizations, publicly protested the use of animals in medicine, in agriculture, and in entertainment, and they sought to educate the public about animals' treatment, generally with the assistance of the media.

In the past decade, dozens of new laws have been passed at the state and local level giving animals additional layers of protection. Although the US federal government has been slow to take up animal legislation, in 2009, animal-friendly congressional members formed the bipartisan Congressional Animal Protection Caucus, a group of more than eighty Representatives who work to pass animal welfare legislation. In addition, in 1999, Humane USA was formed, the nation's first political action committee devoted to electing animal-friendly candidates to state and federal office. Through the actions of Humane USA, the Congressional Animal Protection Caucus (and its predecessor, the Friends of Animals Caucus), and animal rights organizations, the United States has seen the passage of a number of new laws passed in recent years.

Thanks to the maturation of the animal protection movement, nonhuman animals around the world now have more protection than ever before.

See also: Animal Welfare Act; Laws and Regulations; Rescue Groups; Sanctuaries; Shelters

References and Further Reading

Bargheer, S. (2006). "The Fools of the Leisure Class. Honor, Ridicule, and the Emergence of Animal Protection Legislation in England, 1740–1840." *European Journal of Sociology/Archives Européennes de Sociologie*, 47 (1): 3–3.

Beers, D. L. (2006). *For the Prevention of Cruelty: The History and Legacy of Animal Rights Activism in the United States*. Athens: Ohio University Press.

Buettinger, C. (1997). "Women and Antivivisection in Late Nineteenth-Century America." *Journal of Social History*, 30 (4): 857–72.

Feuerstein, A. (2015). "I Promise to Protect Dumb Creatures." *Society & Animals*, 23 (2): 148–65.

Finsen, L. and S. Finsen (1994). *The Animal Rights Movement in America*. New York: Twayne.

Gaarder, E. (2011). "Where the Boys Aren't: The Predominance of Women in Animal Rights Activism." *Feminist Formations*, 23 (2): 54–76.

Herman, E. (2012). *The Rights of the Defenseless: Protecting Animals and Children in Gilded Age America*. By Susan J. Pearson. Chicago, IL: University of Chicago Press, 2011.

Kean, H. (1998). *Animal Rights: Political and Social Change in Britain Since 1800*. London: Reaktion Books.

Munro, L. (2001). "Caring About Blood, Flesh, and Pain: Women's Standing in the Animal Protection Movement." *Society & Animals*, 9 (1): 43–61.

Regan, T. (2004). *The Case for Animal Rights*. Berkeley: University of California Press.

Singer, P. (1975). *Animal Liberation*. New York: Harper Collins.

ANIMAL WELFARE ACT

The Animal Welfare Act is the most important piece of federal legislation in the United States that protects animals. American President Lyndon Johnson signed the Laboratory Animal Welfare Act (LAWA) into law in 1966 to appease public outrage over the theft of pets for animal research. Now named the Animal Welfare Act (AWA), this law governs the transportation, housing, feeding, and veterinary care of warm-blooded animals in laboratories, as well as animals bred and transported for the pet industry and for those in zoos or circuses and other entertainment establishments. However, the US Department of Agriculture's (USDA) definition of "warm-blooded animal" excludes mice and rats, the two animals most commonly used in research, as well as birds. (Animals raised for food have been excluded since the earliest days of the law.) The Act provides a set of standards that cover the handling, care, housing, feeding, ventilation, and veterinary care—including euthanasia—of the animals used for research, entertainment, or pets. Even though the AWA is the United States' most important piece of legislation protecting animals in this country, it is very limited in scope, and even in the three areas that the law covers, it is full of a variety of exceptions, leaving many animals—both warm- and cold-blooded—unprotected.

The act was passed thanks to the love that one family had for their pet dog and the public outrage generated by the theft and death of that dog. Pepper was a Dalmatian owned by the Lakavage family of Pennsylvania. One day in 1965, Pepper was stolen from their yard, and, they would ultimately come to find out, sold to a research facility in New York where he was used for research and then killed.

Prior to that time, there were no formal avenues to acquire animals for medical research or product testing, so laboratories largely used "random source animals," that is, stray dogs and cats as well as pets from animal shelters. The dealers who sold such animals to the laboratories were not required to disclose from where they received the animals, so pet theft became one of the most common methods by which to procure animals. (Interestingly, this parallels the procurement of human corpses for medical education in the nineteenth century. Then, as in the 1960s, there was no formal system to provide bodies to medical schools, so doctors relied upon "body snatchers" to steal newly interred bodies from graves.)

After Pepper was stolen, the Lakavages contacted their congressional representative, Joseph Resnick, who helped them find out what happened to Pepper and who introduced the Laboratory Animal Welfare Bill that year to provide some sort of regulation surrounding the procurement of animals for research, as well as some basic standards for care. Originally, the Act included the following language, demonstrating once again how important pets were to the signing of this legislation:

> to protect the owners of dogs and cats from theft of such pets, to prevent the sale or use of dogs and cats which have been stolen, and to ensure that certain animals intended for use in research facilities are provided humane care and treatment.

Originally, the LAWA covered six species of animals, all classified at that time as pets: dogs, cats, nonhuman primates, guinea pigs, hamsters, and rabbits. The bill has been amended nine times since 1966. The first was in 1970 and expanded the covered animals to include all warm-blooded animals, except for farmed animals, and added two additional venues besides animal research: the pet industry and the animal entertainment industry. In 1976, a new amendment expanded the protections for the transportation of animals and also included animal fighting.

It was not until 1985 that the first amendment was passed to include requirements for some kind of enrichment for nonhuman primates as well as dogs. It also required that universities which undertake animal research form Institutional Animal Care and Use Committees (IACUCs) to provide another layer of oversight in those facilities.

In 1990, another amendment was passed that targeted the use of pets in research. Even after the passage of the AWA, which outlawed pet theft, pets were still being used in research—they were taken from animal shelters, a practice known as "pound seizure." Not only was this legal, in many states it was mandated that animal shelters turn over unclaimed animals to dealers. With the 1990 amendment, shelters were now required to keep stray or surrendered animals for a minimum of five days to allow time for those animals to be claimed by their owners or rehomed. Today, many states have banned pound seizure entirely, while Oklahoma is the only state left that mandates pound seizure.

Most of the amendments to the AWA have expanded protections for the animals used in the industries covered by the law. But the 2002 amendment excluded, for the first time, all rodents from protection. Today, rats and mice make up 99 percent of all of the animals used in research, but they lack any protections whatsoever.

The most recent amendments, passed in 2007, 2008, 2013, and 2019, added more requirements regarding animal fighting (which was ultimately banned), the international transportation of dogs, and which excluded 4-H and other similar organizations from the requirements of the AWA (another example of animals getting protections removed with an amendment).

Facilities that use what are called "covered animals," that is, animals who the USDA includes in the legislation, fall under the jurisdiction of the USDA, which handles licensing, inspection, and all other compliance and enforcement measures. Individuals, businesses, and organizations that fall under the licensing requirements of the AWA must register for a USDA APHIS Animal Care License. Licensing fees are $120 for three years.

Broadly speaking, the USDA provides three types of licenses: Class A, which is for those who breed animals; Class B, for those who transport animals; and class C, for those who keep animals for research or entertainment. Within the pet world, this breaks down into the following categories:

- Pet Wholesalers—Anyone buying, selling, transferring, or trading pets wholesale—that is, they do not sell to the public directly but to retailers who sell to the public; this requires a Class A license.
- Pet Breeders—Anyone breeding pets with annual sales over $500, and do not sell directly to the public; this requires a Class A license.
- Non-native (Exotic) Animal Dealers—Anyone buying, selling, or trading animals that are non-native to the United States for research, teaching, testing, experimentation, or exhibition purposes, or as a pet; this requires a Class B license.
- Wild Animal Dealers—A business or individual selling wild animals for regulated purposes; this requires a Class B license.
- Suppliers of Specimens—Anyone who sells dead animals that are regulated or who sells the blood, serum, or parts of these animals for regulated purposes; this requires a Class B license.
- Carriers: Any enterprise transporting regulated animals for hire as a common carrier must be registered as a carrier; this requires a Class B license.

So for pets, they are covered under the AWA if they were bred in a licensed Class A facility and if they were transported by a licensed Class B carrier. If they were bred by any other formal or informal breeder, they have no coverage. In addition, pet stores that sell animals to the public are not covered, so those animals are not protected under the AWA, nor are pets covered at one's residence, at places like dog shows, or at any other venue. That is where the patchwork of state anti-cruelty laws can step in to provide some protections for pets once they have been sold.

Each year, the USDA releases a number of reports regarding the number of licensees, annual inspections, complaints, violations, and enforcement actions. Most of those relate to research facilities, but the agency does release an annual listing of active licensees. In the latest report, 2,709 facilities are listed as Class A licensees and another 703 are Class B licensees. There are fewer than 100 USDA inspectors for all licensed pet breeders, dealers, and transporters, as well as every research facility or animal exhibit in the country, with a total of more than 9,000 facilities in all.

See also: Animal Protection Movement; Laws and Regulations; Presidential Pets; Puppy Mills; Shelters

References and Further Reading

Cardon, A. D., M. R. Bailey, and B. T. Bennett (2012). "The Animal Welfare Act: From Enactment to Enforcement." *Journal of the American Association for Laboratory Animal Science*, 51 (3): 301–5.

Cohen, H. (2006). "The Animal Welfare Act." *J. Animal l.*, 2: 13.

Phinizy, C. (1965, November 29). "The Lost Pets That Stray to the Labs." *Sports Illustrated*.

USDA. (2022). *Research Facility Annual Usage Summary Report* (FY2019–FY2021). https://www.aphis.usda.gov/awa/research-facility-report/annual-summary (accessed May 21, 2024).

ANTHROPOMORPHISM

Anthropomorphism refers to the attribution of human traits onto nonhuman animals. While the basic definition of anthropomorphism is relatively simple, in practice, it is far from simple, in part because what constitutes a "human" trait is both changing constantly, and subject to a great deal of interpretation.

Research in ethology and animal behavior has increasingly revisited the role of anthropomorphism—not as a whimsical imposition of human traits onto nonhuman subjects but as a carefully modulated tool that can open the doors to the complex inner lives of animals when applied critically. Historically, anthropomorphism was often dismissed as a naïve projection that hindered objective inquiry; early behaviorists, in particular, cautioned against attributing human motives to animals, arguing that such projections were unscientific (Lulka, 2008). Yet, as the field evolved, scholars began to recognize that anthropomorphic language, when used as a tool of "careful" or "critical" anthropomorphism, could generate testable hypotheses about animal cognition and social behavior. This reappraisal has been influenced by cognitive psychology and neuroscience, which have demonstrated that humans naturally and unconsciously attribute human-like features to nonhuman agents—a tendency that, if acknowledged and critically framed, can provide valuable insights rather than misleading conclusions (Varella, 2018).

A central development in contemporary thought has been the articulation of "careful" or "critical anthropoorphism," a perspective that employs anthropomorphic inferences cautiously and contextually. Advocates like Marc Bekoff have argued that many social animals display behaviors, emotions, and even moral-like tendencies that render such comparisons more than mere projection—it becomes a means of understanding complex affective and social processes in nonhumans (Deane-Drummond, 2009). In a similar vein, proponents of licensed anthropomorphizing, as discussed by Ågren and Patten, contend that integrating agential thinking with rigorous quantitative modeling may enhance our understanding of evolutionary pressures without overstepping empirical boundaries (Ågren and Patten, 2022).

Moreover, the term "anthropomorphism" itself is often misunderstood. Popular discourse tends to conflate crude human projections with the nuanced interpretations now advocated within scientific circles. While lay usage may imply an uncritical imposition of human qualities onto animals, modern ethologists and cognitive scientists emphasize that careful anthropomorphism allows researchers to explore animal behavior in a way that respects species-specific behaviors while still drawing on familiar analogies to uncover underlying emotional or cognitive processes (Deane-Drummond, 2009; Ågren and Patten, 2022).

Today, at least in some academic fields, anthropomorphism is no longer viewed solely as a methodological flaw but as a potentially powerful interpretive strategy when used judiciously. Historical misunderstandings of the term have given way to frameworks of careful and critical anthropomorphism—a perspective supported by the work of scholars such as Marc Bekoff and, by extension, Robert Mitchell—which seeks to reconcile human intuitive understanding with the scientific study of animal behavior (Lulka, 2008; Deane-Drummond, 2009; Ågren and Patten, 2022).

See also: Ethics of Pet Ownership

References and Further Reading

Ågren, J. and M. Patten (2022). "Genetic Conflicts and the Case for Licensed Anthropomorphizing." *Behavioral Ecology and Sociobiology*, 76 (12): 166.

Deane-Drummond, C. (2009). "Are Animals Moral? A Theological Appraisal of the Evolution of Vice and Virtue." Zygon, 44 (4): 932–50.

Lulka, D. (2008). "Embodying Anthropomorphism: Contextualizing Commonality In the Material Landscape." *Anthrozoös*, 21 (2): 181–96.

Mitchell, R. W., N. S. Thompson, and H. L. Miles, eds. (1997). *Anthropomorphism, Anecdotes, and Animals*. Albany: SUNY Press.

Varella, M. A. C. (2018). "The Biology and Evolution of the Three Psychological Tendencies to Anthropomorphize Biology and Evolution." *Frontiers in Psychology*, 9: 1839.

ART, PETS IN

The significance of animals, including pets, in shaping human cultures is profoundly reflected in their artistic representations throughout history. For thousands of years, animals have been depicted in various forms of human art, serving as critical symbols through which humans interpret the world and their own identities. Biologist Edward

Wilson articulated this notion by stating that animals are "agents of nature translated into the symbols of culture" (Wilson, 1984: 97). The manner in which animals are represented in art profoundly influences the ways in which they are treated by humans.

Animals are among the most ancient figures featured in art, and their symbolic characteristics render them ideal subjects for artistic expression. More intriguingly, given that nonhuman animals share similarities with humans while remaining distinct, their inclusion in art and literature enables profound narrative explorations of human experience.

A focused examination of pets in artistic representations reveals their evolving status and the deepening bond between humans and these animals. Artists have utilized animals as decorative elements, to authentically depict the societal roles of animals, and symbolically to convey the characteristics they embody within culture (Werness, 2006). The genesis of artistic representation featuring animals can be traced back to the Paleolithic era, where cave paintings from 30,000 years ago showcased animals of substantial significance to human life—often more so than humans themselves (Morphy, 2014; Mithen, 1988).

With the Neolithic revolution and the domestication of animals, the portrayal of domesticated species became more prevalent in art, gradually superseding that of wild animals. For instance, cattle, critical for labor, sustenance, and religious practices, emerged in adornments such as jewelry, pottery, and the walls of significant structures in classical civilizations. It was during the nineteenth century, coinciding with the rise of the concept of "pets," that animals began to take on novel artistic connotations.

In Medieval European art, animals were represented with diverse meanings. They adorned jewelry, paintings, and functional artifacts, as well as illustrated religious texts. Lambs, for instance, symbolized purity and Christ's sacrificial role for humanity, while doves depicted love, and griffins served as guardians of the deceased. Coins from Medieval England frequently featured animals whose meanings were readily understood within the cultural context. Artistic depictions included both mythical creatures, like lions and peacocks—often derived from fables—and commonly encountered species such as chickens and doves. Hens, for example, were emblematic of the Mother Church, while the peacock symbolized immortality and the resurrection of Christ. Shockingly fantastical representations of lions, which had not been directly observed by artists, conveyed messages related to Christ and divine vigilance, while snakes, seen in the Christian narrative as symbols of evil, were paradoxically included for their perceived protective qualities (Klingender, 2019).

The Renaissance marked a pivotal shift, as animals began to be rendered with greater realism and less reliance on symbolism compared to the preceding Medieval period. Artists including Leonardo da Vinci and Albrecht Dürer created detailed portraits of animals, while other works portrayed deceased animals alongside sustenance such as fruits, vegetables, and bread. Hunting remained a prominent theme during the Renaissance (Cohen, 2008, 2014). Furthermore, this period heralded what could be construed as early representations of pets.

Although the essence of pet ownership as we understand it commenced later, elite individuals of the Renaissance did keep pets, and these companion animals often featured prominently in portraits of noble families. Women and children were frequently illustrated with small pets, such as lapdogs, underscoring the societal connection between femininity, childhood, and companion animals, while men were typically depicted with significant hunting dogs or horses. In these instances, the animals served not to sentimentalize the subjects but to signify their elevated social and economic ranks.

In contemporary art, the portrayal of animals often diverges drastically from historical representations, with a noticeable scarcity of symbolic or sentimental depictions. A notable example is found in the work of British artist Damien Hirst, who in the 1990s created installations not merely representing animals but constructed from the remains of deceased creatures (Huang, 2015). His early installation, A Thousand Years (1990), featured a rotting cow's head encased in glass, inhabited by flies and maggots. Another notable work, Mother and Child Divided (1993), involved the preservation of a bisected cow alongside her calf. Hirst's provocative approach has invited controversy amid concerns from gallery owners regarding audience reactions and criticism from art critics who deemed his work overrated, as well as protests from animal rights advocates condemning the use of deceased animals.

Other contemporary artists have employed, and at times killed, live animals to elicit public reactions. The shock value derived from the use of pets frequently surpasses that of more common species. For instance, in 2003, Chilean artist Marco Evarisitti exhibited live goldfish swimming in blenders filled with water, allowing attendees the opportunity to engage with the blenders at the expense of the fish's lives, with Evarisitti framing the experience as a challenge to the audience's conscience.

Similarly, Swedish photographer Nathalia Edenmont's work entailed the explicit slaughtering of rabbits, mice, chickens, and cats, with the aftermath photographed alongside flowers and other objects, anchoring her intention to provoke public reflection on societal ethics concerning animal treatment (Wallis, 2012).

In a particularly controversial exhibit in 2007, Costa Rican artist Guillermo Vargas showcased a starving dog tethered to a gallery wall in a piece titled Expo Exposición N° 1. The wall bore the phrase "eres lo que lees" (you are what you read), formed from dog food. Allegations arose (though unsubstantiated) that the dog eventually starved during the exhibition. Vargas contended that his artwork aimed to illuminate the public's indifference to the plight of starving dogs on the streets.

American artist Eduardo Kac sparked debate in 2000 when he commissioned the National Institute of Agronomic Research in Paris to create a genetically modified rabbit capable of emitting a phosphorescent glow by integrating jellyfish DNA. Although an agreement purportedly existed for the rabbit to live with Kac in Chicago, the institution disputed this claim. The resulting imagery, despite not being physically housed by Kac, emerged as one of his most recognized works (Kac, 2003).

As pet ownership has proliferated across the Western world and beyond, the presence of pets in artistic representation has become increasingly commonplace. Their likenesses have appeared in paintings and photographs of families, thereby normalizing the existence of multispecies households. This phenomenon has also led to the emergence of new art forms, such as pet portraits, further popularized by the accessibility of platforms like Etsy. Furthermore, advancements in technology in the twenty-first century have enabled "ordinary" individuals to engage in artistic creations featuring their pets.

Marie Emilie Coignet de Courson (1716–1806) with a Dog, by French artist Jean Honoré Fragonard, c. 1769. Women and children from aristocratic families were often portrayed in portraits holding small lapdogs. (Metropolitan Museum of Art, Fletcher Fund, 1937)

Contemporary art focusing on dogs, cats, and companion animals marks a significant evolution from abstract symbolism to tangible representations of real, individually named animals. Such modern "pet art" aptly conveys the complexity and affection inherent in human–animal relationships, even when exploring themes of mortality. For instance, artist Julia Schlosser's photographic series "Claire Last Days" (2016) poignantly captured her personal bond with her rabbit Claire before and after her death, intertwining images of Claire with the remnants of her life—her toys, food, medications, and other tokens of a love-filled existence culminating in a tender farewell.

Art critic John Berger has noted the fading presence of animals within contemporary society, suggesting that current artistic engagements with animals serve as memorials to their loss. Works such as "Claire Last Days" poignantly encapsulate the affection shared between humans and their companions, even as they navigate the emotional terrain of loss.

See also: Anthropomorphism; Film, Pets In; Literature, Pets In; Television, Pets On

References and Further Reading

Aloi, G. (2012). *Art and Animals*. London: I. B. Tauris.

Baker, S. (1993). *Picturing the Beast: Animals, Identity and Representation*. Manchester: Manchester University Press.
Berger, J. (1980). "Why Look at Animals?" In *About Looking*. New York: Pantheon.
Broglio, R. (2011). *Surface Encounters: Thinking with Animals and Art*. Minneapolis: University of Minnesota Press.
Cohen, S. (2008). Animals as Disguised Symbols in Renaissance Art (Vol. 169). Leiden: Brill.
Cohen, S. (2014). "Animal Imagery in Renaissance Art." *Renaissance Quarterly*, 67 (1): 164–80.
DeMello, M. (Ed.). (2016). Mourning Animals: Rituals and Practices Surrounding Animal Death. East Lansing: MSU Press.
Donald, D. (2008). *Picturing Animals in Britain, 1750–1850*. New Haven, CT: Yale University Press.
Huang, T. H. (2015). "On the Use of Animals in Contemporary Art: Damien Hirst." *Concentric: Literary and Cultural Studies*, 41 (1): 87–118.
Kac, E. (2003). "GFP Bunny." *Leonardo*, 36 (2): 97–102.
Klingender, F. (2019). *Animals in Art and Thought: To the End of the Middle Ages*. Routledge Library Editions; Medieval World 28. London: Routledge.
Malamud, R. (2012). *An Introduction to Animals and Visual Culture*. London: Palgrave Macmillan.
Mithen, S. J. (1988). "To Hunt or to Paint: Animals and Art in the Upper Palaeolithic." Man, 23: 671–95.
Morphy, H. (2014). *Animals Into Art*. London: Routledge.
Wallis, J. (2012). "Behaving Badly: Animals and the Ethics of Participatory Art." *Journal of Curatorial Studies*, 1 (3): 315–27.
Werness, H. B. (2006). Continuum Encyclopedia of Animal Symbolism in World Art. A&C Black.

ARTHROPODS

Arthropods are invertebrate animals and include insects, the largest group, as well as crustaceans (such as crabs or lobsters) and spiders. Research on arthropods as pets reflects a growing yet historically nuanced field that intersects with cultural interests, scientific inquiry, and contemporary debates on exotic pet ownership.

The practice of keeping arthropods—ranging from insects to arachnids—can be traced back to the Victorian era when naturalists collected and observed insects for both scientific study and leisure. This early interest in arthropods laid the foundation for later trends wherein species such as tarantulas, scorpions, and stick insects were maintained as companion animals, not just as curiosities but as subjects of sustained hobbyist engagement.

While conventional pets such as dogs and cats receive extensive regulatory attention, arthropods have been marginalized in both policy and academic discourse. However, the increasing popularity of keeping arthropods—as part of a larger trend toward exotic pets—has attracted scholarly interest in their species-specific husbandry needs and welfare concerns. Research emphasizes that, despite their small size and relatively simple physiologies, many arthropods require carefully controlled environmental conditions, including specific humidity, temperature, and dietary provisions, to ensure their well-being. These husbandry nuances highlight that arthropods, although markedly different from traditional companion animals, demand specialized care protocols that are frequently underappreciated (Reaser et al., 2008).

Moreover, the emergence of arthropods as pets has raised important regulatory and ethical questions. Scholars such as Reaser et al. (2008) have pointed to the zoonotic potential associated with nontraditional pets and cautioned that the trade in wild-caught arthropods could inadvertently facilitate the spread of novel pathogens. This discussion extends to concerns over conservation, as the collection of wild populations for the pet trade may disturb local ecological balances (Grant et al., 2017) underscore that the broader category of exotic pets—which increasingly includes arthropods—necessitates improved regulation and clearer guidelines to safeguard both animal welfare and public health. These perspectives urge the incorporation of arthropods into growing discussions about ethical pet ownership, calling for a reevaluation of how companion animal status is defined in legal and social terms.

One culture with a history of arthropod keeping is Japan. Research on insect pet keeping in Japan, encapsulated in the ethnocategory known as mushi, reveals a long historical and cultural tradition that markedly differs from Western perspectives on insects. In Japan, mushi is not limited to what Western paradigms typically classify as insects; rather, it encompasses insects, larvae, and other small animals and is imbued with nuanced cultural and symbolic meanings (Laurent, 1995). Early naturalistic observations and traditional practices promoted by rural and educational settings encouraged children to collect, observe, and even "keep" insects as a means to understand nature. This engagement, integral to childhood learning, reflects an appreciation of mushi as gentle and integral components of the natural world, forming a significant part of Japan's historical practice of nurturing an empathetic relationship with nature (Laurent, 1995).

Age and gender differences shape the perception and practice of mushi engagement among Japanese children. Fieldwork research indicates that children's interpretations of mushi evolve with age, reflecting a deeper, more empathetic understanding of animal life as they mature. Notably, gender differences emerge in that girls and boys may adopt subtly distinct stances toward these small creatures. Girls are often observed to approach mushi with a greater sense of care and aesthetic sensitivity, attributing delicate and nurturing characteristics, whereas boys might be drawn to the dynamic and robust features of these animals. Such differences highlight how cultural, social, and developmental factors interweave in the formation of attitudes toward nature and animal encounters, a nuance that is often absent in Western paradigms that tend to homogenize perceptions of insects (Laurent, 1995; Prokop and Tunnicliffe, 2008).

From their origins in early scientific and cultural explorations to contemporary debates on animal welfare and regulatory oversight, arthropods occupy a unique niche within the companion animal landscape. This niche is characterized by distinct care requirements, potential zoonotic implications, and a need for more targeted research and policy interventions to ensure that the increasing trend of arthropod pet ownership is managed in an ethically and ecologically responsible manner (Reaser et al., 2008; Grant et al., 2017).

See also: Exotic Pets; Japan, Pets In

References and Further Reading

Grant, E. H. C., E. Muths, R. A. Katz, S. Canessa, M. J. Adams, J. R. Ballard, ... and C. L. White (2017). "Using Decision Analysis to Support Proactive Management of Emerging Infectious Wildlife Diseases." Frontiers in Ecology and the Environment, 15 (4): 214–21.

Laurent, É. (1995). "Definition and Cultural Representation of the Category Mushi in Japanese Culture." *Society and Animals*, 3 (1): 61–77.

Prokop, P., and S. Tunnicliffe (2008). "'Disgusting' Animals: Primary School Children's Attitudes and Myths of Bats and Spiders." *Eurasia Journal of Mathematics Science and Technology Education*, 4 (2): 87–97.

Reaser, J., E. Clark, and N. Meyers (2008). "All Creatures Great and Minute: A Public Policy Primer for Companion Animal Zoonoses." Zoonoses and Public Health, 55 (8–10): 385–401.

ATTACHMENT

Attachment refers to the emotional connections that are formed by and between living beings like humans and nonhuman animals. Most of what we know about attachment originally came from the world of child psychology and childhood development, but today, the field of anthrozoology has produced a great deal of literature on attachment between humans and nonhumans.

The history of attachment studies in children has evolved from early theoretical formulations to a nuanced, multi-method research field. Early work, influenced by the foundational ideas of child psychologists John Bowlby and Mary Ainsworth, laid the groundwork by emphasizing the critical nature of the early child–caregiver relationship. Studies initially focused on characterizing attachment patterns through structured observational measures, notably the Strange Situation Procedure, which enabled researchers to identify secure, insecure, and disorganized patterns of attachment (Fraley and Spieker, 2003).

Comparative studies have extended attachment research to diverse child populations. For instance, Zeanah et al. (2005) investigated attachment in institutionalized children in Romania, revealing that the absence of consistent caregiver relationships in these settings often leads to high incidences of disorganized or unclassifiable attachment behaviors. Such studies underscore the sensitivity of the attachment system to variations in caregiving environments and the consequential impact of early adverse experiences on later social and emotional functioning.

Research on attachment between humans and pets has begun to extend traditional attachment theory beyond the parent–child relationship. Foundational studies, such as Topál et al. (1998), demonstrated that pet dogs exhibit attachment behaviors similar to those observed in children during Ainsworth's Strange Situation Test, thereby suggesting that many human–dog relationships function as secure bases. Building on these insights, subsequent investigations have confirmed that attachment bonds with pets—especially dogs—can serve as effective sources of emotional support and stress buffering, paralleling the role of human attachment figures (Kurdek, 2009; Beck and Madresh, 2008).

Contemporary research has further explored the psychological mechanisms underlying pet attachment. For example, Lass-Hennemann et al. (Lass-Hennemann et al., 2022; and Faner et al., 2024) have developed a number of tools to measure attachment strength and to see what attachment to an animal might correlate with. Their work suggests that strong bonds with pets might develop as a compensatory response to adverse human interpersonal experiences (Lass-Hennemann et al., 2022; Faner et al., 2024). With the development of validated scales to measure attachment, such as the first and most well-used of those, the Lexington Attachment to Pets Scale, we are beginning to see these correlations; strong pet attachment appears to be linked with better mental health outcomes, particularly among individuals with limited social support Antonacopoulos and Pychyl, 2010; Sung and Han, 2023).

Moreover, investigations into the human–pet attachment bond have addressed variability across demographic groups and contexts. Özen et al. (2024) and Rockett and Carr (2014) highlight that sociodemographic factors, personality traits, and life circumstances mediate the formation and quality of such bonds. For instance, pet attachment has been shown to be particularly salient among older adults coping with loneliness or depression (Miltiades and Shearer, 2011), while research involving both children and adults has noted that a secure attachment to pets can promote individual well-being and broader empathy and social cognitive skills (Groenewoud et al., 2023; Maruyama, 2005). In the domain of human–pet interactions, attachment theory has been extended to consider how pets might buffer negative effects of human social stressors, thereby contributing to better quality of life and overall psychological health (Peacock, Chur-Hansen, and Winefield, 2012; Riggs et al., 2024).

Pet attachment, whether examined through behavioral tests such as the Strange Situation or through multidimensional questionnaires like the Lexington scale, provides both a source of emotional security and a potential avenue for enhancing human well-being, particularly when human social networks are lacking.

See also: Adults and Pets; Bond, Human–Animal; Children and Pets; Human Health, Pets And; Mental Health, Pets And

References and Further Reading

Antonacopoulos, N. M. D. and T. A. Pychyl (2010). "An Examination of the Potential Role of Pet Ownership, Human Social Support and Pet Attachment in the Psychological Health of Individuals Living Alone." Anthrozoös, 23 (1): 37–54.

Beck, L. and E. A. Madresh (2008). "Romantic Partners and Four-Legged Friends: An Extension of Attachment Theory to Relationships with Pets." Anthrozoös, 21 (1): 43–56.

Birmingham, R., K. Bub, and B. Vaughn (2016). "Parenting in Infancy and Self-Regulation in Preschool: An Investigation of the Role of Attachment History." *Attachment & Human Development*, 19 (2): 107–29.

Duschinsky, R. (2020). *Cornerstones of Attachment Research*. Oxford: Oxford University Press.

Faner, J. M. V., E. A. R. Dalangin, L. A. T. C. De Leon, L. D. Francisco, Y. O. Sahagun, and E. F.Acoba (2024). "Pet Attachment and Prosocial Attitude toward Humans: The Mediating Role of Empathy to Animals." Frontiers in Psychology, 15: 1391606.

Fraley, R., and S. Spieker (2003). "Are Infant Attachment Patterns Continuously or Categorically Distributed? A Taxometric Analysis of Strange Situation Behavior." *Developmental Psychology*, 39 (3): 387–404.
Groenewoud, D., M. J. Enders-Slegers, R. Leontjevas, A. Van Dijke, T. de Winkel, and K. Hediger (2023). "Children's Bond with Companion Animals and Associations with Psychosocial Health: A Systematic Review." Frontiers in Psychology, 14: 1120000.
IJzendoorn, M., C. Schuengel, and M. Bakermans-Kranenburg (1999). "Disorganized Attachment in Early Childhood: Meta-Analysis of Precursors, Concomitants, and Sequelae." *Development and Psychopathology*, 11 (2): 225–50.
Kurdek, L. A. (2009). "Pet Dogs as Attachment Figures for Adult Owners." Journal of Family Psychology, 23 (4): 439.
Lass-Hennemann, J., S. K. Schäfer, M. R. Sopp, and T. Michael (2022). "The Relationship between Attachment to Pets and Mental Health: The Shared Link via Attachment to Humans." BMC Psychiatry, 22 (1): 586.
Maruyama, M. (2005). *Humane Education: The Effects of Animals in the Classroom on Children's Empathy in Japanese Elementary Schools*. Dissertations and thesis, Portland State University.
Miltiades, H. and J. Shearer (2011). "Attachment to Pet Dogs and Depression in Rural Older Adults." Anthrozoös, 24 (2): 147–54.
Özen, D., N. Bıçakcı, and Y. Demirbaş (2024). "Exploring the Dynamics of Human-Pet Attachment: An In-Depth Analysis of Socio-Demographic Factors and Relationships." *Ankara Üniversitesi Veteriner Fakültesi Dergisi*, 71 (4): 385–93.
Peacock, J., A. Chur-Hansen, and H. Winefield (2012). "Mental Health Implications of Human Attachment to Companion Animals." *Journal of Clinical Psychology*, 68 (3): 292–303.
Riggs, D., H. Fraser., N. Taylor, and E. Peel (2024). "Exploring the Contribution of Animal Companionship to Human Wellbeing: A Three-Country Study." *International Journal of Wellbeing*, 14 (1): 1–17.
Rockett, B., and S. Carr (2014). "Animals and Attachment Theory." *Society and Animals*, 22 (4): 415–33.
Sung, J. Y. and J. S. Han (2023). "Exploring the Role of Empathy as a Dual Mediator in the Relationship between Human–Pet Attachment and Quality of Life: A Survey Study among Adult Dog Owners." Animals, 13 (13): 2220.
Topál, J., Á. Miklósi, V. Csányi, and A. Dóka (1998). "Attachment Behavior in Dogs (canis familiaris): A New Application of Ainsworth's (1969) Strange Situation Test." *Journal of Comparative Psychology*, 112 (3): 219–29.
Zeanah, C., A. Smyke, S. Koga, and E. Carlson (2005). "Attachment in Institutionalized and Community Children in Romania." *Child Development*, 76 (5): 1015–28.

AUSTRALIA AND NEW ZEALAND, PETS IN

Australia and New Zealand both, like the United States, owe much of their cultural beliefs and practices to their origins as an English colony, and as such, their pet keeping practices largely follow the English pattern. It should not surprise us to learn that, as in the United States, a majority of households in both countries own pets—most commonly dogs and cats—and that these relationships are incredibly significant. But because of some of the very unique factors found in both nations—especially the unique flora and fauna—pet keeping has an impact on wildlife biodiversity that is relatively unique to this region.

Many Australian pet owners, for example, are not only aware of the ecological implications of keeping pets—such as the impact of free-roaming cats on native wildlife—

but actively engage in practices like neutering and spaying their pets to mitigate these effects (Hall et al., 2016). Another example has to do with rabbits. Because of the import of European rabbits to Australia in the nineteenth century, and their subsequent spread across the continent, rabbits as pets has been prohibited in the state of Queensland, where the damage caused by the wild rabbits is most severe.

As in other Western nations, the pet industry is a major economic force in Australia and New Zealand. In Australia, the pet industry is valued at several billion Australian dollars annually, driven by expenditures in veterinary care, specialized pet foods, grooming, and accessories. New Zealand, while smaller in scale, exhibits similar growth trends as pet ownership increases and consumers demand higher-quality and more personalized pet care products and services. Such economic indicators underscore the importance of the sector within the overall economy, as the increasing trend toward pets as family members not only enhances owners' quality of life but simultaneously stimulates commercial activity in pet-related industries.

As in other Western nations, and particularly those with roots in England, pet keeping in Australia and New Zealand is marked by high household ownership rates, evolving conservation concerns, and a robust economic impact.

See also: United Kingdom, Pets In; United States, Pets In

References and Further Reading

Birchall, M. (2020). "History, Sovereignty, Capital: Company Colonization in South Australia and New Zealand." *Journal of Global History*, 16 (1): 141–57.

Feinleib, M., and H. Ritvo (1988). "The Animal Estate: The English and Other Creatures in the Victorian Age." *The Journal of Interdisciplinary History*, 19 (2): 322.

Hall, C. M., N. A. Adams, J. S. Bradley, K. A. Bryant, A. A. Davis, C. R. Dickman, ... and M. C. Calver (2016). "Community Attitudes and Practices of Urban Residents Regarding Predation by Pet Cats on Wildlife: An International Comparison." PLoS One, 11 (4): e0151962.

Johnston, L., J. Szczepanski, and P. McDonagh (2017). "Demographics, Lifestyle and Veterinary Care of Cats in Australia and New Zealand." *Journal of Feline Medicine and Surgery*, 19 (12): 1199–205.

Rudolph, S., and M. Hogarth (2020). "Contested Citizenship Education in Settler Colonies on First Nations Land." In *The Palgrave Handbook of Citizenship and Education*, 537–48. Cham: Springer International Publishing.

B

BIRDS

The history of keeping birds as pets spans millennia and reflects a complex interplay of indigenous practices, international trade, and status. Early indigenous cultures across the globe maintained intimate and complicated relationships with birds which reflect their status as not just animals but as creatures with a great deal of spiritual significance, and this conception of birds extended into the earliest of the ancient civilizations.

For instance, in pre-Columbian MesoAmerica, early civilizations such as the Aztecs and Maya not only kept birds as pets but also integrated them into elaborate social and ceremonial practices, where feathers as well as whole birds were used as symbols of prestige, power, and spiritual connection Sousa and Caplan, 2020). Similarly, traditional practices in South America reveal that indigenous communities valued birds both for aesthetic enjoyment and as trade items, with feathers serving as potent indicators of social status and identity.

In Southeast Asia, particularly Indonesia, bird keeping constitutes an important cultural practice with significant economic and social dimensions. Research has shown that species such as the Straw-headed Bulbul are highly prized, with wild-caught individuals often sought after by wealthier households as markers of exclusivity and status (Bergin et al., 2017). This tradition coexists with indigenous approaches in which the collection and maintenance of birds are embedded in local knowledge systems and ecological practices. The act of capturing and keeping birds is intertwined with ritual and everyday life, demonstrating an integrated relationship with nature that reflects a broader cultural worldview.

Indigenous perspectives on birds are also vividly illustrated in other parts of the world. The Nuer of western Ethiopia, for example, saw birds as sacred entities imbued with spiritual qualities who were thought to act as intermediaries between the human and supernatural realms (Aticho et al., 2024). This belief is found throughout the world and makes intuitive sense, as birds are one of the only creatures on the planet that can fly, and as such, are thought to have access to spiritual realms that humans are shut out of. Such beliefs have contributed to community-driven conservation practices, ensuring that both cultural and ecological values of birds are preserved.

Additionally, historical accounts from Medieval Europe further tell us more about the history of bird keeping. During this period, exotic birds and their feathers were not only kept as pets but also utilized to display wealth and power, echoing many of the status-related dynamics observed in the pre-Columbian Americas. These diverse historical and cultural threads demonstrate that the practice of keeping birds is neither purely utilitarian nor solely ornamental; it has served as a vehicle through which communities express cultural identity, negotiate social status, and engage in long-distance trade.

How do birds—given that very few birds have ever been domesticated—become pets? Historically, all birds kept as pets were taken from the wild, while in the past few years, captive breeding of a variety of birds, most commonly psittacines, has replaced the wild bird trade in some countries and for some species.

In tropical regions such as Brazil and Southeast Asia, indigenous traditions and both local and international demand for birds have resulted in greater pressure on those bird species, and in some regions, local ceremonial or spiritual use of birds has been supplanted by capturing birds for the international trade. For example, in Indonesia, bird-keeping has evolved from a culturally rooted practice to a significant commercial activity, with wild birds being captured in large numbers for domestic consumption and export despite the known conservation impacts and limited regulatory capacity (Jepson and Ladle, 2005). In these systems, birds are prized not only for their vocal abilities and visual ornaments but also for their cultural symbolism, which drives both local and international demand. Ethnographic studies in Mexico further illustrate how specialized local knowledge regarding avian biology and behavior underpins the capture and maintenance of wild birds for trade, thereby reinforcing long-standing sociocultural attachments to these species (Roldán-Clarà et al., 2017).

International responses to the wild bird trade, including trade bans and stricter enforcement of CITES (the Convention on International Trade in Endangered Species of Wild Fauna and Flora) regulations, have been shaped by both conservation and public health concerns.

When birds are kept as pets, there are significant welfare challenges that emerge that are not found within the more traditional pets such as dogs and cats. Because pet birds are wild, and because most people are relatively uninformed about bird behavior, bird cognition, bird emotions, or anything else about these complicated, intelligent creatures, pet birds are often kept in circumstances that are far from ideal. In addition, because psittacines live such a long time, many birds are passed from person to person throughout their lifetimes, increasing the chances of their needs not being met.

Pet birds, which are estimated to be owned by millions of households in the United States, form a significant yet often overlooked segment of the companion animal market (Burmeister et al., 2020). Researchers have developed new instruments such as the Owner-Bird Relationship Scale (OBRS) to quantify human–bird attachment, revealing that many bird owners display strong emotional bonds with their pets that, while sometimes differing qualitatively from those observed with dogs or cats, still contribute to owners' overall psychological well-being (Burmeister et al., 2020; Teo and Thomas, 2019).

Among the challenges associated with owning pet birds, significant attention has been directed toward the health risks and management practices within this population. For example, investigations into antimicrobial treatment practices have shown that some pet bird owners resort to short-term or even prophylactic antibiotic usage—often obtained off-label over the Internet—which not only poses risks of treatment failure but also increases the potential for the emergence of zoonotic pathogens that may be transmitted to humans (Meyst et al., 2022).

Alongside these challenges, the benefits of keeping pet birds have been documented in relation to owner well-being and social connectedness. Research has demonstrated that high levels of attachment to pet birds can function as an emotional buffer, contributing to reduced feelings of loneliness and stress among owners. Although the nature of the relationship between bird owners and their avian companions often involves less tactile interaction compared to other pet species, the bond formed can nonetheless be profound and is associated with positive mental health outcomes (Burmeister et al., 2020; Teo and Thomas, 2019). Additionally, studies utilizing attachment theory frameworks have emphasized that these bonds are bidirectional, with the personality traits of both owners and birds influencing the quality of the relationship and overall life satisfaction (Wells and Treacy, 2024).

Keeping birds as pets poses both unique challenges—such as health and welfare management and zoonotic risks—and important psychological benefits linked to strong owner attachment. But the keeping of pet birds is also directly linked to questions of biodiversity, making them an even more complicated pet.

See also: Elite Pet Keeping; Exotic Pets; Indigenous Pet Keeping; Status Pets

References and Further Reading

Aloysius, S., D. Yong, J. Lee, and A. Jain (2019). "Flying into Extinction: Understanding the Role of Singapore's International Parrot Trade in Growing Domestic Demand." *Bird Conservation International*, 30 (1): 139–55.

Alves, R., J. Lima, and H. Araújo (2012). "The Live Bird Trade in Brazil and its Conservation Implications: An Overview." *Bird Conservation International*, 23 (1): 53–65.

Aticho, A., A. Beyene, S. Zelelew, C. Puok, A. Gosler, N. Stenseth, and D. Chala (2024). "Birds As Sacred 'Children of God', Beholders of Spirit and Symbols of Twinhood: Human–Bird Interaction Among the Nuer in Western Ethiopia." *People and Nature*, 6 (6): 2418–30.

Bergin, Daniel, Serene CL Chng, James A. Eaton, and Chris R. Shepherd (2018). "The final straw? An overview of Straw-headed Bulbul Pycnonotus zeylanicus trade in Indonesia." *Bird Conservation International* 28(1): 126–32.

Burmeister, A. K., K. Drasch, M. Rinder, S. Prechsl, A. Peschel, R. Korbel, and N. J. Saam (2020). "Development and Application of the Owner-Bird Relationship Scale (OBRS) to Assess the Relation of Humans to Their Pet Birds." Frontiers in Veterinary Science, 7: 575221.

De Meyst, A., R. Aaziz, J. Pex, L. Braeckman, M. Livingstone, D. Longbottom, ... and D. Vanrompay (2022). "Prevalence of New and Established Avian Chlamydial Species in Humans and Their Psittacine Pet Birds in Belgium." Microorganisms, 10 (9): 1758.

Jepson, P., and R. Ladle (2005). "Bird-Keeping in Indonesia: Conservation Impacts and the Potential for Substitution-Based Conservation Responses." *Oryx*, 39 (4): 442.

Roldán-Clarà, Blanca, Víctor M. Toledo, and Ileana Espejel (2017). "The use of birds as pets in Mexico." *Journal of ethnobiology and ethnomedicine* 13(1): 35

Sousa, L., and A. Caplan (2020). "Introduction: Knowledge of Birds and Feathers in the Ancient and Colonial Meso-American World." *Ethnohistory*, 67 (3): 345–54.
Souto, W., M. Torres, B. Sousa, K. Lima, L. Vieira, G. Pereira, and B. Pralon (2017). "Singing for Cages." *Tropical Conservation Science*, 10: 1–19.
Teo, J. T. and S. J. Thomas (2019). "Psychological Mechanisms Predicting Wellbeing in Pet Owners: Rogers' Core Conditions versus Bowlby's Attachment." Anthrozoös, 32 (3): 399–417.
Wells, D. L. and K. R. Treacy (2024). "Pet Attachment and Owner Personality." Frontiers in Psychiatry, 15: 1406590.
Wright, T., C. Toft, E. Enkerlin-Hoeflich, J. Gonzalez-Elizondo, M. Albornoz, A. Rodríguez-Ferraro, ... J. Wiley (2001). "Nest Poaching in Neotropical Parrots." *Conservation Biology*, 15 (3): 710–20.

BOND, HUMAN–ANIMAL

The term human–animal bond refers to the idea that having a relationship with other species can be beneficial to humans, and more recently, to the animals as well. Without this recognition—that humans are not only drawn to animals but benefit from a connection with them—there would be no basis for animal-assisted therapies, service animals, or indeed our entire understanding of the health benefits of living with another species.

Early conceptualizations of the human–animal bond were influenced by pioneers such as Konrad Lorenz and Boris Levinson, whose work highlighted the emotional and therapeutic significance of interactions with animals. This early foundation was later expanded by veterinary professionals and social scientists who organized conferences and published seminal works during the 1970s and 1980s, thereby framing the concept as a legitimate object of scholarly inquiry and practical application (Hines 2003). Indeed, interest in the human–animal bond is directly responsible for the rise of anthrozoology as a new field of inquiry.

In the 1970s, research on animal-assisted activity and therapy began to gain momentum, and the importance of human–animal relationships for mental health and well-being became more widely recognized. Early studies were largely descriptive and correlational in nature, relying on observational data and anecdotal evidence. Over time, however, researchers began developing more rigorous empirical methodologies.

Current research on the human–animal bond has shifted toward integration with public health frameworks. Hoy-Gerlach and Townsend (2023) advocate viewing bond support as a primary, secondary, and tertiary public health intervention. The expansion of the One Health approach, as championed by organizations like the One Health Commission and incorporated into medical education (Rabinowitz et al., 2017), underscores the bidirectional benefits of strong human–animal bonds for both human and animal well-being. More recent systematic reviews also point toward the need for a sound theoretical basis and physiological measurements to validate outcomes in areas such as animal-assisted interventions (Holder et al., 2020, 2020). Research by Carver (2022) demonstrates that the strength of the human–animal bond is a significant predictor of well-being, with many studies using standardized instruments like the 10-item Pet Attachment Scale (PAS) to capture variances in attachment strength.

Leaders in the field, including researchers such as Aubrey Fine and Alan Beck, have been influential not only in advancing academic literature but also in shaping policy and practice. These figures, along with organizations such as the Human Animal Bond Research Institute (HABRI) and the International Society for Anthrozoology (ISAZ), have contributed to elevating the field from anecdotal reporting to a scientifically recognized discipline. Their work has promoted interdisciplinary approaches that combine rigorous empirical research with the ethics of caring, suggesting that traditional moral theories must be supplemented with an ethics of care to fully address the responsibilities engendered by human–animal bonds (Russow, 2002).

Recent critiques, such as those by Macauley and Chur-Hansen (2022), note that much of the existing literature still relies on descriptive and correlational methodologies. Nonetheless, the evolution of the field highlights an ongoing transition toward more controlled, interdisciplinary research frameworks that incorporate psychological, physiological, and sociocultural dimensions. This evolution has deepened our understanding of the benefits—such as enhanced mental health, improved social connectedness, and reduced stress—and clarified the challenges inherent in translating observational bonds into clinical and public health interventions.

Through continued interdisciplinary research and the development of robust theoretical frameworks and measurement tools, the field is better positioned to influence policy, clinical practice, and educational initiatives that recognize the mutual benefits between humans and animals.

See also: Attachment; Human Health, Pets And; Mental Health, Pets And

References and Further Reading

Carver, L. F. (2022). "One Health, the Human-Animal Bond and Well-Being During the COVID-19 Pandemic." One Health Innovation, 1 (1). https://doi.org/10.24908/ohi.v1i1.16190

Hines, L. M. (2003). "Historical Perspectives on the Human-Animal Bond." American Behavioral Scientist, 47 (1): 7–15.

Holder, T. R., M. E. Gruen, D. L. Roberts, T. Somers, and A. Bozkurt (2020a). "A Systematic Literature Review of Animal-Assisted Interventions in Oncology (Part I): Methods and Results." Integrative Cancer Therapies, 19: 1534735420943278.

Holder, T. R., M. E. Gruen, D. L. Roberts, T. Somers, and A. Bozkurt (2020b). "A Systematic Literature Review of Animal-Assisted Interventions in Oncology (Part II): Theoretical Mechanisms and Frameworks." Integrative Cancer Therapies, 19: 1534735420943269.

Hosey, G., and V. Melfi (2014). "Human-Animal Interactions, Relationships and Bonds: A Review and Analysis of the Literature." *International Journal of Comparative Psychology,* 27 (1): 117–42.

Hoy-Gerlach, J. and L. Townsend (2023). "Reimagining Healthcare: Human–Animal Bond Support as a Primary, Secondary, and Tertiary Public Health Intervention." International Journal of Environmental Research and Public Health, 20 (7): 5272.

Macauley, L. and A. Chur-Hansen (2022). "Human Health Benefits of Non-Conventional Companion Animals: A Narrative Review." Animals, 13 (1): 28.

Rabinowitz, P. M., B. J. Natterson-Horowitz, L. H. Kahn, R. Kock, and M. Pappaioanou (2017). "Incorporating One Health into Medical Education." BMC Medical Education, 17 (1): 45.

Russow, L. M. (2002). "Ethical Implications of the Human-Animal Bond in the Laboratory." ILAR Journal, 43 (1): 33–7.

BREEDS, POPULARITY OF

Dog breeds, and to a lesser extent cat breeds, rise and fall in terms of public sentiment and popularity as pets. Research on dog breeds and their popularity has yielded insights into both intrinsic breed characteristics and the influence of external sociocultural factors, such as media exposure, on changes in public preferences over time. Wilson et al. (2018), for example, demonstrated that fluctuations in American Kennel Club registrations from 1926 to 2005 correlate with specific clusters of canine behavioral traits. Their work indicates that shifts in the demand for breeds are not random but can be partly explained by the emergence or decline of particular behavioral profiles over historical periods. This study implies that as older breeds fall out of favor, new ones—often with differing behavioral and physical characteristics—rise to prominence, suggesting a dynamic interplay between breed traits and owner expectations in the marketplace.

Media representations have played a crucial role in this process, as evidenced by the work of Ghirlanda, Acerbi, and Herzog (2014), who investigated the influence of movies on breed popularity. Their findings show that film releases featuring dog movie stars can lead to a significant increase in the demand for the featured breeds, with the effect lasting up to ten years after a movie's release. The impact of a movie on breed popularity was found to correlate with its opening weekend viewership, thereby underlining the power of mass media to transform public perceptions and influence breed choice on a national scale (Ghirlanda, Acerbi, and Herzog, 2014). Such trends underscore the notion that popular movies and television shows act as cultural catalysts, rapidly altering market preferences and contributing to the "fad" nature of certain breeds.

Additional research has examined how breed preferences change over time due to a combination of media portrayal and evolving consumer values. For example, studies on morphological trends (Teng et al., 2016), have highlighted shifts in the desired physical attributes of purebred dogs in Australia, which can mirror broader global trends. These changing preferences are influenced not only by media portrayals but are also intertwined with contemporary concerns over dog health and welfare; certain physical characteristics targeted by popular media may predispose breeds to genetic disorders (Packer, Murphy, and Farnworth, 2017; Menor-Campos, 2024). Thus, while the allure of a breed may be temporarily boosted by its association with popular culture, long-term shifts in breed popularity can also be driven by growing awareness of the potential health and behavioral consequences inherent to selective breeding.

Unlike dog breeds, which were originally based on functional traits selected for in breeding, cat breeds were primarily developed in the past 150 years and are mostly based on selecting for aesthetic traits rather than functional ones (Montague et al., 2014).

Consumer research tells us that preferences for certain cat breeds are strongly influenced by early life experiences and media exposure. Studies in consumer psychology suggest that nostalgic associations formed during adolescence and early adulthood can have a lingering effect on aesthetic tastes. Such early exposure to specific images of cats in movies and on television can drive long-term preferences and even lead to shifts in breed popularity over

A woman holds a toy Boston Terrier, c. 1905. Sociocultural factors play a role in changes in public preference for certain breeds over time. (Library of Congress)

time Schindler and Holbrook (2003). For instance, when popular films or television shows feature a particular breed, those breeds may experience surges in demand due to their established emotional resonance with viewers' formative experiences.

In addition, recent investigations into owner preferences have focused on phenotypic traits such as skull morphology and overall facial structure as significant factors in breed selection. This dynamic suggests that cultural fads and nostalgia—fueled by media portrayals—can precipitate cyclical booms and busts in the popularity of specific cat breeds. The interplay between aesthetic selection by breeders and the impact of cultural forces thus shapes not only breed characteristics but also trends in the pet industry over time.

In summary, the current body of research on dog and cat breeds and their popularity demonstrates that both historical fluctuations and modern cultural influences, particularly through movies and television, shape breed preferences over time. These findings suggest that breed preferences are dynamic and are continually redefined by a confluence of inherent canine traits, societal trends, and media portrayals.

See also: Advertising, Pets In; Cats; Dogs; Television, Pets On

References and Further Reading

Ghirlanda, S., A. Acerbi, and H. Herzog (2014). "Dog Movie Stars and Dog Breed Popularity: A Case Study in Media Influence on Choice." *Plos One*, 9 (9): e106565.

Herzog, H. (2006). "Forty-Two Thousand and One Dalmatians: Fads, Social Contagion, and Dog Breed Popularity." *Society & Animals*, 14 (4): 383–97.

Menor-Campos, D. (2024). "Ethical Concerns About Fashionable Dog Breeding." *Animals*, 14 (5): 756.

Montague, M. J., G. Li, B. Gandolfi, R. Khan, B. L. Aken, S. M. Searle, … and W. C. Warren (2014). "Comparative Analysis of the Domestic Cat Genome Reveals Genetic Signatures Underlying Feline Biology and Domestication." *Proceedings of the National Academy of Sciences*, 111 (48): 17230–5.

Packer, R., D. Murphy, and M. Farnworth (2017). "Purchasing Popular Purebreds: Investigating the Influence of Breed-Type on the Pre-Purchase Motivations and Behaviour of Dog Owners." *Animal Welfare*, 26 (2): 191–201.

Schindler, R. M. and M. B. Holbrook (2003). "Nostalgia for Early Experience as a Determinant of Consumer Preferences." Psychology & Marketing, 20 (4): 275–302.

Teng, K. T., P. D. McGreevy, J. A. L. Toribio, and N. K. Dhand (2016). "Trends in Popularity of Some Morphological Traits of Purebred Dogs in Australia." Canine Genetics and Epidemiology, 3 (1): 2.

Wilson, B., J. Serpell, H. Herzog, and P. McGreevy (2018). "Prevailing Clusters of Canine Behavioural Traits in Historical us Demand for Dog Breeds (1926–2005)." *Animals*, 8 (11): 197.

C

CANADA, PETS IN

Pet keeping in Canada is largely parallel to what we know of pet keeping in the United States. Not only do both countries share a similar colonial history, but because of the geographic proximity of the United States to Canada, and the ability of citizens of both nations to access the same products and services, including pets and pet care, we do not find any significant differences between the two countries.

Market analyses indicate that evolving demographic trends in Canada, as in the United States, also shape pet ownership statistics. As in the United States changing household compositions—such as the increase in smaller households, delayed family formation, and more individuals living in rented accommodations where pet policies may be more restrictive—substantially impact pet keeping numbers. These demographic shifts, coupled with the rising interest in pet companionship, are expected to influence both the level of pet ownership and the financial dynamics within the pet services market.

The literature on animal rescue, shelters, and pet relinquishment in Canada reflects a complex interplay of socioeconomic, behavioral, and systemic factors that influence both the intake and outcomes for companion animals. Research conducted during the Covid-19 pandemic in cities such as Vancouver has shown that while shelters experienced surges in pet adoptions due to increased isolation and the human need for companionship (Schor and Protopopova, 2021), there remains a challenge in supporting vulnerable populations who face difficulties in sustaining pet ownership over time. Such studies point to the importance of understanding the local context, which may include economic stressors and shifts in service provision, thereby highlighting the necessity for adaptable policies and services in animal shelters. While this is no different for the United States, in Canada there is a greater social safety net which may result in different expectations and practices surrounding vulnerable human populations and their needs.

Multiple studies, albeit from international settings, provide insight into reasons for pet relinquishment that are relevant for Canadian shelters. For instance, research has consistently identified behavioral problems as a major factor influencing owners to give up their pets. Powell et al. (2021) showed that standardized behavioral questionnaires can help identify issues at the time of relinquishment, suggesting that early intervention and targeted support might reduce unnecessary shelter intakes. Similarly, Salman et al. (2000) reported that aggression and other behavioral challenges are frequently cited as reasons for relinquishment, emphasizing that overcoming owner misperceptions about pet behavior may serve as a critical strategy for shelters seeking to reduce returns. Additionally, a systematic content analysis indicated that the reasons behind relinquishment may evolve, particularly during periods of societal stress such as the Covid-19 pandemic (Kisley et al., 2024).

See also: Australia and New Zealand, Pets In; United States, Pets In

References and Further Reading

Hunter, S., and R. A. Brisbin (2016). *Pet Politics: The Political and Legal Lives of Cats, Dogs, and Horses in Canada and the United States*. West Lafayette: Purdue University Press.

Kisley, M., E. Chung, and H. Levitt (2024). "Investigating the Reasons behind Companion Animal Relinquishment: A Systematic Content Analysis of Shelter Records for Cats and Dogs, 2018–2023." Animals, 14 (17): 2606.

Powell, L., C. Houlihan, M. Stone, I. Gitlin, X. Ji, C. L. Reinhard, and B. Watson (2021). "Animal Shelters' Response to the COVID-19 Pandemic: A Pilot Survey of 14 Shelters in the Northeastern United States." Animals, 11 (9): 2669.

Salman, M. D., J. Hutchison, R. Ruch-Gallie, L. Kogan, J. C. New Jr, P. H. Kass, and J. M. Scarlett (2000). "Behavioral Reasons for Relinquishment of Dogs and Cats to 12 Shelters." Journal of Applied Animal Welfare Science, 3 (2): 93–106.

Schor, M. and A. Protopopova (2021). "Effect of COVID-19 on Pet Food Bank Servicing: Quantifying Numbers of Clients Serviced in the Vancouver Downtown Eastside, British Columbia, Canada." Frontiers in Veterinary Science, 8: 730390.

Stull, J. W., A. S. Peregrine, J. M. Sargeant, and J. S. Weese (2013). "Pet Husbandry and Infection Control Practices Related to Zoonotic Disease Risks in Ontario, Canada." *BMC Public Health*, 13: 1–15.

CAPITALISM, PETS AND

Capitalism is an economic system, found today throughout much of the world, in which private parties (rather than the government) own the means of production, and where the goal of the system is to produce profit for those owners. Capitalism is linked to pet keeping in a variety of ways. First, we know that contemporary pet ownership is deeply embedded within broader socioeconomic paradigms and consumer cultures (Beroš and Mandić, 2023). Second, pet keeping began to expand in Western societies only after the development of industrial capitalism and the subsequent rise of a new middle class with the means to own "wasteful" things such as pets. And finally, pets—at least in most countries—are both living beings but also property. As such, they are commoditized, and the production of them is geared toward producing profit rather than, say, the happiest or healthiest pets. This perspective argues that the transformation of the animal-human relationship is not solely a matter of emotional attachment but is also significantly shaped by profit motives and market expansions inherent to capitalist systems (Beroš and Mandić, 2023).

Historically, pet keeping has been associated with middle-class privilege, a shift that Soares (2021) traces back to the late nineteenth and early twentieth centuries, when maintaining pets was less a matter of necessity and more one of social differentiation. This historical perspective provides a baseline for understanding contemporary trends, wherein significant segments of the middle class allocate increased financial resources toward pet care. In recent years, even during economic or public health crises, such as the Covid-19 pandemic, research by Adams et al. (2021) indicates that households with more robust economic resources engage in careful pet care planning, which reflects the complex relationship between financial stability and pet welfare.

Moreover, from a sociological standpoint, analyses by Redmalm (2020) apply Foucauldian frameworks that view pet keeping as a site where disciplinary structures and biopower are at play. This research emphasizes that the domestic practices surrounding pet keeping—ranging from behavioral regulation to the crafting of pet-friendly environments—are also mechanisms through which capitalist values are reproduced within households. The tension between freedom and regulation in pet ownership mirrors the broader dynamics in capitalist societies, where individual choices are simultaneously enabled and constrained by socioeconomic forces. Thus, pet keeping, underpinned by market-driven imperatives, becomes a lens through which one can observe the convergence of commodification, consumer behavior, and regulatory power in modern society (Redmalm, 2020).

Multiple studies have documented that as disposable incomes grow and the middle-class population expands, pet ownership evolves from a purely emotional or utilitarian relationship into a marker of social status and a gateway into a burgeoning consumer market (Halonen and Laihonen, 2019). Specifically, Halonen and Laihonen (2019) argue that pets have come to signify social capital; owning a pet, especially a dog, is often perceived as indicative of upper or middle-class status. This delineation underscores a changing cultural attitude toward animals and aligns pet keeping with modern consumer culture.

The convergence of increasing consumer spending, new business models in pet services, and the symbolic capital of pet ownership strengthens the view that pet keeping is deeply embedded in contemporary consumer culture, where monetary resources and social aspirations are intricately linked (Halonen and Laihonen, 2019).

Overall, research on pets and economic status, coupled with studies on social change, reveals that the expanding middle class plays a pivotal role in reshaping pet ownership. This phenomenon is marked by increased spending on pet care and welfare, evolving norms around pet-related consumer practices, and clear signals of how economic growth and social stratification continue to influence the human–pet relationship.

See also: Communism, Pets And; Elite Pet Keeping; Status Pets; United States, Pets In

References and Further Reading

Adams, B. L., J. W. Applebaum, M. N. Eliasson, S. E. McDonald, and B. A. Zsembik (2021). "Child and Pet Care-Planning During COVID-19: Considerations for the Evolving Family Unit." Family Relations, 70 (3): 705–16.

Beroš, M., and M. Mandić (2023). "Overview of Animal Dehumanization in the Context of Changing Capitalism." *Jahr*, 14 (1): 145–56.

Halonen, M., and P. Laihonen (2019). "From 'No Dogs Here!' to 'Beware of the Dog!': Restricting Dog Signs as a Reflection of Social Norms." Visual Communication, 20 (4): 501–26.

Redmalm, D. (2020). "Discipline and Puppies: The Powers of Pet Keeping." *International Journal of Sociology and Social Policy*, 41 (3/4): 440–54.

Soares, C. (2021). "'The Many Lessons Which the Care of Some Gentle, Loveable Animal Would Give': Animals, Pets, and Emotions in Children's Welfare Institutions, 1870–1920." The History of the Family, 26 (2): 236–65.

CAT SHOWS

Cat shows are competitions whereby cat lovers enter their cats who compete with each other on conformation to breed standards, as well as other characteristics. The current research specifically addressing cat shows is relatively limited; however, insights can be gleaned from studies on feline demographics, human–cat interaction styles, and broader demographic shifts in pet ownership. Although few studies directly focus on the phenomenon of cat shows, researchers have explored related themes that can be informative for understanding the demographics, motivations of participants, types of cats exhibited, and the evolving nature of these events.

The history of the cat fancy and cat shows is rooted in the broader domestication of cats and has evolved rapidly in the modern era as a reflection of both genetic selection and sociocultural transformations. Early domestication of cats, descending from the African wildcat (Felis silvestris), initially served an agricultural purpose before transitioning into symbols of cultural significance. Subsequently, cats became objects of aesthetic refinement and competition O'Brien (2004). This foundational narrative paved the way for the emergence of the cat fancy, which has developed significantly in the past seventy-five years, with over 85 percent of currently recognized cat breeds being established during this period through deliberate selection for aesthetic traits (Kurushima et al., 2012). Such rapid breed development underscores the dynamic interplay between human intervention through breeding practices and the emergent cultural ideals that now define what constitutes a "purebred" cat.

Existing work on feline demographics provides a partial context for cat shows by detailing the profile of cat owners and the characteristics of the cats themselves. For instance, Johnston, Szczepanski, and McDonagh (2017) highlight that a substantial proportion of cat owners in Australia and New Zealand access regular veterinary care and maintain structured lifestyles for their cats. Although this study was not centered on cat shows, its findings suggest that participants in specialized events such as cat shows may represent a segment of the population with a higher attention to feline health, breed standards, and the overall care necessary to compete successfully in a show environment.

Parallel to animal health studies, Finka et al. (2022) examined individual differences in human behavior during interactions with cats and suggested that personal traits and attitudes toward cats may predict how people manage and showcase their pets. These findings imply that motivations for participating in cat shows may extend beyond aesthetic appreciation to include a desire for social recognition and personal fulfillment through the demonstration of "best practice" in cat handling and care. In the context of cat shows, such motivations are likely intertwined with the pursuit of prestige, the desire to promote high breeding standards, and the opportunity to network with other enthusiasts.

Furthermore, broader research on demographic changes provides an overarching framework that can help explain shifts in the profile of cat show participants over time. While the reference by Macera (2025) on demographic changes is not specifically focused on pet ownership or cat shows, it addresses the evolving socioeconomic landscapes which can impact traditional pet ownership patterns. These changes may translate into a growing and more diverse base of cat show participants who are motivated not only by tradition or competitive success but who also view cat showing as a social and cultural activity that reinforces community identity and belonging. Additionally, such shifts might encourage organizers to adapt cat show formats to engage younger and more diverse audiences.

The formation of formal organizations and clubs dedicated to the cat fancy provided the necessary framework for systematic shows and competitions, where stringent breed standards and judging criteria were established (Comben, 1961). These events, originally conceived as leisure activities, have over time adopted characteristics of both a competitive sport and a cultural exhibition. Research by Stone explores how cat shows have grown into complex social events that promote breed excellence while also serving as forums for community building among enthusiasts (Stone, 2022). Cat shows thus represent a dual role: they are simultaneously venues for public display and competition, as well as spaces where participants engage in cultivating shared meanings surrounding aesthetics, pedigree, and even identity within the feline community.

More recent scholarly work reinforces the idea that cat shows continue to evolve in response to changes in human–cat interrelatedness. For instance, Stone's work illustrates how personal and cultural dimensions shape participation in these events, suggesting that the practice is as much about celebrating the human–cat bond as it is about showcasing breeding success (Stone, 2022). Historical analyses, such as that by Lloyd, provide further context by reflecting on how societal changes have influenced the symbolic and practical dimensions of the cat fancy, illustrating a progression from ornamental hobby to a structured, competitive pursuit (Lloyd, 1987).

In summary, while direct research on the specifics of cat shows remains sparse, the available literature on feline demographics, human–cat interactions, and sociodemographic shifts provides important context. It indicates that cat shows are likely influenced by a confluence of factors—ranging from the health practices and care standards of pedigree cats Johnston, Szczepanski, and McDonagh (2017) to the individual behavioral traits and motivations of their owners Finka et al. (2022)—as well as by overarching social and demographic changes. This evolving landscape reflects a dynamic interplay between scientific understanding of feline genetics, the aesthetics of breed standards, and the social practices that underpin the organized display of cats.

See also: Breeds, Popularity Of; Cats; Dog Shows; Fancy Clubs

References and Further Reading

Comben, N. (1961). "The Cat Fancy." *Journal of Small Animal Practice*, 2 (1–4): 48–50.
Finka, L. R., L. Ripari, L. Quinlan, C. Haywood, J. Puzzo, A. Jordan, and M. L. Brennan (2022). "Investigation of Humans' Individual Differences as Predictors of Their Animal Interaction Styles, Focused on the Domestic Cat." *Scientific Reports*, 12 (1): 12128.
Johnston, L., J. Szczepanski, and P. McDonagh (2017). "Demographics, Lifestyle and Veterinary Care of Cats in Australia and New Zealand." *Journal of Feline Medicine and Surgery*, 19 (12): 1199–205.
Kurushima, J., M. Lipinski, B. Gandolfi, L. Froenicke, J. Grahn, R. Grahn, and L. Lyons (2012). "Variation of Cats Under Domestication: Genetic Assignment of Domestic Cats to Breeds and Worldwide Random-Bred Populations." *Animal Genetics*, 44 (3): 311–24.
Lloyd, A. T. (1987). "Cats from History and History from Cats." *Endeavour*, 11 (3): 112–15.
O'Brien, S. (2004). "Cats." *Current Biology*, 14 (23): R988–9.
Stone, E. (2022). *Cat People: Human–Cat Interrelatedness in the Cat Fancy*. London: Routledge.

CATS

Cats are one of the earliest domesticated animals, and, like dogs, represent one of the only domesticated species that were not domesticated for food. In addition, cats, more than any other animal, most likely domesticated themselves, making cats one of the most interesting animals to have been made into a pet.

Historically, the human–cat relationship has transitioned from utilitarian roots—where cats were originally valued for pest control—to an emotionally rich bond embodying companionship and personal identity (Ovenden, Bassett, and Sumner, 2024). This historical shift is not only reflected in changing care practices but has also been reinterpreted as part of broader cultural narratives that have gradually embraced cats as valued family members.

Modern trends in global cat ownership have been closely linked to economic development and expanding disposable incomes. As nations experience rapid urbanization and rising middle-class populations, there has been a notable increase in the number of households that can support pets (Knight, 2023), including, of course, cats. But because of cats' unique history with humans, and their early role as "pest control," they have retained much of their "wildness," which has been reflected in systems of care and control for domestic cats—in particular, the shift in recent years from cats living indoor-outdoor lives with relative independence to living primarily indoors, in a more pampered environment.

The current debate regarding the management of domestic cats revolves around several complex factors, including the status of free-roaming and community cats, concerns about wildlife conservation and public health, and the contrasting welfare outcomes between indoor and outdoor lifestyles. Advocates for indoor-only cat care argue that indoor cats live longer, have better welfare, and do not have a negative impact on wildlife—particularly birds. In contrast, many people continue to argue that cats should live, or at least have access to, the outdoors, as hunting, climbing, interacting with other cats, and establishing a territory are all important aspects of a cat's natural behavior, even when living outdoors results in negative welfare for the cat and an increased impact on the environment and wildlife (Grant and George, 2014; Reed, 2022).

To complicate matters further, studies exploring the welfare and behavioral aspects of cats indicate that indoor confinement can lead to an increase in certain behavioral problems, even as it protects cats from external hazards such as vehicle strikes, disease, and predation (Sandøe et al., 2017). For example, research comparing confined cats with their free-roaming counterparts has found that while indoor cats may experience higher incidences of behavioral issues possibly due to restrictive environments, they are shielded from many external dangers (Sandøe et al., 2017). Conversely, outdoor access allows for the exercise of species-typical behaviors, yet it exposes cats to environmental dangers and risks to local biodiversity (Reed, 2022). This dichotomy underpins much of the current debate, where animal welfare and conservation objectives sometimes appear to be in conflict.

Parallel to these concerns are evolving owner attitudes and public perceptions that increasingly favor cat containment. Recent surveys and behavioral studies indicate that a growing number of cat guardians now prefer to keep their pets indoors or within controlled outdoor boundaries, driven by worries over wildlife predation, public nuisance issues, and overall cat safety (Gemma and McLeod, 2023; Siracusa and Provoost, 2017). However, there remains a contingent of advocates—backed by arguments centered on animal well-being and natural behavioral expression—who oppose strict indoor confinement, contending that providing outdoor access is critical for a cat's psychological health and fulfillment (Abbate, 2019).

Finally, research examining the environmental sustainability of cat ownership highlights an emerging concern: the cumulative impact of feeding and caring for a rising cat population. Environmental studies indicate that pet diets, including those for cats, contribute a measurable share to resource consumption and greenhouse gas emissions, forecasting increased environmental pressures as global pet populations continue to expand (Okin, 2017). These findings present a dual narrative whereby the benefits of companionship and improving animal welfare are accompanied by new challenges in reconciling pet ownership with sustainability goals.

See also: Cat Shows; Fancy Clubs; Free-Roaming Cats

References and Further Reading

Abbate, C. (2019). "A Defense of Free-Roaming Cats From a Hedonist Account of Feline Well-Being." *Acta Analytica*, 35 (3): 439–61.

Gemma, C., and L. McLeod (2023). "Understanding the Factors Influencing Cat Containment: Identifying Opportunities for Behaviour Change." *Animals*, 13 (10): 1630.

Grant, C., and E. George (2014). "Domestic Cats: Management of a Charismatic Non-Native Species." *Proceedings of the Vertebrate Pest Conference*, 26.

Johnston, L., J. Szczepanski, and P. McDonagh (2017). "Demographics, Lifestyle and Veterinary Care of Cats in Australia and New Zealand." *Journal of Feline Medicine and Surgery*, 19 (12): 1199–205.
Knight, A. (2023). "The Relative Benefits for Environmental Sustainability of Vegan Diets for Dogs, Cats and People." *PLOS One*, 18 (10): e0291791.
Lloyd, A. (1987). "Cats from History and History from Cats." *Endeavour*, 11 (3): 112–15.
Lyons, L. (2012). "Variation of Cats Under Domestication: Genetic Assignment of Domestic Cats to Breeds and Worldwide Random-Bred Populations." *Animal Genetics*, 44 (3): 311–24.
Mitsui, K., S. Sato, and Y. Kakuma (2020). "Effects of the Community Cats Program on Population Control, Migration and Welfare Status of Free-Roaming Cats in Tokyo, Japan." *Animals*, 10 (3): 461.
O'Brien, S. (2004). "Cats." *Current Biology*, 14 (23): R988–R989.
Okin, G. S. (2017). "Environmental Impacts of Food Consumption by Dogs and Cats." *PLOS One*, 12 (8): e0181301.
Ovenden, K., I. Bassett, and C. Sumner (2024). "'I Want You to Want Me': How Owners Value Cats' Choices Has Implications for Cat Containment." *People and Nature*, 6 (2): 548–61.
Reed, L. (2022). "The Effects of Free-Roaming Cats on Native Wildlife Populations." *Wildlife Rehabilitation Bulletin*, 40 (1): 17–21.
Sandøe, P., A. Nørspang, B. Forkman, C. Bjørnvad, S. Kondrup, and T. Lund (2017). "The Burden of Domestication: A Representative Study of Welfare in Privately Owned Cats in Denmark." *Animal Welfare*, 26 (1): 1–10.
Siracusa, C., and L. Provoost (2017). "The Advantages and Disadvantages of Confining Cats Indoors." *Cab Reviews Perspectives in Agriculture Veterinary Science Nutrition and Natural Resources*: 1–6.
Stone, E. (2022a). *Cat People: Human–Cat Interrelatedness In the Cat Fancy*. London: Routledge.
Wolf, P., and J. Schaffner (2019). "The Road to TNR: Examining Trap-Neuter-Return Through the Lens of Our Evolving Ethics." *Frontiers in Veterinary Science*, 5: 341.

CEMETERIES, PET

Pet cemeteries, first established in the late nineteenth century, are dedicated tracts of land, privately owned, that offer pet owners a place to both bury their companion animals after death, as well as a place to visit their remains.

According to Stanley Brandes (2009), there are over 600 six hundred pet cemeteries in the United States alone, while according to Ambrós (2010) there are over 900 in Japan. There is even a small but growing movement to allow "whole family cemeteries" in the United States and elsewhere—cemeteries that would allow pets and humans to be buried together in the same grave (something that had, at one time, been practiced as far back as the late Paleolithic).

Historical and archaeological research on ancient pet burials (that predate the development of the pet cemetery) reveals the variety of ways in which deceased animals can appear either as independent burials, as parts of co-burials with humans, or as ritual food or sacrificial offerings. It is often difficult to interpret these archaeological remains and to understand whether the animal buried represents a pet, a ritual object, food, or something else. It was not until the rise of the pet cemetery that pet animals began to receive different treatment from other animals, not just in life, but also in death.

Historically, pet cemeteries emerged alongside modern shifts in mortuary practices and urbanization during the Industrial era, when pets began to be recognized as integral members of the family, worthy of their own dedicated memorial spaces. Early pet cemeteries in post-industrial societies reflected not only the personalization of grief but also the burgeoning economic segmentation within pet culture, where the care, commemoration, and disposal of pet remains became specialized services.

Comparative studies highlight significant regional variations in pet memorialization practices. In Japan, for instance, pet cemeteries are interwoven with Buddhist rituals and have become common features in urban landscapes, where pets are memorialized through cremation and ritualized burials that sometimes include joint human–pet ceremonies (Ambrós, 2010, 2012). These practices illustrate a transformation in how nonhuman animals are viewed in relation to human society—positioned in a liminal space that both honors their emotional significance and negotiates their difference from human kin. Similar trends are observed in other parts of the world; for example, research in post-socialist Estonia documents an evolution toward individualized memorialization within pet cemeteries, reflecting a broader societal trend in the commodification and professionalization of death care services (Kõivupuu, 2017).

Recent quantitative analyses of pet cemetery markers reinforce the notion that the language and symbols used in these spaces have evolved over time, and reflect and are

A historic family pet cemetery located in the gardens of Powerscourt Estate in Enniskerry, County Wicklow, Ireland. (Cshirk2/Dreamstime.com)

drawn from human funerary practices. These inscriptions not only document the farewell to beloved pets but also mirror a cultural shift in which pet death is increasingly ritualized and individualized, marking a transition from a utilitarian approach to animal loss toward one steeped in personal and emotional significance (Auster, Auster-Gussman, and Carlson, 2020). In tandem, broader discussions in the pet death industry emphasize trends toward eco-friendly memorial practices and a diversification of services—which include not only traditional in-ground burials but also creative alternatives such as biodegradable markers—underscoring how consumer expectations and sustainability concerns are reshaping the field (Spiegelman and Kastenbaum, 1990).

Taken together, this body of research shows that pet cemeteries serve as critical sites for the expression of grief, identity, and cultural change. They encapsulate the evolving human–animal bond across time and geographical contexts, illustrating how memorial practices adapt to contemporary social values while simultaneously fostering a specialized, global pet death industry that responds to both economic and ethical challenges.

See also: Grief and Pet Loss; Pet Death Industry

References and Further Reading

Ambrós, B. (2010). "The Necrogeography of Pet Memorial Spaces: Pets as Liminal Family Members in Contemporary Japan." *Material Religion*, 6 (3): 304–35.

Ambrós, B. (2012). "Petto Kuyō: Changing Views of Animal Spirits in Contemporary Japan." In *Handbook of Contemporary Japanese Religions*, 487–507. Leiden and Boston: Brill.

Auster, C., L. Auster-Gussman, and E. Carlson (2020). "Lancaster Pet Cemetery Memorial Plaques 1951–2018: An Analysis of Inscriptions." *Anthrozoös*, 33 (2): 261–83.

Brandes, S. (2009). "The Meaning of American Pet Cemetery Gravestones." *Ethnology: An International Journal of Cultural and Social Anthropology*, 48 (2): 99–118.

Dixon, H. (2018). "Late 1st-Millennium B.C.E. Levantine Dog Burials as an Extension of Human Mortuary Behavior." *Bulletin of the American Schools of Oriental Research*, 379: 19–41.

Laffranchi, Z., S. Zingale, U. Tecchiati, A. Amato, V. Coia, A. Paladin, M. Milella, et al. (2024). "'Until Death Do Us Part'. A Multidisciplinary Study on Human- Animal Co- Burials From the Late Iron Age Necropolis of Seminario Vescovile in Verona (Northern Italy, 3rd-1st c. BCE)." *Plos One*, 19 (2): e0293434.

Pręgowski, M. P. (2016). "Pet Cemeteries in Poland and Beyond: Their Histories, Meanings, and Symbolism." In M. P. Pręgowski and J. Wlodarczyk (eds), *Free Market Dogs: The Human-Canine Bond in Post-Communist Poland* (p. 157). West Lafayette: Purdue University Press.

Spiegelman, V., and R. Kastenbaum (1990). "Pet Rest Memorial: Is Eternity Running Out of Time?" *Omega – Journal of Death and Dying*, 21 (1): 1–13.

Tourigny, E. (2020). "Do All Dogs Go To Heaven? Tracking Human-Animal Relationships Through the Archaeological Survey of Pet Cemeteries." *Antiquity*, 94 (378): 1614–29.

CHILDREN AND PETS

Children have been symbolically associated with nonhuman animals for much of our history. But as animals moved closer and closer to people, as pet keeping expanded, this association became much greater. We can see this in the rise of whole industries catering to children since the nineteenth century: children's literature, toys, children's clothing, and decor, etc. The nineteenth century was the era in which pet keeping expanded into the middle class, and it was also the era when children became an important demographic and started to receive some very limited rights. These two developments crystallized the connection between not just animals, but pets, and human children.

As we have noted, modern pet keeping took off in the nineteenth and twentieth centuries, with the publication of popular magazines and books, especially those aimed at children, full of poetry, advice, fiction, and pictures about pets. According to Katherine Grier, whose work focuses on the history of pet keeping, the turn of the century was also the period in which the commercialization of pet keeping began, as booklets produced by pet supply companies were distributed offering pet care advice.

According to Grier (2010), Victorian parents and moralists essentially invented the child-pet relationship as a means of inculcating middle-class virtues like kindness, gentility, and self-control into their children, and as a way to counteract boys' "natural" tendency toward violence. She wrote, "Pet keeping, an activity long interpreted and tolerated as a personal indulgence, was transformed into a morally purposive act."

Nowhere is this more clear than by looking at children's literature. Much of children's literature derives from folktales and fairy tales, which themselves featured animals, many talking, so it's no surprise to find that children's literature, which often evokes fantasyland settings and creatures, would also heavily feature animals.

Animals are used in children's books in one of two primary ways. They are either stand-ins for humans, representing a number of characteristics that humans have or that the author wants to teach to the reader, or they are animals—simply bears, foxes, or rabbits. In children's literature, animals are far more commonly used as human models, and usually, as substitute children. As such, they are either presented realistically, as animals who can convey love, or fear, or loyalty, or they are fully anthropomorphized, with human clothing, mannerisms, or language. Olivia the pig, the White Rabbit, the Cat in the Hat, Peter Rabbit, and Winnie the Pooh represent the latter: fully realized human characters in animal guise. Examples of the former—realistically portrayed animals who demonstrate important moral qualities—are the two dogs and a cat in *The Incredible Journey* (1960), who taught readers about love and perseverance and how to overcome hardship, and the title character in *Sounder* (1969), a dog who demonstrates loyalty and love, are both examples of the former—realistically portrayed animals who demonstrate important moral qualities. While ancient literature featuring animals—think about Aesop's Fables for example—featured primarily wild animals or working animals, contemporary children's literature almost exclusively includes pets. These are the animals with whom children have relationships, and those relationships are often at the root of such books.

Some books are a blend of the realistic portrayal of animals, with a bit of anthropomorphic portrayal as well. Richard Adams' *Watership Down* (1972), which was based on Ronald Lockley's *The Private Life of the Rabbits* (1976), shows many realities of rabbits' lives, including their social structure, the way they dig warrens, their mating behavior, their fighting, the relationships between rabbits both wild and domestic, and their daily rhythms of feeding, resting, and playing. In addition, the book is also quite explicit about the ways in which rabbits die at human hands—including being shot, gassed inside their warrens, and caught in snares. Rabbits in this book are not a symbol of anything, nor are they simply stand-ins for humans. But the rabbits do talk—a necessary device to give the readers insight into the minds and motivations of the characters.

Some children's book writers use animals as characters because it allows them to show children diversity without relying on ethnic stereotypes. In addition, animal characters can also do things that human characters cannot, because those things would be dangerous or odd. Of course, animal characters do things in children's books that animals cannot do either, but the reader's disbelief is suspended because the character is not human. In fact, many animal characters in children's books behave neither like real children nor like real animals.

Animals might play such an important role in children's literature because children seem to naturally love animals (Melson, 2001; Cole and Stewart, 2016). Children in all cultures are drawn to animals from a very young age, forming attachments to them and making animals central in their lives. This may be a result of the biophilia hypothesis, created by E. O. Wilson (1995), which suggests that humans have an innate interest in watching and trying to understand animals. Children, too, anthropomorphize animals: scholars say that children have not reached the point that so many adults have where the animal and human worlds have become separate; animals are, to many children, playmates, parents, friends, and teachers. As a result, children project "human" characteristics and traits onto animals. And while scientists are just now coming around to the idea that feelings like love, anger, or jealousy are most definitely shared by both humans and many animals, children seem to intuitively believe this, using their own bodily experiences to relate to the experiences of animals—again, primarily pets.

In his classic work on children's fairy tales, *The Uses of Enchantment* (2010), Bruno Bettelheim writes that the line between humans and animals is much less sharply drawn for children than for adults, so the idea that animals can be children or can turn into humans seems quite possible. In addition, children are drawn to animals that seem more like little people: bears, for example, though fierce in real life, with their fat bodies, big heads, short limbs, and upright gait, seem like little people—and thus was born the teddy bear (Geerdts, 2016).

Children relate to animals, and, since at least the Victorian age, adults have understood that they could use the natural affinity between children and animals to teach children valuable social skills (Grier, 2010). Through watching animals, children learn empathy, relationship skills, kindness, and compassion (Ascione, 1992). First, children identify with animals, and from that identification comes empathy. If children empathize with animals, then it's much easier for them to empathize with other humans as well and to learn the difference between treating someone right and treating someone wrong (Beierl, 2008).

In addition, children also use their relationships with animals to help develop their own identities as people. Finally, children use animals as an emotional safety net—retreating to them when feeling sad, or upset, or scared. This is another factor in the link between animal abuse and violence to humans. Because many children develop such close relationships with their pets, if there is abuse in the home, that relationship becomes even more important. Unfortunately, the abuser recognizes this as well, often harming or removing the pet in order to better control the child.

Because of children's connection to real animals, representations of animals are also popular with children, explaining to them why animals so heavily populate their toys, books, clothing, and decor.

If animals are "good for children to think," the corollary is that as children develop the skills that they need to be adults, that they will (or should) no longer need animals at all. As children grow up, they are expected to shrink from animals, and if they don't, they are thought to be immature, to be hanging onto childhood. Many children's books, while emphasizing the closeness between child and animal, end with the child growing up and, sometimes, the animal's death. The most obvious example of this was found in Fred Gipson's *Old Yeller* (1956), which details the love between a boy named Travis and his dog, Yeller. But after Yeller contracts rabies, the boy must shoot him. The book ends with the replacement of Yeller by Yeller's puppy and, as Travis grows up, with the gift of a horse.

Ultimately, animals are useful vehicles for both educating and entertaining children because of their ability to be both like us and not like us. The make-believe world in which fantasy creatures live and animals talk to each other (and to humans) is the ideal world in which to include lessons on friendship, morality, kindness, bravery, or perseverance.

See also: Adults and Pets; Empathy; Gender, Pets And; Literature, Pets In; Socialization, Pets And

References and Further Reading

Adams, R. (1972). *Watership Down*. London: Rex Collings.

Armstrong, W. H. (1969). *Sounder*. New York: Harper & Row.

Ascione, F. R. (1992). "Enhancing Children's Attitudes About the Humane Treatment of Animals: Generalization to Human-Directed Empathy." *Anthrozoös*, 5 (3): 176–91.

Beierl, B. H. (2008). "The Sympathetic Imagination and the Human–Animal Bond: Fostering Empathy Through Reading Imaginative Literature." *Anthrozoös*, 21 (3): 213–20.

Bettelheim, B. (2010). *The Uses of Enchantment: The Meaning and Importance of Fairy Tales*. New York: Vintage.

Burnford, S. E. (1960). *The Incredible Journey*. London: Hodder & Stoughton.

Cole, M., and K. Stewart (2016). *Our Children and Other Animals: The Cultural Construction of Human-Animal Relations in Childhood*. London: Routledge.

Geerdts, M. S. (2016). "(Un)Real Animals: Anthropomorphism and Early Learning About Animals." *Child Development Perspectives*, 10 (1): 10–14.

Gipson, F., and J. Weiss (1956). *Old Yeller*. New York: Harper & Row.

Grier, K. C. (2010). *Pets in America: A History*. Chapel Hill: University of North Carolina Press.

Kellert, S. R., and E. O. Wilson (1995). *The Biophilia Hypothesis*. Washington, DC: Island Press.

Lockley, R. M. (1976). *Private Life of the Rabbit: An Account of the Life History and Social Behaviour of the Wild Rabbit*. London: Corgi.

Melson, G. F. (2001). *Why the Wild Things Are: Animals in the Lives of Children*. Cambridge, MA: Harvard University Press.

CHINA, PETS IN

Research on pet keeping in China indicates that the country's companion animal culture has undergone a marked transformation over recent decades, driven by rapid urbanization, economic growth, and shifting cultural values. Historical perspectives reveal that traditional Chinese attitudes toward animals, where they were often kept for utilitarian or symbolic purposes, have gradually evolved into a modern pet culture that emphasizes companionship and quality of life. Recent analyses highlight that China's pet keeping trends mirror broader societal changes, where rising disposable incomes and an expanding middle class have fostered a more consumer-oriented and emotionally grounded approach to pet ownership.

Contemporary research demonstrates that, in the post-Covid context, attitudes toward pets in China have undergone further change. The isolation and social challenges imposed by the pandemic have led to increased demand for animal companionship, as individuals have increasingly turned to pets for emotional support and well-being. In this scenario, pet keeping is not only seen as a marker of a modern urban lifestyle but also as an essential component of personal and social welfare. Studies focusing on the impact of pet companionship on personal development have shown that such relationships contribute positively to mental health and social outcomes, particularly among younger demographics—for example, among students Li et al. (2023). This dimension of pet keeping underscores the cultural shift toward viewing pets as integral family members and supportive agents in times of stress and social disruption.

It is also clear that pet keeping's expansion in China has much to do with the country's economic changes over the past few decades. Pets, like other things associated with affluence, are on the rise in China because they demonstrate, in part, that China, and by extension, Chinese citizens, have reached a similar level of consumption practices to Westerners, with pets carrying an extra level of status.

See also: Communism, Pets And; Japan, Pets In

References and Further Reading

Cheang, S. (2006). "Women, Pets, and Imperialism: The British Pekingese Dog and Nostalgia for Old China." *Journal of British Studies*, 45 (2): 359–87.

Gong, X., H. Qi, J. Xie, and N. Qi (2020). "From Livestock to Families: Taking Pets to Tourism Activities in China." *Journal of Tourism and Cultural Change*, 18 (6): 728–42.

Huang, C., Y. Hu, W. Liu, Z. Zou, and J. Sundell (2013). "Pet-Keeping and its Impact on Asthma and Allergies Among Preschool Children in Shanghai, China." *Chinese Science Bulletin*, 58: 4203–10.

Li, P., J. Yao, Y. Xu, and F. Zhou (2023). "Impact of Pet Companionship on Student Development: A Meta-Analysis." *Best Evidence of Chinese Education*, 14 (1): 1727–43.

CLASSROOM PETS

Classroom pets are animals—usually "pocket pets" like guinea pigs, hamsters, or rabbits—kept in the classroom as part of a teacher's educational curriculum. Sometimes those animals live full time in the classroom, only going home (with either the teacher or a student) on weekends and holidays, and at other times, the classroom pet is really the pet of the teacher, who travels back and forth to school each day. Current research on pets in education—and classroom pets in particular—reveals a multifaceted picture that spans educational benefits, student socialization and behavior, as well as animal welfare and ethical controversies. In educational settings, classroom pets are employed as pedagogical tools to foster empathy, responsibility, and social-emotional learning. Several studies (Williams et al., 2021; Hawkins et al., 2017) have demonstrated that school-based interventions involving animals can enhance children's understanding of animal welfare and promote positive attitudes toward nonhuman animals. These programs, designed as targeted educational activities, contribute to children's knowledge acquisition and can potentially lead to lasting changes in how young learners interact with animals, thereby fostering compassion and responsible pet ownership.

However, these benefits are counterbalanced by potential challenges and controversies. Research by Chiu et al. (2013) provides evidence that the presence of a classroom pet may inadvertently support a "teacher's pet" phenomenon among students, where certain children develop preferential relationships with the animal. This dynamic can contribute to group divisions, with students who do not develop a close bond with the classroom animal feeling alienated or less favored. Such findings raise concerns about the social implications of classroom pet integration and underscore the need for teachers to manage these interactions carefully to ensure inclusive classroom dynamics.

From the perspective of animal welfare, there is ongoing debate about the suitability of the traditional classroom environment for living animals. Concerns include the possibility that high noise levels, constant handling, and an unpredictable routine might contribute

Schoolchildren playing with a pet rabbit in the classroom. The use of pets in a school setting can bring both benefits and challenges. (Shannon Fagan/Dreamstime.com)

to stress or compromise the health of the animal. Ensuring that classroom pets receive proper care, rest, and enrichment is critical, and researchers argue that the responsibilities of maintaining these welfare standards should be clearly defined as part of the program design (Williams et al., 2021; Hawkins et al., 2017). Moreover, the ethical use of animals as educational tools invites controversy related to the potential for long-term negative impacts on the animal's quality of life if the classroom setting does not meet their behavioral and physical needs.

While classroom pets offer a promising avenue for enhancing educational outcomes—especially in terms of socialization and empathy building among young learners—the research simultaneously highlights risks associated with group dynamics and animal welfare. Addressing these concerns requires the development of best practices and clear protocols to manage both human and animal needs, ensuring that the benefits of classroom pet programs do not come at the expense of social inequity or compromised welfare standards.

See also: Children and Pets; Pocket Pets; Socialization, Pets And

References and Further Reading

Chiu, S. I., J. Lee, and T. Liang (2013). "Does the Teacher's Pet Phenomenon Inevitably Cause Classroom Conflict? Comparative Viewpoints of Three Pet-Student Groups." *School Psychology International*, 34 (1): 3–16.

Gee, N. R., A. H. Fine, and P. McCardle, eds. (2017). *How Animals Help Students Learn: Research and Practice for Educators and Mental-Health Professionals.* New York: Routledge.

Gee, N. R., J. M. Rawlings, M. E. O'Haire, P. C. Bennett, D. Snellgrove, and J. M. Peralta (2017). "Caring for Classroom Pets." In N. R. Gee, A. H. Fine, and P. McCardle (eds), *How Animals Help Students Learn*, 197–211. New York: Routledge.

Hawkins, R. D., J. M. Williams, and Scottish Society for the Prevention of Cruelty to Animals (Scottish SPCA) (2017). "Childhood Attachment to Pets: Associations Between Pet Attachment, Attitudes to Animals, Compassion, and Humane Behaviour." *International Journal of Environmental Research and Public Health*, 14 (5): 490.

McCullough, A., A. Ruehrdanz, R. Garthe, C. Hellman, and M. O'Haire (2021). "Measuring the Social, Behavioral, and Academic Effects of Classroom Pets on Third and Fourth-Grade Students." *Human-Animal Interaction Bulletin*, 9: 1–21.

Meadan, H., and B. Jegatheesan (2010). "Classroom Pets and Young Children." *YC Young Children*, 65 (3): 70.

Williams, J., M. Cardoso, S. Zumaglini, A. Finney, and M. Knoll (2021). "'Rabbit Rescuers': A School-Based Animal Welfare Education Intervention for Young Children." *Anthrozoös*, 35 (1): 55–73.

COMMUNICATION, HUMAN-PET

Human-pet communication refers to the ways in which humans communicate with nonhuman animals, as well as the study of such communication. The research on human communication with companion animals spans multiple disciplines, integrating cognitive science, developmental psychology, and sociocultural analysis to explore the nuanced dynamics of interspecies interaction.

One of the most important criteria for being a pet is having a name, because having a name symbolically and literally incorporates that animal into the human domestic sphere. But with respect to communication, having a name also allows for human–animal communication: we can talk to animals, even if they cannot (yet) respond to us in kind. While nonhuman animals do not possess human verbal language, we can and do still talk to them, and many companion animals understand much of what we say, based on our tone, inflection, body language, and facial expressions.

Sociologists who have studied human–animal communication have shown that, like baby talk, human-pet communication has a clear structure and, as well, a distinctive tone, set of bodily gestures, and comportment. Beyond immediate communication, this talk serves as a sort of glue in human–animal relationships and, more broadly, enhances the social lubricative function of companion animals in human-human relationships. Sociologist Clinton Sanders (1999) has studied the communication between humans and dogs and maintains that language enables human and canine interactants to construct and share a mutually defined reality. Animals, because they lack human language, are normally excluded from social exchange with humans, but in the domestic realm, guardians of pets have made a number of allowances for that lack of language. Ask any caretaker, and they will not only admit that they talk to their animals, but they will maintain that their animals understand what they are saying. In addition, we speak "*for*" our animals—to friends, to family, to the veterinarian. We also speak *through* them; sometimes people use their pet dog or cat as a sort of mediator to communicate information to another person. Pet owners see cross-species communication as real and possible, and this possibility itself allows for that communication and for the reciprocal relationships that we have developed with our companions.

By opening up the door to cross-species communication, and by including (some) animals in our own worlds, we humanize those animals, giving them a "person-like" status. Sanders, for example, describes a form of social exchange involving his dog asking (through body language or barking) to be let out, and Sanders acquiescing and letting him out. Both partners played an equal role in that exchange, and both were able to anticipate and acknowledge the needs and interests of the other.

Another area of study looks at how and whether communication is related to the level of attachment. Castillo and Chen (2022) found that attachment to companion animals among future human services professionals reflects broader socioemotional frameworks, indicating that human–animal communication is embedded in cultural practices of care and empathy.

Another research stream has explored the specificity of interspecies communication behaviors. Nawroth (2017) demonstrated that even species traditionally not regarded as companion animals, such as goats, exhibit intentional, audience-dependent gazing behavior similar to that observed in dogs and horses. This finding suggests that domestication processes have broadly influenced communication strategies across species, facilitating complex and meaningful exchanges between humans and animals. In parallel, Saito et al. (2019) examined the communicative differences between cats and dogs, revealing that despite their relatively independent nature, cats have developed unique communicative mechanisms that allow them to engage effectively with humans, albeit in ways that differ in frequency and style from the more overt social behaviors of dogs.

López-Cepero and Español (2025) argues that the richness of these relationships, whether characterized as friendship or mutual caregiving, underscores the importance of viewing such bonds through multiple conceptual lenses—from the guardian-dependent model to the companion friend model. This narrative is echoed in multispecies family contexts, where Solhjoo (2024) details how communication processes contribute to the shared knowledge and emotional lives of human–animal households through an information experience approach. In addition, research such as Melson's (2003) review on child development emphasizes that early interactions with companion animals foster communication skills and social-emotional growth, further demonstrating how these bonds evolve and enrich human development.

Research into how humans communicate with their pets has increasingly focused on the development and application of augmentative interspecies communication (AIC) devices to supplement traditional verbal and nonverbal cues. A central theme in this body of work is the use of button-based soundboards, which allow humans to "speak" to their pets through pre-recorded words or phrases that are activated by the press of a button. Studies have shown that dogs, such as Bunny, a standard poodle with a massive social media following, can learn to associate these button presses with specific meanings and can even coordinate their own button-pressing behavior in cooperative food-reward tasks McGetrick et al. (2024).

In addition to facilitating one-way communication, these devices also afford bidirectional interactions. The use of soundboard systems, reviewed by Smith et al. (Smith et al., 2023), illustrates that through repeated training, dogs can learn to respond to button-triggered vocalizations and, in some cases, initiate communication using similar devices. Although most research has focused on dogs, these findings have sparked interest in exploring whether similar approaches can be applied to other companion animals, including cats, even though studies specifically addressing cats with AIC devices remain sparse—even as more cats using buttons appear on social media.

See also: Anthropomorphism; Attachment; Bond, Human–Animal

References and Further Reading

Castillo, Y. A., and R. Chen (2022). "Assessing Companion Animal Attachment Among Future Latino Human Services Professionals." *Journal of Social, Behavioral, and Health Sciences*, 16 (1): 178–94.

López-Cepero, J., and A. Español (2025). "Negotiating Relationships Within Multispecies Families: Analysing Dilemmas on Humanisation of the Bond." *Human Arenas*. https://doi.org/10.1007/s42087-025-00504-y

McGetrick, J., L. Fux, J. Schullern-Schrattenhofen, J. L. Rault, and F. Range (2024). "Do Pet Dogs Reciprocate the Receipt of Food From Familiar and Unfamiliar Conspecifics?" *Ethology*, 130 (4): e13430.

Melson, G. F. (2003). "Child Development and the Human–Companion Animal Bond." *American Behavioral Scientist*, 47 (1): 31–9.

Nawroth, C. (2017). "Invited Review: Socio-Cognitive Capacities of Goats and Their Impact on Human–Animal Interactions." *Small Ruminant Research*, 150: 70–5.

Saito, A., K. Shinozuka, Y. Ito, and T. Hasegawa (2019). "Domestic Cats (*Felis catus*) Discriminate Their Names from Other Words." *Scientific Reports*, 9 (1): 5394.
Sanders, C. (1999). *Understanding Dogs*. Philadelphia, PA: Temple University Press.
Smith, G., A. Bastos, A. Evenson, L. Trottier, and F. Rossano (2023). "Use of Augmentative Interspecies Communication Devices in Animal Language Studies: A Review." *Wiley Interdisciplinary Reviews Cognitive Science*, 14 (4): e1647.
Solhjoo, N. (2024). "Knowing Within Multispecies Families: An Information Experience Study." *Journal of Information Science*. https://doi.org/10.1177/01655515241268845

COMMUNISM, PETS AND

Communism is an economic system whereby the people (either directly or via the government) own the means of production. Today, only five nations currently can be described as Communist (China, Cuba, Laos, Vietnam, and North Korea), with post-Soviet Russia no longer seen as a Communist country, even though its 1917 revolution was the first communist revolution in the world. Communism is historically antithetical to pet keeping, as it was, and still is to some extent, seen as a symbol of the wastefulness of the bourgeoisie.

Research on pet keeping in communist countries reveals a complicated landscape in which official ideology, state policies, and everyday practices intersected to shape the human–animal relationship. Under communist regimes, pet keeping was viewed negatively: official narratives emphasized utilitarian and collective values over what was seen as bourgeois individual indulgence, while many citizens maintained informal practices of caring for companion animals despite limited state support.

In Russia, for example, street dogs have historically been a visible component of urban environments under communism. These animals often navigated unregulated public spaces amid constrained municipal services such as the Moscow subway, reflecting both neglect and the resilience of community responses in caring for stray animals. The presence of street dogs was not only a marker of challenges in public infrastructure and urban management but also highlighted informal networks of care that emerged in response to state shortcomings, providing rudimentary forms of animal welfare, even as official policy largely overlooked the issue.

Following the collapse of the Iron Curtain, Eastern European societies that were once controlled by the Soviet Union under communism experienced rapid social and economic transformations that profoundly affected pet ownership. The transition from centrally planned to market economies and the rising influence of Western consumer culture contributed to a marked increase in pet keeping among emerging middle classes (Christiaens, Mark, and Jarillo, 2017). New economic opportunities and greater disposable incomes facilitated a shift in attitudes: pets increasingly came to be seen as family members and symbols of modernity. In this perspective, the rise in pet ownership after 1989 in Eastern Europe can be understood as a reaction against previous constraints—a reassertion of individual agency and an embrace of practices once considered luxuries. Moreover, the increasing professionalization of pet-related services and the expansion of pet industries in these regions further attest to the rapid cultural shift that has redefined the role of companion animals in post-communist societies. This research suggests that the rise in pet ownership can be seen as both a symbolic and practical adaptation to newfound socioeconomic opportunities and the subsequent redefinition of personal and public spaces—a cultural shift that mirrors trends observed in Western societies.

See also: Capitalism, Pets And; Europe, Pets In; Russia, Pets In

References and Further Reading

Applebaum, A. (2013). *Iron Curtain: The Crushing of Eastern Europe, 1944–1956. Choice Reviews Online*, 50 (11): 50-6346-50–6346.

Cherkaev, X. (2024). "Russian Late-Liberalism: Notes Hierarchy, Responsibility, and Legally Ownerless Dogs." *General Anthropology*, 31 (1–2): 19–22.

Christiaens, K., J. Mark, and J. Jarillo (2017). "Entangled Transitions: Eastern and Southern European Convergence or Alternative Europes? 1960s–2000s." *Contemporary European History*, 26 (4): 577–99.

Hoffmann, E. (1972). "Social Science and Soviet Administrative Behavior." *World Politics*, 24 (3): 444–71.

Keen, M., and J. Mucha (2004). "Sociology in Central and Eastern Europe in the 1990s: A Decade of Reconstruction." *European Societies*, 6 (2): 123–47.

Zhulenko, A. S., and G. V. Polynova (2016). "Controlling Street Dog Population in Moscow." *Вестник Российского университета дружбы народов. Серия: Экология и безопасность жизнедеятельности*, (4): 7–17.

COMMUNITY, PETS AND

The sociological literature on pets and community examines how companion animals—most notably dogs—serve as social actors that mediate community interactions, influence perceptions of neighborhood safety, and reflect broader issues related to gentrification and racial/ethnic dynamics.

Researchers have found that activities such as dog walking and visits to dog parks do more than simply provide exercise for pets; they help build social capital among residents. For example, Wood et al. (2017) demonstrate that dog walking in public open spaces facilitates spontaneous social interactions between pet owners and contributes to perceptions of community friendliness and collective surveillance, which can enhance neighborhood safety. This everyday practice is viewed as a form of informal social networking that can bridge social divides, although the benefits are not distributed equally across different sociodemographic groups.

At the same time, sociologists are increasingly attentive to the role pets play in broader processes of urban transformation and social stratification. Žažar and Clausen (2024) argue that pets perform multifaceted social functions in contemporary societies, acting as both mediators of human interaction and as symbols in contested spaces. In rapidly gentrifying neighborhoods, the proliferation of "pet-friendly" amenities such as upscale dog parks and relaxed leash laws can signal shifting social norms and market values that sometimes lead to the marginalization of long-term residents from lower-income or minority backgrounds. These public spaces can also become arenas for racial and ethnic conflict, where differing cultural expectations around pet management and use of communal areas highlight and, at times, exacerbate existing social boundaries.

Complementing these perspectives, Christian et al. (2016) provide quantitative evidence that dog walking is positively associated with owners' perceptions of neighborhood safety. Their findings suggest that by transforming public spaces into sites for casual interaction and mutual monitoring, pets help create an environment where residents feel more secure and connected. However, the same processes may also reinforce social inequalities if the benefits of increased safety and community interaction are primarily accessible to individuals from more affluent or racially dominant groups. Leash laws and regulations around dog parks, while seemingly neutral, are often enforced in ways that reflect and reproduce these underlying social tensions.

Together, this body of literature reveals that pets are integral to understanding contemporary community dynamics. The everyday practices surrounding pet ownership and the use of communal spaces not only contribute to enhanced social capital and perceptions of safety but also intersect with processes of gentrification and racial/ethnic conflict, reflecting the complex, layered nature of urban modernity.

See also: Ethnicity, Pets And; Social Capital, Pets And; Socioeconomic Class, Pets And

References and Further Reading

Christian, H., L. Wood, A. Nathan, I. Kawachi, S. Houghton, K. Martin, and S. McCune (2016). "The Association Between Dog Walking, Physical Activity and Owner's Perceptions of Safety: Cross-Sectional Evidence from the US and Australia." *BMC Public Health*, 16 (1): 1010.

Wood, L., K. Martin, H. Christian, S. Houghton, I. Kawachi, S. Vallesi, and S. McCune (2017). "Social Capital and Pet Ownership – A Tale of Four Cities." *SSM – Population Health*, 3: 442–7.

Žažar, K., and L. Clausen (2024). "Towards the Sociology of Pets—Social Functions of Animals in Contemporary Societies Observed from Systems Theoretical Perspective." *Behavioral Science*, 42 (2): 434–46.

COVID-19

Covid-19 is a viral disease that, starting in early 2020, caused a global pandemic, influencing every country on the planet and changing our lives for, most likely, decades to come. The pandemic's impact extended to our relationship with pets, demonstrating, yet again, the ways in which pets are integrated into every facet of our lives today.

The research on pets during the Covid-19 pandemic reveals multiple impacts on pet adoption patterns, pet behavior and socialization, as well as fears related to zoonotic transmission of Covid as well as other diseases.

Early in the pandemic, social isolation and the accompanying psychological stress led many individuals and families to adopt companion animals. King et al. (2024) reported that families with children experienced unique pressures—balancing remote schooling and childcare—that influenced patterns of both pet adoption and relinquishment. This trend was echoed globally, as Protopopova (2021) documented surprising increases in pet adoptions from shelters, attributing these trends to the heightened need for companionship and emotional support during periods of severe social distancing.

Simultaneously, research has shown that increased engagement with pets helped mitigate the emotional distress associated with the pandemic. Gradidge et al. (2024) demonstrated a "pet effect" whereby enhanced play and interaction contributed to improved subjective well-being and reduced anxiety among pet owners. However, the extended period of isolation also resulted in challenges. Bennetts and Crawford (2022) described how disruptions in routine and reduced socialization opportunities led to behavioral issues, such as separation anxiety in dogs, when owners eventually returned to work. This reflects a broader pattern: while pets provided comfort during lockdown, their altered behavioral habits sometimes required additional training and adaptation as the restrictions eased.

In parallel with these changes in adoption and behavior, there was a notable public fear regarding the transmission of Covid-19 via pets. Although the scientific consensus has largely dispelled significant evidence for pets acting as major vectors, instances were reported wherein pet abandonment, or even the killing of pets, was spurred by such fears (Sun et al. 2020).

As the initial surge in pet adoption began to level off with the rollout of Covid vaccines and the gradual return to pre-pandemic routines, such as going back to the workplace, animal advocates raised concerns about the long-term fate of these newly adopted animals. Ho, Hussain, and Sparagano (2021) warned that despite early reports indicating a high retention rate for these animals, the transition back to normalcy might trigger behavioral challenges and possibly increased rates of pet relinquishment if owners were unprepared for the post-lockdown realities.

Together, this body of research underscores that the pandemic not only spurred a dramatic increase in pet adoption driven by a need for emotional support but also brought to the fore challenges in pet socialization and behavior management as lifestyles changed. Moreover, lingering fears around zoonotic transmission—despite limited scientific backing—highlight the importance of clear public health communication to protect both animal welfare and public sentiment during global crises.

See also: China, Pets In; Human Health, Pets And; Zoonoses

References and Further Reading

Adams, B., J. Applebaum, M. Eliasson, S. McDonald, and B. Zsembik (2021). "Child and Pet Care-Planning During COVID-19: Considerations For the Evolving Family Unit." *Family Relations*, 70 (3): 705–16.

Bennetts, S. K., S. B. Crawford, T. Howell, B. Ignacio, F. Burgemeister, K. Burke, and J. M. Nicholson (2022). "Companionship and Worries in Uncertain Times: Australian Parents' Experiences of Children and Pets During COVID-19." *Anthrozoös*, 35 (6): 833–46.

Carroll, G., A. Torjussen, and C. Reeve (2023). "Companion Animal Adoption and Relinquishment During the COVID-19 Pandemic: Households with Children at Greatest Risk of Relinquishing a Cat or Dog." *Animal Welfare*, 32: e56.

Gradidge, S., S. Loughnan, and N. Gibson (2024). "Exploring the 'Pet Effect': Does Playing With Pets Contribute to Owner Wellbeing?" *Pets*, 1 (3): 328–39.
Ho, J., S. Hussain, and O. Sparagano (2021). "Did the COVID-19 Pandemic Spark a Public Interest in Pet Adoption?" *Frontiers in Veterinary Science*, 8: 647308.
King, E., S. Dowling-Guyer, E. McCobb, and M. Mueller (2024). "Longitudinal Patterns of Companion Animals in Families with Children During the COVID-19 Pandemic: Findings From the Adolescent Brain Cognitive Development (ABCD) Study®." *Frontiers in Veterinary Science*, 11: 1364718.
Protopopova, A. (2021). "10 The Impact of Global Disasters on Our Pets: Lessons from COVID-19." *Journal of Animal Science*, 99 (Supplement_3): 5–6.
Sun, A., Y. Zhao, G. Zhuang, and G. Zhang (2020). "Pets Are Irresponsible for COVID-19." *Authorea Preprints*.

CRUELTY

Psychologist Frank Ascione characterizes animal cruelty as "socially unacceptable behavior that deliberately inflicts unnecessary pain, suffering, or distress on, or causes the death of, an animal" (1993: 228). Importantly, it is not the act itself that is deemed cruel; rather, it is determined by its classification as "socially unacceptable" or "unnecessary." In the context of the United States, animal cruelty is not a fixed phenomenon; it is situational, influenced by historical contexts, and constrained by cultural, economic, and political considerations. For example, while raising, killing, and consuming dogs is entirely unacceptable in North America and Europe, these practices are permitted and legal in countries such as China and Korea. In many regions globally, companion animals are considered members of the family, and individuals would go to great lengths to safeguard them. Nevertheless, millions of cats and dogs, including those in nations where pets are celebrated, are less fortunate, and incidents of cruelty and neglect frequently appear in media outlets. Although some of these notable cases draw public attention to a grave issue, a significant majority of instances of animal cruelty go unnoticed. In the United States, the predominant reports of abuse involve companion animals, yet there is a lack of reliable statistics documenting the prevalence of animal cruelty and neglect. A study conducted by the Humane Society of the United States (HSUS) in 2003 (HSUS First Strike Campaign 2003 Report of Animal Cruelty Cases) indicated that 57 percent of documented cases comprised intentional cruelty, while 43 percent involved severe neglect. The analysis demonstrated that the most frequently reported forms of cruelty included shooting, animal fighting, torture, and beating. Among the neglected animals, 70 percent suffered from malnourishment, and 30 percent endured starvation, with others experiencing drowning, stabbing, or even being set on fire. However, these sensational cases reflect only a small fraction of the millions of animal cruelty and neglect instances occurring annually. Due to the limited tracking of these figures in various countries, ascertaining the total number of affected animals remains challenging. A 2016 study found that approximately 55,000 to 60,000 cases of animal cruelty are reported annually to the RSPCA in Australia. However, the authors cautioned that these figures are likely an underrepresentation given the infrequency with which cruelty incidents are reported (Alberthsen et al., 2016). Animal cruelty occurs universally across both urban and rural settings and transcends socioeconomic and racial/ethnic lines. The HSUS indicates that dogs are unequivocally the most frequent victims of animal cruelty, comprising over 60 percent of all cruelty cases reported in the United States, with cats following closely behind. Notably, pit bull-type dogs represent an increasingly significant portion of victimized animals (Iliopoulou, Carleton, and Reese, 2019). Cases of livestock abuse typically surface only through undercover investigations, creating significant difficulties in gauging the extent of abuse occurring beyond the routine maltreatment observed in stockyards and slaughterhouses. Indeed, numerous animal cruelty statutes globally often exempt livestock or recognized agricultural practices, and even where existing laws are robust, persuading law enforcement to make arrests or seize farm animals subjected to neglect or abuse can prove challenging. Nonetheless, arrests do occasionally occur. In May 2010, the animal rights organization Mercy for Animals released an undercover video depicting employees at Conklin Dairy Farms in Plain City, Ohio, engaging in acts of violence against dairy cows and their calves, including kicking, stomping, stabbing, and beating. Following the video's release, a pertinent question arose: does this represent "deviant" animal cruelty, or is such abuse a widespread issue within the industry? Conklin Dairy responded promptly, asserting that the footage merely showcased a few "bad apples" within the industry and that the behaviors of these individuals should not be attributed to the sector as a whole. In response to the incident, the featured employee was terminated, and retraining was mandated for the rest of the staff, as per the company's statement. In 2017, one of the most significant animal cruelty cases related to farmed animals led Massachusetts authorities to charge the owner of a tenant farm and twenty-five tenant farmers with 151 counts of animal cruelty after discovering hundreds of animals enduring appalling conditions. The farmers pleaded guilty and were sentenced to community service. Nonetheless, the majority of conduct classified as animal cruelty does not occur on farms or involve livestock, regardless of the severity of many farming practices. Instead, such actions predominantly take place behind closed doors, within private residences. In the United States, companion animals not only constitute the type of animals most frequently subjected to human violence, but they also hold a unique legal status, at least to some extent, as they are the only category of animals receiving protection from such violence. Animal cruelty is not comprehensively addressed in most federal laws; rather, it is determined by state regulations that define and stipulate penalties for cruelty, generally excluding all animals except for pets and a small selection of working animals, primarily dogs. Given the predominantly private nature of many of these violent acts, the majority of cruelty incidents remain concealed from public view, are seldom reported, and consequently go unprosecuted. This circumstance highlights a significant challenge in the ongoing battle against animal cruelty, illustrating that even in countries that appear to have well-established animal protection legislation, effective legal recourse is contingent upon various factors being adequately addressed to secure the welfare of the animals in question.

See also: Gender, Pets And; Link between Animal and Human Abuse

References and Further Reading

ArlUKe, A. (2002). "Animal Abuse as Dirty Play." *Symbolic Interaction*, 25: 405–30.
Ascione, F. R., S. E. McDonald, P. Tedeschi, and J. H. Williams (2018). "The Relations Among Animal Abuse, Psychological Disorders, and Crime: Implications for Forensic Assessment." *Behavioral Sciences & the Law*, 36 (6): 717–29.
Bailey, S. K., V. K. Sims, and M. G. Chin (2016). "Predictors of Views About Punishing Animal Abuse." *Anthrozoös*, 29 (1): 21–33.
Beirne, P. (2009). *Confronting Animal Abuse—Law, Criminology, and Human-Animal Relationships*. Lanham, MD: Rowman & Littlefield.
Bonnar-Kidd, K. K. (2010). "Sexual Offender Laws and Prevention of Sexual Violence or Recidivism." *American Journal of Public Health*, 100 (3): 412–19.
Brewster, M. P., and C. Reyes (2012). *Animal Cruelty: A Multidisciplinary Approach to Understanding*. Durham, NC: Carolina Academic Press.
Carlisle-Frank, P., and T. Flanagan (2006). *Silent Victims: Recognizing and Stopping Abuse of the Family Pet*. Lanham, MD: University Press of America.
Iliopoulou, M. A., C. L. Carleton, and L. A. Reese (2019). "Beloved Companion or Problem Animal: The Shifting Meaning of Pit Bull." *Society & Animals*, 27 (3): 327–46.
Ison, J. (2019). "Animal Abuse and Advocating for the Carceral: Critiquing Animal Abuse Registries." *Animal Studies Journal*, 8 (2): 55–80.
Levitt, L., G. Patronek, and T. Grisso, eds. (2015). *Animal Maltreatment: Forensic Mental Health Issues and Evaluations*. Oxford: Oxford University Press.
Maher, J., H. Pierpoint, and P. Beirne (2017). *Palgrave International Handbook of Animal Abuse Studies*. New York: Palgrave Macmillan.
Marceau, J. (2019). *Beyond Cages: Animal Law and Criminal Punishment*. Cambridge: Cambridge University Press.
Richardson, D. A. (2017). *Veterinarians and Their Perception of the Treatment of Animal Abuse Cases in the Criminal Justice System*. Doctoral dissertation, Illinois State University, Normal, IL.

D

DIPLOMACY AND PETS

Diplomacy refers to the work of engaging in international relations. The emerging scholarly work on animals, pets, and international diplomacy positions nonhuman animals not as mere curiosities but as active agents and symbols. Research in this area has sought to reconceptualize diplomacy by incorporating what Leira and Neumann (2017) refer to as "beastly diplomacy." This perspective underscores that animals matter not only for their corporeal presence but for the ways in which they represent nation-states and function as diplomatic subjects, as well as for the ways they are mobilized as tools in diplomatic practice. In this view, animals have been used to convey friendly intentions, construct national images, and facilitate dialogue far beyond the traditional, human-focused concept of diplomacy.

A particularly celebrated example of animal diplomacy is seen in the practice of "Panda Diplomacy," as analyzed by Hartig (2013). Here, the Chinese use of the giant panda as a diplomatic gift and symbol exemplifies how nation-states leverage culturally appealing animals to project "soft power" and foster bilateral ties. This practice not only engenders goodwill among recipient nations but also reinforces China's distinct national identity in the realm of public diplomacy. Similarly, Cushing and Markwell's study of "Platypus Diplomacy" (2009) illustrates how other nations have used native animals as diplomatic gifts to communicate cultural uniqueness and to negotiate mutually beneficial relationships. Both cases underscore that animal diplomacy provides a tangible and visually compelling tool for states to express political intentions and cultivate transnational connections.

Animals—whether as pets or state symbols—play an instrumental role in today's diplomatic practices. They help bridge cultural and political divides by offering nonverbal, affective communication channels that can complement formal negotiation processes and foster goodwill. This reconceptualization of diplomacy not only enriches the theoretical landscape of international relations but also calls for more empirical investigations into the ways nonhuman actors are intertwined with global politics.

See also: Presidential Pets; Status Pets

References and Further Reading

Cushing, N., and K. Markwell (2009). "Platypus Diplomacy: Animal Gifts in International Relations." *Journal of Australian Studies*, 33 (3): 255–71.

Hartig, F. (2013). "Panda diplomacy: The Cutest Part of China's Public Diplomacy." *The Hague Journal of Diplomacy*, 8 (1): 49–78.

Leira, Halvard, and Iver B. Neumann (2017). "Beastly diplomacy." *The Hague Journal of Diplomacy* 12(4): 337–59.

DISABILITY, PETS AND

Research on the role of pets in the lives of individuals with disabilities highlights a dual pathway: on one hand, companion animals provide significant emotional support that enhances life satisfaction, and on the other, service animals perform practical functions to assist with daily tasks and emergency situations.

From the perspective of emotional and psychological well-being, studies have shown that pets offer a unique form of social and emotional support that can buffer against the negative effects of disability. For example, Haron et al. (2018) demonstrated that emotional support significantly moderates the relationship between disability and overall life satisfaction among older adults. In such contexts, the presence of a supportive pet can boost self-efficacy in health-related activities, reduce depression, and promote better overall well-being. Essentially, by fostering a sense of companionship and facilitating healthy self-management behaviors, pets help mitigate some of the psychological distress that often accompanies physical impairment.

In addition to offering companionship, trained service animals have a proven record of providing essential practical assistance to people with disabilities. Dalziel et al. (2003) reviewed the use of seizure-alert dogs and highlighted how these dogs not only warn their owners about impending health crises but also assist with other daily tasks—ranging from guiding individuals with visual impairments to aiding mobility in those with various physical disabilities. Such trained animals provide a level of independence to their handlers that might otherwise be compromised because of physical limitations.

The existing research underscores a twofold benefit of pet ownership for people with disabilities. On the one hand, companion animals contribute to improved emotional welfare and overall life satisfaction by offering consistent emotional support. On the other hand, service animals fulfill practical roles that enhance functionality and independence in daily living. Both roles are instrumental in reducing the isolation often experienced by disabled persons and in empowering them to manage the challenges associated with their conditions.

See also: Human Health, Pets And

References and Further Reading

Carlisle, G. K., and S. M. Craven (2023). "Pet Ownership and Persons with Developmental and Physical Disabilities: Impact on Mental and Physical Health." In A. H. Fine, M. K. Mueller, Z. Y. Ng, A. M. Beck, and J. M. Peralta (eds), *The Routledge International Handbook of Human-Animal Interactions and Anthrozoology*, 259–71. New York: Routledge.

Dalziel, D. J., B. M. Uthman, S. P. McGorray, and R. L. Reep (2003). "Seizure-Alert Dogs: A Review and Preliminary Study." *Seizure*, 12 (2): 115–20.

Hall, S. S., J. MacMichael, A. Turner, and D. S. Mills (2017). "A Survey of the Impact of Owning a Service Dog on Quality of Life for Individuals with Physical and Hearing Disability: A Pilot Study." *Health and Quality of Life Outcomes*, 15: 1–9.

Haron, S. A., H. F. Foong, and T. A. Hamid (2018). "Moderating Effect of Emotional Support on the Relationship Between Disability and Life Satisfaction in Older Adults." *Geriatrics & Gerontology International*, 18 (9): 1361–5.

Taniguchi, Y., S. Seino, B. Headey, T. Hata, T. Ikeuchi, T. Abe, S. Shinkai, and A. Kitamura (2022). "Evidence that Dog Ownership Protects Against the Onset of Disability in an Older Community-Dwelling Japanese Population." *PloS One*, 17 (2): e0263791.

DOG BREEDING AND EUGENICS

Eugenics is an ideology and pseudo-science focused on blood purity. In the United States and a number of European countries in the nineteenth and early twentieth centuries, eugenics became a focus of politicians, scientists, and public figures who wanted to use practices taken from animal breeding—but specifically dog breeding—in order to "improve" the human species by encouraging reproduction among healthy white people and discouraging it among those who were considered to be "unfit," from a racial, biological, and social perspective.

Historical research on dog breeding during the nineteenth and twentieth centuries reveals a complex interplay between the material invention of dog breeds, emerging scientific ideologies—including eugenics—and the institutionalization of breeding practices through the founding of kennel clubs. Early in the Victorian era, dog breeders in Britain reimagined and standardized breeds based on aesthetic and functional criteria, an enterprise that was not only commercial but also deeply intertwined with contemporary scientific discourses on heredity and selection. This period witnessed an explosion of breed formation as dog fanciers began to compile detailed genealogies and physical standards, practices that would later become emblematic of the scientific approach to animal improvement, which included eugenic theories.

The eugenics movement provided a legitimizing ethos for breeding practices that emphasized the purity and superior traits of specific lineages. The notion that desirable traits could be selectively enhanced through controlled mating found resonance both in human and canine contexts. Researchers have highlighted how the scientific fascination with heredity during this period influenced not only the design of breed standards but also the broader cultural perception of dogs as "improved" or idealized animals. This convergence of eugenic thinking and dog breeding meant that early kennel clubs, emerging in the late nineteenth century, played a significant role in disseminating these ideas. By codifying breed standards and maintaining stud books, institutions such as the Kennel Club in Britain provided a formal structure that promoted selective breeding practices aimed at preserving the physical and, by extension, the genetic purity of dog breeds.

The institutionalization of dog breeding through these clubs not only standardized practices but also helped spread these ideals internationally, eventually leading to the recognition of hundreds of breeds. The process of breed stabilization through kennel clubs has since become a central narrative in the history of dog breeding, even as contemporary scholarship questions the inevitability of current breed standards and highlights the contingent nature of breed formation. In sum, the historical trajectory from the Victorian invention of dog breeds, through the adoption of eugenic ideologies, to the establishment of formal kennel clubs marks a pivotal transformation in the way dogs have been bred, regulated, and culturally celebrated.

The scholarly research on the dog fancy increasingly reveals that the evolution of modern pedigree breeding in the nineteenth and twentieth centuries was deeply interwoven with contemporary ideas of selective breeding and eugenics. The dog fancy emerged as both a cultural pastime and a rigorous scientific enterprise, where breeders sought to create idealized dogs by emphasizing specific physical traits and bloodlines. This pursuit of "breed purity" paralleled eugenic ideologies that prized hereditary fitness and genetic "excellence" in humans, influencing public perceptions of both canine quality and status.

Breed standards and institutional structures such as kennel clubs were established, in part, to enforce these ideals and to promote a vision of genetic refinement that has had lasting impacts on the welfare and characteristics of many popular breeds. This historical interplay between the dog fancy world and eugenic thought and practice continues to shape debates over animal welfare, genetic health, and the ethics of selective breeding practices within modern canine culture.

See also: Breeds Popularity Of; Dog Shows; Fancy Clubs; Holocaust, Pets During The; Selective Breeding

References and Further Reading

Howell, P. (2013). "The Dog Fancy at War: Breeds, Breeding, and Britishness, 1914–1918." *Society & Animals*, 21 (6): 546–67.
Huff, C. (2002). "Victorian Exhibitionism and Eugenics: The Case of Francis Galton and the 1899 Crystal Palace Dog Show." *Victorian Review*, 28 (2): 1–20.
Jung, C., and D. Pörtl (2019). "How Old are (Pet) Dog Breeds?" *Pet Behaviour Science*, (7): 29–37.
Lockwood, R. (2019). "In Search of Pedigrees: Why Do We Harm the Dogs We Love?" *Journal of Animal Ethics*, 9 (2): 220–5.
Ritvo, H. (1986). "Pride and Pedigree: The Evolution of the Victorian Dog Fancy." *Victorian Studies*, 29 (2): 227–53.
Vesey-Fitzgerald, B. (2023). *The Domestic Dog: An Introduction to Its History*. London: Routledge.
Zelinger, A. (2019). "Race and Animal-Breeding: A Hybridized Historiography." *History and Theory*, 58 (3): 360–84.

DOG SHOWS

Dog shows are competitions in which people and their dogs compete on the basis of breed standards, agility, or other measures.

In Great Britain, the Victorian era marked the genesis of the dog fancy—a cultural practice that celebrated purebred lineages and specific aesthetic standards. Dog shows emerged as both competitive events and social gatherings, where enthusiasts could display carefully bred animals according to the standards codified by pioneering institutions such as the Kennel Club. These early shows were not only about judging physical conformation; they also underscored the ideals of heredity and refinement prevalent in the period's scientific and social discourses.

Across the Atlantic, American dog shows evolved under the strong cultural influence of British practices. The establishment of the American Kennel Club (AKC) in the late nineteenth century paralleled the British model by introducing organized competitions and standardized breed registries. The AKC recognized a multitude of breeds—many with roots in British breeding traditions—yet over time, American dog shows began to reflect distinctive national characteristics, emphasizing both aesthetic appeal and functional

The largest and smallest dogs at a 1923 dog show in the United States. Dog breed recognition and formalization were points of contention in early American dog shows. (Library of Congress)

traits that resonated with American cultural values. The interplay between the British origins of the dog fancy and the emerging American independent style created a dynamic field where debates over breed standards, animal health, and competitive fairness became central.

Research shows that the process of formalizing breeds—and by extension, dog show competitions—was contingent and subject to ongoing contestation. Disputes over breed recognition illustrate how dog shows have been instrumental in negotiating cultural and scientific notions of "purity" and "fitness" in canines. These events, therefore, stand at the intersection of scientific breeding practices and popular entertainment, maintaining a dual role in both reinforcing and challenging prevailing ideas about breed idealism.

Dog shows serve as a prominent avenue for the exhibition of purebred dogs, highlighting both their aesthetic characteristics and their adherence to breed standards. These events, while primarily focused on the visual appraisal of dogs, also raise important issues pertaining to dog health, and in particular, the detrimental health impacts of many official breed standards.

In addition to health concerns, dog shows also encapsulate the complexity of dog-owner interactions, which can navigate both competitive success and behavioral communication. Research by Henschel et al. explores how shared information and behaviors between owners and dogs influence their performance in shows, suggesting a dynamic relationship that may affect the outcomes of such events (Henschel et al., 2020). This understanding contributes to a broader discourse on the role of human behavior in shaping canine performance and well-being in competitive settings.

Thus, the phenomenon of dog shows is intertwined with critical discussions about canine health, the implications of standardized breed traits, and the human–animal relationship dynamics. As these shows continue to evolve, it is crucial for stakeholders to address these underlying issues to ensure the continued health and welfare of competing dogs.

See also: Breeds, Popularity Of; Dog Breeding and Eugenics; Dogs; Fancy Clubs; Holocaust; Selective Breeding

References and Further Reading

Corbee, R. (2012). "Obesity in Show Dogs." *Journal of Animal Physiology and Animal Nutrition*, 97 (5): 904–10.

Henschel, M., J. Winters, T. Müller, and J. Bräuer (2020). "Effect of Shared Information and Owner Behavior on Showing in Dogs (canis familiaris)." *Animal Cognition*, 23 (5): 1019–34.

Howell, P. (2013). "The Dog Fancy at War: Breeds, Breeding, and Britishness, 1914–1918." *Society & Animals*, 21 (6): 546–67.

Ritvo, H. (1986). "Pride and Pedigree: The Evolution of the Victorian Dog Fancy." *Victorian Studies*, 29 (2): 227–53.

Syrjämaa, T. (2017). "Spectacles of Modern Companionship: Men, Dogs and Early Finnish Dog Shows." In T. Räsänen and T. Syrjämaa (eds), *Shared Lives of Humans and Animals*, 63–78. New York: Routledge.

DOGS

Dogs were the first animals to partner with humans and were the first animals to become domesticated—or, as modern scholars might say, to domesticate themselves. This makes them, arguably, the most important animals from the perspective of humanity and our development as a species.

Archaeologists have long said that dogs were first domesticated at least 15,000 years ago, based on evidence from burials in which dogs were buried along with humans—the oldest of these dates from 14,000 years ago in Bonn, Germany. However, that date is being challenged on at least two fronts. Archaeologists have located what they believe is a domesticated dog skull in a cave called Goyet in Belgium that they date to 31,700 years ago (Ovodov et al., 2011). Another cave, called Chauvet, in France, contains fossilized dog skeletons dating to 26,000 years ago (Galibert et al., 2011). In addition, DNA evidence using mitochondrial DNA shows that the split between wolves and dogs could go as far back as 100,000 years (Vilà et al., 1997), meaning that humans may have been living with dogs for much longer than we thought.

All of the early breeds of domesticated dogs were bred for hunting, such as the pointer, the hound, and the retriever. As humans migrated around the planet, a variety of dog breeds migrated with them (Wang et al., 2016; Leathlobhair et al., 2018). With agriculture came new specializations for the dog—from hunting to herding and protection, for racing and for pulling loads, and to keep rats away from the food supply. Today, there are more than 400 breeds of dog, all with specialized appearances and behavior, and all derived from the wolf.

Some researchers have speculated that the benefits that dogs offer to humans might have contributed to the rapid expansion of humans into the New World. Archaeological and historical evidence shows that the earliest inhabitants of North America were selectively breeding dogs for traits that would be useful for hunting. That evidence also indicates that the people used dogs as beasts of burden, tying wooden sleds called *travois* to them, on which they loaded firewood and other necessities. The Clallam Indians of Puget Sound, Washington, kept long-haired dogs whose hair they spun the hair into clothing and blankets, and other native groups most likely used dogs for protection as well.

The research on dogs as pets offers a comprehensive picture that spans from their ancient domestication to the multiple roles they now play in families and communities across the globe, as well as the challenges associated with dog ownership.

Early studies widely concur that dogs were the first species to be domesticated, diverging from wild wolf ancestors during the late Pleistocene. Although the exact timing remains debated, it is generally accepted that this process involved both natural and artificial selection for traits that improved human–dog communication and cooperation. This long history of co-evolution has laid the groundwork for the multifaceted relationship between humans and dogs, where dogs have come to embody roles that are far more than utilitarian; they serve as companions, aides, therapists, and even cultural symbols.

Because of their historical closeness with humans, dogs tend to be represented positively in religions around the world. In Native American traditions, dogs symbolize friendship and loyalty, for example. But dogs are often seen as guardians of the dead as well. In Greek and Roman mythology, Cerberus, the three-headed dog, guards the gates of the underworld. In the ancient Zoroastrian religion of Persia, dogs were intermediaries between the profane and the sacred realms and were especially important in rituals surrounding death. If a person died, for example, dogs were brought in to witness the body, and they played a role in funerals. After death, the soul is accompanied to the afterlife by two dogs. Many of the cultures of ancient MesoAmerica, like the Olmec, Toltec, Maya, and Aztecs, saw the hairless dog as the guardian of the dead, and some of these animals were kept as sacrificial food, whereas others were used to protect homes from evil spirits and to heal the sick. In addition, the dog was thought to accompany the souls of the dead to the underworld; for this reason, mummified dogs were buried in the tombs of the deceased throughout MesoAmerica. Even today, small dogs are thought to ward off ailments in parts of Central America. In addition, because dogs are scavengers and often eat the bodies of the dead, they are seen as filthy in some traditions, such as in Islam.

Today, the roles that dogs fulfill vary significantly around the world. In many households, dogs provide essential emotional support and companionship, with research showing that even untrained family pets can improve overall family functioning and reduce anxiety, particularly among children and individuals with special needs (Wright et al., 2015; Adkins et al., 2023). Studies have documented that dogs contribute to enhanced social capital through routines such as dog walking—a practice that not only benefits physical health but also promotes social interactions and neighborhood connectedness (Koohsari et al., 2021). In marginalized communities, dog ownership has been linked to improved social acceptance among individuals facing stigma (Kerr-Little et al., 2025). Moreover, therapeutic applications, such as the use of assistance dogs to support families of children with autism, underscore the significant psychosocial and practical roles that dogs play (Hellings et al., 2022).

The popularity of dogs as pets has grown steadily over time, driven partly by shifting cultural values and the influence of media. For instance, research indicates that popular culture—such as films featuring canine stars—can dramatically influence breed popularity, independent of considerations related to temperament or health (Ghirlanda et al., 2014). Surveys and demographic studies reveal that rising economic prosperity and evolving lifestyle choices have contributed to increased dog ownership across different cultures, even as selection criteria increasingly emphasize aesthetics, as observed in trends toward brachycephalic and smaller breed types (Packer et al., 2023; Teng et al., 2022). However, this surge in popularity has also led to a host of challenges.

Problems around dog ownership include both practical and ethical dimensions. Many prospective and current owners face challenges in managing the daily responsibilities associated with dog care. Research suggests that unrealistic expectations—often resulting from media hype or inadequate preadoption counseling—can lead to issues such as improper training, owner dissatisfaction, and even eventual relinquishment of the pet (Powell et al., 2021). Health problems are common, particularly among breeds that have been intensively selected for specific physical traits, as seen in studies on brachycephalic dogs where the pursuit of aesthetic appeal has been linked to substantial welfare issues (Packer et al., 2019). Additionally, broader societal concerns, such as the environmental impacts of pet ownership and the difficulties in managing free-roaming or untrained dogs, are emerging as important areas for further study (Westgarth et al., 2015).

See also: Breeds, Popularity Of; Dog Shows; Domestication; Fancy Clubs; Selective Breeding

References and Further Reading

Adkins, J. R., C. M. Mulé, D. E. Linder, A. Must, S. B. Cash, and S. C. Folta (2023). "Exploring Health Behaviors and the Role of Pet Dogs in Households with Autistic Children: The DANE Study." *Frontiers in Pediatrics*, 11: 1153124.

Galibert, F., P. Quignon, C. Hitte, and C. André (2011). "Toward Understanding Dog Evolutionary and Domestication History." *Comptes Rendus Biologies*, 334 (3): 190–6.

Ghirlanda, S., A. Acerbi, and H. Herzog (2014). "Dog Movie Stars and Dog Breed Popularity: A Case Study in Media Influence on Choice." *PLOS One*, 9 (9): e106565.

Halonen, M., and P. Laihonen (2019). "From 'No Dogs Here!' To 'Beware of the Dog!': Restricting Dog Signs As A Reflection of Social Norms." *Visual Communication*, 20 (4): 501–26.

Hellings, D., A. Joosten, M. Hatfield, and J. Netto (2022). "Benefits and Challenges of Assistance Dogs for Families of Children on the Autism Spectrum: Mothers' Perspectives." *Qualitative Health Research*, 32 (11): 1648–56.

Howell, P. (2013). "The Dog Fancy at War: Breeds, Breeding, and Britishness, 1914–1918." *Society & Animals*, 21 (6): 546–67.

Jung, C., and D. Pörtl (2019). "How Old Are (Pet) Dog Breeds?" *Pet Behaviour Science*, 7: 29–37.

Kisley, M., E. Chung, and H. Levitt (2024). "Investigating the Reasons Behind Companion Animal Relinquishment: A Systematic Content Analysis of Shelter Records for Cats and Dogs, 2018–2023." *Animals*, 14 (17): 2606.

Kerr-Little, A., J. G. Bramness, R. C. Newberry, and S. Biong (2025). "Dog Ownership for People with Substance Use Disorder: Self-Reported Influence on Substance Use and Mental Health." *Substance Abuse Treatment, Prevention, and Policy*, 20 (1): 27.

Koohsari, M. J., A. Yasunaga, A. Shibata, K. Ishii, R. Miyawaki, K. Araki, … K. Oka (2021). "Dog Ownership, Dog Walking, and Social Capital." *Humanities and Social Sciences Communications*, 8 (1): 1–6.

Larson, G., E. K. Karlsson, A. Perri, M. T. Webster, S. Y. W. Ho, J. Peters, P. W. Stahl, K. E. Comstock, et al. (2012). "Rethinking Dog Domestication by Integrating Genetics, Archeology, and Biogeography." *Proceedings of the National Academy of Sciences*, 109 (23): 8878–83.

Leathlobhair, M. N., A. R. Perri, E. K. Irving-Pease, K. E. Witt, A. Linderholm, J. Haile, S. Brace, et al. (2018). "The Evolutionary History of Dogs in the Americas." *Science*, 361 (6397): 81–5.

Lockwood, R. (2019). "In Search of Pedigrees: Why do We Harm the Dogs We Love?" *Journal of Animal Ethics*, 9 (2): 220–5.

Losey, R. J., V. I. Bazaliiskii, S. Garvie-Lok, M. Germonpré, J. A. Leonard, A. L. Allen, M. A. Katzenberg, and M. V. Sablin (2011). "Canids as Persons: Early Neolithic Dog and Wolf Burials, Cis-Baikal, Siberia." *Journal of Anthropological Archaeology*, 30 (2): 174–89.

McGetrick, J., L. Fux, J. Schullern-Schrattenhofen, J. Rault, and F. Range (2024). "Do Pet Dogs Reciprocate the Receipt of Food from Familiar and Unfamiliar Conspecifics?" *Ethology*, 130 (4).

Okin, G. (2017). "Environmental Impacts of Food Consumption by Dogs and Cats." *Plos One*, 12 (8): e0181301.

Ovodov, N. D., S. J. Crockford, Y. V. Kuzmin, T. F. Higham, G. W. Hodgins, and J. van der Plicht (2011). "A 33,000-Year-Old Incipient Dog from the Altai Mountains of Siberia: Evidence of the Earliest Domestication Disrupted by the Last Glacial Maximum." *PLOS One*, 6 (7): e22821.

Packer, R. M., A. M. Davies, H. A. Volk, H. L. Puckett, S. L. Hobbs, and R. C. Fowkes (2019). "What Can We Learn from the Hair of the Dog? Complex Effects of Endogenous and Exogenous Stressors on Canine Hair Cortisol." PLOS One, 14 (5): e0216000.

Packer, R. M., C. L. Brand, Z. Belshaw, C. L. Pegram, F. Dale, K. B. Stevens, and D. G. O'Neill (2023). "Is UK Puppy Purchasing Suffering a Long COVID Effect? Ongoing Negative Impacts of the COVID-19 Pandemic upon Puppy Purchase Motivations and Behaviours in 2021." *Animals*, 13 (13): 2186.

Pollinger, J. P., K. E. Lohmueller, E. Han, H. G. Parker, P. Quignon, J. D. Degenhardt, K. Bryc, et al. (2010). "Genome-Wide SNP and Haplotype Analyses Reveal a Rich History Underlying Dog Domestication." *Nature*, 464 (7290): 898–902.

Powell, L., D. Duffy, K. Kruger, B. Watson, and J. Serpell (2021). "Relinquishing Owners Underestimate their Dog's Behavioral Problems: Deception or Lack of Knowledge?" *Frontiers in Veterinary Science*, 8: 734973.

Ritvo, H. (1986). "Pride and Pedigree: The Evolution of the Victorian Dog Fancy." *Victorian Studies*, 29 (2): 227–53.

Teng, L., S. Liao, X. Zhou, C. Jia, M. Feng, H. Pan, ... M. Yue (2022). "Prevalence and Genomic Investigation of Multidrug-Resistant *Salmonella* Isolates from Companion Animals in Hangzhou, China." *Antibiotics*, 11 (5): 625.

Vesey-Fitzgerald, B. (2023). *The Domestic Dog: An Introduction to Its History*. London: Taylor & Francis.

Vilà, C., P. Savolainen, J. E. Maldonado, I. R. Amorim, J. E. Rice, R. L. Honeycutt, ... R. K. Wayne (1997). "Multiple and Ancient Origins of the Domestic Dog." *Science*, 276 (5319): 1687–9.

Wang, G. D., W. Zhai, H. C. Yang, L. U. Wang, L. Zhong, Y. H. Liu, ... Y. P. Zhang (2016). "Out of Southern East Asia: The Natural History of Domestic Dogs Across the World." *Cell Research*, 26 (1): 21–33.

Westgarth, C., H. E. Christian, and R. M. Christley (2015). "Factors Associated with Daily Walking of Dogs." *BMC Veterinary Research*, 11 (1): 116.

Wright, H. F., S. Hall, A. Hames, J. Hardiman, R. Mills, Paws Team, and D. S. Mills (2015). "Acquiring a Pet Dog Significantly Reduces Stress of Primary Carers for Children with Autism Spectrum Disorder: A Prospective Case Control Study." *Journal of Autism and Developmental Disorders*, 45 (8): 2531–40.

DOMESTICATION

Domestication refers to the process by which an animal's genetic makeup has been altered, via selective breeding, such that it results in a new species of animal. As has long been noted by archaeologists and historians (Clutton-Brock, 1987; Zeuner, 1963; Zeder, 2012), animal domestication first occurred during either the Mesolithic Era or the late Paleolithic Era with the domestication of the dog. Even though the dog was domesticated as a partner to humans—to help with hunting, guarding, and other functions—the fact that the first animal to be domesticated ended up being the animal with the closest relationship with humans is interesting to note.

Domesticating animals involves more than simply taming wild animals. Domestication refers to the process of maintaining animals in captivity, selectively breeding them for human purposes, and controlling their food supply, reproduction, and other aspects of life, thus creating a dependency on humans for survival and a marked alteration in appearance and behavior.

Dogs were domesticated as hunting partners to humans at least 15,000 years ago, according to the fossil record, but perhaps as early as 34,000 years ago based on genetic evidence (see Skoglund et al., 2015; Pollinger et al., 2011; Larson et al., 2012), with some scientists suggesting that dogs may have been domesticated twice, in both eastern Asia and western Eurasia (Frantz et al., 2016). During this period, humans were still hunting and gathering and had no real control over the production of their food. Evidence suggests that dogs were first domesticated in East Asia and perhaps in as many as three other independent locations, and from there they might have spread around the world. Dogs have been found in human burials from around the world dating to as much as 14,000 years ago (Losey et al., 2018), demonstrating how important they were to Mesolithic-Era humans.

With the domestication of the "livestock" animals as well as the domestication of plants in the Neolithic period, starting about 10,000 years ago, our ancestors moved from food collection to food production as their primary economic activity. This change is among the most monumental changes in human—and animal—history.

After the dog, the first animals to be domesticated were the goat, sheep, pig, and the cow; all were domesticated for their milk, meat, and wool or skin. Horses, which were first hunted as food animals, were the first true beasts of burden, being used to pull plows in farming communities perhaps 5,000 years ago; another 2,000 years later, horse riding developed. Horses were tremendously important to humans, allowing for travel, warfare, trade, and work, and, later, entertainment through horse racing and sports like the rodeo. Horses' importance to humans declined only with the development of the internal combustion engine in the early nineteenth century, but by that time, they were so essential to human society that today they serve as an important source of companionship in many cultures.

Horses were followed by cats, chickens, llamas, alpacas, and camels, and finally, less than 2,000 years ago, smaller animals such as rabbits and guinea pigs were domesticated.

As has been demonstrated by a number of scholars, of the fourteen or so large animals to have been *truly* domesticated, most (the cat is a notable exception) can be defined by a number of behavioral traits, including a tendency toward scavenging, a rapid maturity rate, reasonable size, a calm disposition, the ability to be bred in captivity, a gregarious nature and willingness to live with others in close quarters, and a hierarchical social life, all of which make animals such as the dog and horse amenable to living with humans, in exchange for feeding and care (Clutton-Brock, 1987, and 2014; Diamond, 1999; Zeuner, 1963). Of course, the animals themselves must have had something to offer to humans as well, such as food, clothing, an ability to work as hunters or beasts of burden, and, later, the promise of companionship.

Although the species that were ultimately domesticated fit the aforementioned criteria, making them natural choices for domestication, the process of domestication itself was a result of both natural and cultural evolution. First, specific behavioral and physical traits of individual animals that scavenged or hung around human encampments were favored by natural selection in the process of domestication. For example, those individuals who demonstrated less fear of and more curiosity about humans would be among the first individuals to approach human societies; after they reproduced, those traits would continue in their offspring (Budiansky, 1992). As each generation is born, the "flight distance" of those animals will be shorter than in the preceding generation, until the domesticates become comfortable living around humans. Humans would have made choices as well, selecting woolly animals from among wild sheep (who are not typically woolly), thus acquiring livestock better suited to lowland heat and from which to obtain wool. Animals would have been selected for smaller size, so that they could be more easily handled, or larger size, in order to produce more meat, or for different colors or textures of wool, fur, or hair.

Next, humans most likely adapted their own behavior to that of the animals, incorporating them into human social and economic structures and, later, manipulating the physiology and behavior of the animals themselves. To be a domesticated animal is to be, at the very minimum, owned and controlled by humans in a human cultural environment. This is a profoundly different human–animal relationship than was seen in non-agricultural societies, in which humans could not conceive of "owning" animals in this way (see Ingold, 1994). As we will see, domestication in the twenty-first century has moved from natural selection to artificial selection and has been shaped almost entirely by human hands.

The domestication of animals was a truly revolutionary stage in the development of human civilization. It allowed humans to have a steady food supply, have new sources of labor, and have new forms of companionship and protection, and it provided resources for new forms of religious worship. Without domestication, it is unlikely that humans and animals could co-exist in the types of multispecies households we see across the world today.

For humans, once the domestication of large ruminants was combined with plant domestication, this led to the rise of agricultural societies which would become the first empires on the planet. It allowed people to live in one place, which then allows for notions like private property and the ownership of land, animals, and other humans. The domestication of the horse provided long-distance transportation, which enabled trade networks, allowing for the exchange of cultural knowledge among peoples. It also paved the way for new kinds of war, which would be used to support the new empires that were developing.

But the results of domestication for the human–animal relationship, and for the animals themselves, were no less revolutionary. As anthropologist Tim Ingold points out (1994), the relationship between animals and humans among traditional hunter-gatherers was often one of mutual trust in which the environment and its resources are shared by both animals and people; animals that were hunted by humans were seen as humans' equals. Domestication changed this relationship into one of dominance and control, as humans took on the role of master and animals that of property, becoming items to be owned and exchanged.

Domestication also had long-range consequences for the animals themselves, as the very nature of the animals changed throughout the process, typically not in the animals' favor. Through domestication, once-wild animals become increasingly more dependent on humans, both physically and emotionally (Price, 1984). As humans became more proficient in breeding animals for characteristics useful to people, domesticated animals grew ever more dependent on humans for their care. One of the results of domestication is the domestication syndrome: the tendency for domestication to result in behavioral and physical features associated with the juveniles of the species such as shorter faces, excess fat, smaller brains, smaller teeth, and bigger eyes, as well as higher-pitched voices, playfulness, and curiosity. The result of hundreds, and even thousands, of years of selective breeding are our modern pets, most of whom are physically and behaviorally unable to live independently, and who are, in fact, perpetual juveniles (a condition known as neoteny; see Coppinger and Smith, 1983; Clutton-Brock, 1992).

Of course, once humans began selectively breeding their animal charges (and killing both those offspring who did not fit the bill as well as wild competitors) in order to emphasize or discourage certain physical or behavioral traits, the animals changed even further. We see this especially in pets, where the goal of breeding is not to produce more meat, better eggs, or higher-quality wool, but to create animals that are best suited to living with humans. Our pets today are, for the most part, smaller (yet fleshier) than their wild cousins, more brightly colored, with shorter faces, rounder skulls, and more variations in fur and hair type as well as ear and tail appearance. In addition, domestication has resulted as well in a permanent loss of genetic diversity within the species.

While it seems clear that many pets today benefit from living with humans, there is no doubt that there has been a loss to those animals of some of their former traits. Researchers believe that years of domestication have led to dogs losing the problem-solving skills they once had in the wild. In behavioral studies, pet dogs fail basic intelligence tests that wolves and wild dogs pass with ease, according to the research. The findings of one such study (Smith and Litchfield, 2010) suggest dogs are now so dependent on people that they are simply "stupid" versions of their forefathers. Another way to look at this, though, is to say that as dogs became dependent on people, they got better at communicating with humans and using us as tools. Rather than opening doors themselves, domesticated dogs just wait until we open them for them.

According to some scholars (e.g., see Budiansky, 1997), the process of domestication was one of mutual benefit in which both the early domesticates and our human ancestors had something to gain from joining forces. The question we must ask ourselves now, 10,000 years later, is this: if nonhuman animals did, in fact, benefit from aligning themselves with human societies, gaining protection from predators, easy access to food, and shelter from the weather, do modern domesticated animals, and in particular our pets, experience the same benefits?

As people learned more about the inheritance of traits, animal breeders began selectively breeding their animals for more specific characteristics, such as overall size, fur color or texture, ear and tail shape, and more. Termed artificial selection by Darwin, selective breeding has led to the creation of hundreds of breeds of dogs, one of the most intensively bred animals in the world. Using dogs as an example, breeds were created in order to fulfill human desires. Some breeds were created to retrieve ducks during a hunt, others were created to herd sheep, and still others were created to race.

Today, the pet industry has borrowed some of the more contemporary methods of artificial selection from the livestock industry. For example, the pet industry has adopted the use of artificial selection as well as artificial insemination to create breeds of animals with favorable (to humans) traits. Recent years have seen an escalation in the development of the varieties of dogs, cats, and other companion animals being developed in order to appeal to discriminating consumers.

And although the early breeds of dogs were created to highlight working traits, recent breeds have been geared more toward aesthetics. On the other hand, given that cats are not working animals, most cat breeds have been created for aesthetic purposes, with an eye toward color, size, fur type, and tail, ear, and body type. The result is hundreds of breeds of dogs and dozens of breeds of cats, rabbits, and other species, all bred by large and small breeders to sell through the pet industry.

Another result is a whole host of health problems associated with these breeds. Dogs, in particular, are at risk of problems associated with the odd proportions in body, legs, and head that are bred into many of the breeds (Ackerman, 2011). Another form of artificial selection refers to breeders emphasizing deleterious traits in the breeding process. Japanese Bobtails (cats with a genetic mutation resulting in a bobbed tail), hairless cats, and Scottish Folds (whose ears are folded down) are examples of this type of breeding.

Genetic manipulation of animals represents a new scientific development that has irreversibly changed animal bodies. Genetic engineering has even found its way into the pet world, with the production of a new so-called "hypoallergenic cat" (selling for $12,000–$28,000), created by manipulating the genes that produce allergens.

In terms of reproduction, cloning animals is the wave of the future, allowing humans the greatest level of control over animal bodies. Thus far, the livestock industry has been most active in the use of cloning, but it is found in the vivisection and pet industries as well. Laboratory scientists are also cloning mice, rabbits, and other laboratory animals in order to ensure that the animals used in research are genetically identical and to control for any so-called "imperfections."

In the pet world, cloning has been less successful (in part because of the enormous number of animals who are either born with terrible deformities or who are "sacrificed" in order to produce the cloned animal), but a handful of companies today either offer companion animal cloning or tissue-freezing services for those animals who cannot yet be cloned (Lee et al., 2005).

Another way in which animal bodies have been changed is through surgical procedures. Because the control of animal reproduction is critical for those who breed animals, castration has been used for thousands of years to ensure that undesirable animals cannot breed, or to increase the size or control the temperament of certain animals. In the twenty-first century, with companion animals rising in popularity, surgical techniques to remove the uterus and ovaries of female animals were developed, and spaying is now an extremely common surgery for companion animals, although it is very rarely performed on livestock.

In the pet-breeding world, companion animals undergo surgical procedures in order to make them conform to the artificial requirements of the breed. Breed standards demand that certain dogs, for example, must have their tails docked, their ears cropped, or both. In addition, many companion animals today experience surgical procedures that are used to control unwanted (by humans) behavior. Some people, for example, have their dogs debarked (by cutting their vocal cords) in order to reduce barking, and many cat owners elect to have their cats declawed (which involves amputating the front portions of a cat's toes) in order to prevent harm to their furniture.

All of these uses of the animal body can be illustrated with French philosopher Michel Foucault's (1990) notion of biopower: the ways in which the modern state controls and regulates its citizens' bodies.

Human civilization would not be what it is today were it not for animal domestication. Without the assistance of working animals who pull plows, carts, sleds, and carriages, who carry goods or people, who assist humans with hunting, and who herd other animals, and without the use of animals as food and fiber, it is difficult to imagine where human societies would be today.

Certainly, many companion animals gain a great deal by living with humans, and many working animals might live better lives with humans than without. But even these points can be debated. According to ecologist Paul Shepard, "the benefit to animals of being domestic is fictitious, for they are slaves, however coddled, becoming more demented and attenuated as the years pass" (1996: 267). In any case, the point is moot because with the possible exception of the mixed-breed cat, today's domesticated animals are so highly bred and engineered for human benefit that they could never again survive on their own. In short, we care for them because they could not live without our care, and they live with and obey us because they no longer have a choice in the matter.

See also: Affection and Control; Dogs; Fancy Clubs; Selective Breeding

References and Further Reading

Ackerman, L. J. (2011). *The Genetic Connection: A Guide to Health Problems in Purebred Dogs*. Lakewood, CO: American Animal Hospital Association.

Budiansky, Steven (1992). *The Covenant of the Wild: Why Animals Chose Domestication*. New Haven, CT: Yale University Press.

Clutton-Brock, J. (1987). *A Natural History of Domesticated Mammals*. Cambridge: Cambridge University Press.

Clutton-Brock, J. (1992). "The Process of Domestication." *Mammal Review*, 22 (2): 79–85.

Coppinger, R. P., and C. K. Smith (1983). "The Domestication of Evolution." *Environmental Conservation*, 10 (4): 283–92.

Diamond, Jared (1999). *Guns, Germs and Steel: The Fate of Human Societies*. New York: W. W. Norton & Co.

Frantz, L. A. F., V. E. Mullin, M. Pionnier-Capitan, O. Lebrasseur, M. Ollivier, A. Perri, A. Tresset, et al. (2016). "Genomic and Archaeological Evidence Suggest a Dual Origin of Domestic Dogs." *Science*, 352 (6290): 1228–31.

Foucault, M. (1990). *The History of Sexuality: An Introduction, Volume I*. Translated by R. Hurley. New York: Vintage.

Ingold, T. (1994). "From Trust to Domination: An Alternative History of Human–Animal Relations." In A. Manning and J. Serpell (eds), *Animals and Human Society*. New York: Routledge.

Larson, G., E. K. Karlsson, A. Perri, M. T. Webster, S. Y. W. Ho, J. Peters, ... K. Lindblad-Toh (2012). "Rethinking Dog Domestication by Integrating Genetics, Archeology, and Biogeography." *Proceedings of the National Academy of Sciences*, 109 (23): 8878–83.

Lee, B. C., M. K. Kim, G. Jang, H. J. Oh, F. Yuda, H. J. Kim, ... W. S. Hwang (2005). "Dogs Cloned from Adult Somatic Cells." *Nature*, 436 (7051): 641.

Losey, R. J., T. Nomokonova, L. Fleming, K. Latham, and L. Harrington (2018). "Domestication and the Embodied Human–Dog Relationship: Archaeological Perspectives from Siberia." In R. J. Losey, R. P. Wishart, and J. P. Laurens Loovers (eds), *Dogs in the North*, 8–27. London: Routledge.

Pollinger, J. P., D. A. Earl, J. C. Knowles, A. R. Boyko, H. Parker, E. Geffen, ... R. K. Wayne (2011). "A Genome-Wide Perspective on the Evolutionary History of Enigmatic Wolf-Like Canids." *Genome Research*, 21 (8): 1294–305.

Price, E. O. (1984). "Behavioral Aspects of Animal Domestication." *Quarterly Review of Biology*, 59 (1): 1–32.

Shepard, P. (1996). *The Others: How Animals Made Us Human*. Washington, DC: Island Press.

Skoglund, P., E. Ersmark, E. Palkopoulou, and L. Dalén (2015). "Ancient Wolf Genome Reveals an Early Divergence of Domestic Dog Ancestors and Admixture into High-Latitude Breeds." *Current Biology*, 25 (11): 1515–19.

Smith, B. P., and C. A. Litchfield (2010). "Dingoes (*Canis dingo*) Can Use Human Social Cues to Locate Hidden Food." *Animal Cognition*, 13 (2): 367–76.

Wang, G. D., W. Zhai, H. C. Yang, L. Wang, L. Zhong, Y. H. Liu, D. M. Irwin, et al. (2016). "Out of Southern East Asia: The Natural History of Domestic Dogs Across the World." *Cell Research*, 26 (1): 21–33.

Zeder, M. A. (2012). "The Domestication of Animals." *Journal of Anthropological Research*, 68 (2): 161–90.

Zeuner, F. E. (1963). *A History of Domesticated Animals*. London: Hutchinson.

E

ELITE PET KEEPING

Prior to the modern era, pet keeping was not a mainstream phenomenon in any culture. In fact, if we look at both the ancient nation-states as well as the European and Asian imperial states leading up to the Renaissance, pets were primarily kept by the wealthy who could afford them.

In Europe and Asia at this time, only landowners, royalty, the church, and other wealthy classes could afford to care for animals that did not earn their keep at this time. In Europe, for example, hunting dogs were highly prized by the nobility and probably served both hunting and companionship functions (Ritvo, 1987). In China, the Pekingese and the Shih Tzu were both bred originally as companion animals, but they also had secondary functions. The Shih Tzu, even though small, was used as a watchdog in the Imperial Palace, and the Pekingese were considered to be sacred because they looked so much like the Chinese lions that were sacred in Buddhism. In these cultures, purebred dogs did not become popular among commoners until the mid-twentieth century.

During the Middle Ages, the Catholic Church viewed the practice of pet keeping as a form of heresy, in that it challenged the human/animal boundary, even though many nuns and monks kept pets (Maehle, 2002). Another reason that pet keeping had been seen as trivial—and still is today, though to a much lesser extent—is the fact that historically, as today, women are the primary caretakers of companion animals. Because of women's association with pets, and in particular, with lap dogs, pets were seen as feminizing and symbolized women's inferiority. Even worse, a woman in Medieval Europe could be accused of being a witch if she had too close of a relationship with an animal. European elites were often satirized in the press for their keeping of pets, although noblemen who kept hunting dogs were exempt from this stigma.

The poor, on the other hand, were prohibited from hunting in England (as lands were owned either by the church, the crown, or the nobility); consequently, they were prohibited from keeping hunting dogs. Although people from all socioeconomic statuses keep pets today, high income remains correlated with pet keeping—the higher one's income, the greater the likelihood of living with a pet. In addition, purebred pets are associated with elites, whereas the poor kept animals that the wealthy saw as dirty and diseased. Even today, purebred animals are still linked to elites.

Research on pet keeping among the wealthy reveals that owning companion animals has long been an expression of luxury, social refinement, and status. Historical studies demonstrate that royal courts and aristocratic households in European societies traditionally maintained elaborate collections of pets and exotic animals as a means of displaying wealth and taste. In the eighteenth century, for example, elite individuals frequently kept lapdogs and exotic species not only for companionship but also to signal their elevated social position and refined cultural sensibilities Brown (2011). Such practices were often criticized, as detractors viewed extravagant pet keeping as a wasteful indulgence—a display of wealth that highlighted the divide between the material excesses of the rich and the more utilitarian approaches to animal companionship found in other social classes (Serpell, 1987).

This duality of pet keeping among the affluent—the simultaneous celebration of luxurious lifestyle elements and the criticism of wastefulness—was further reinforced during the Victorian and Edwardian eras. Research by Strange (2021) illustrates how the practices and ideals surrounding pet keeping in late nineteenth-century Britain emphasized class distinctions. The types of pets kept (often pedigreed or exotic) and their display

Prince and Princess August Wilhelm of Prussia with their two bulldogs in 1911. The dogs won prizes at the Berlin annual dog show in Germany. (Library of Congress)

in domestic settings served as markers of exclusive identity and taste. Over time, these historical precedents have contributed to a modern cultural landscape in which luxury pet products and specialized services cater to the wealthy while perpetuating the notion that pet keeping is an emblem of affluence and social prestige.

Overall, the scholarly literature indicates that among the wealthy, pet ownership has evolved from a royal tradition of exotic and highly curated animal companionship into a contemporary marker of elite lifestyle, even as it attracts criticism for representing wastefulness and ostentatious consumption. This complex history highlights how pets have functioned as both cherished companions and potent symbols within stratified societies.

See also: Capitalism, Pets And; Socioeconomic Class, Pets And; Status Pets

References and Further Reading

Brown, L. (2011). "The Lady, the Lapdog, and Literary Alterity." *The Eighteenth Century*, 52 (1): 31–45.

Maehle, A. H. (2002). "Cruelty and Kindness to the 'Brute Creation': Stability and Change in the Ethics of the Man–Animal Relationship, 1600–1850." In A. Manning and J. Serpell (eds), *Animals and Human Society*, 113–37. London: Routledge.

Ritvo, H. (1987). "The Emergence of Modern Pet-Keeping." *Anthrozoös*, 1 (3): 158–65.

Serpell, J. A. (1987). "Pet-keeping in Non-Western Societies: Some Popular Misconceptions." *Anthrozoös*, 1 (3): 166–74.

Strange, J. M. (2021). "When John Met Benny: Class, Pets and Family Life in Late Victorian and Edwardian Britain." *The History of the Family*, 26 (2): 214–35.

EMPATHY

A number of recent studies point to a correlation between positive attitudes toward companion animals and a more humane attitude toward other animals, and even some very preliminary studies are showing there is a link between positive attitudes toward animals and a more compassionate attitude toward people.

Elizabeth Paul and James Serpell (1993) trace the evolution of animal-keeping in the West and its association with attitudes toward fellow humans. They point out, for example, that starting in the seventeenth century, many of the most enlightened humanitarians had an affinity for animals, and that scholars and ancient Greek philosophers, dating back to the ancient Greeks, thought that eliminating violence toward animals would make humans more peaceful. We also know that abolitionists and animal rights activists were often the same people, so there appears to be at least a small correlation between affinity for animals and social justice. In fact, as we have seen earlier, during the nineteenth century, when the commercial pet industry began to develop, many saw animal companionship as a way to cultivate virtues such as kindness and self-control in young people. In addition, the humane education movement at that time took as its central premise the idea that childhood pet keeping could be used as a springboard to teach children empathy toward other animals.

In 1882, George Angell, founder of the Massachusetts SPCA, created the Bands of Mercy, which were after-school clubs where children met, learned about animals, prayed, and testified about the animals that they had helped. Currently, thirteen US states have laws mandating that public schools teach some form of humane education, acknowledging the importance of having young people learn about kindness toward animals.

Today, some scholars (Cloutier and Peetz, 2016; Paul, 2000; Daly and Suggs, 2010; Hawkins et al., 2017) think that living with animals might, in fact, teach empathy and compassion—toward both animals and people. However, other studies challenge this notion, finding evidence that, for example, children with pets might not show higher empathy, and that living with cats is, in fact, negatively correlated with empathy (Daly and Morton, 2006). Another point to consider is that pets occupy a unique position in human society—as cultural theorist Erica Fudge points out, "a pet is a pet first, an animal second" (2014: 32). In other words, we can love pets, but still not think too much about other animals, because pets aren't *really* animals, after all.

See also: Adults, and Pets; Children and Pets; Socialization, Pets And

References and Further Reading

Cloutier, A., and J. Peetz (2016). "Relationships' Best Friend: Links Between Pet Ownership, Empathy, and Romantic Relationship Outcomes." *Anthrozoös*, 29 (3): 395–408.

Daly, B., and L. L. Morton (2006). "An Investigation of Human-Animal Interactions and Empathy as Related to Pet Preference, Ownership, Attachment, and Attitudes in Children." *Anthrozoös*, 19 (2): 113–27.

Daly, B., and S. Suggs (2010). "Teachers' Experiences with Humane Education and Animals in the Elementary Classroom: Implications for Empathy Development." *Journal of Moral Education*, 39 (1): 101–12.

Ellingsen, K., A. J. Zanella, E. Bjerkås, and A. Indrebø (2010). "The Relationship Between Empathy, Perception of Pain and Attitudes Toward Pets Among Norwegian Dog Owners." *Anthrozoös*, 23 (3): 231–43.

Fudge, E. (2014). *Pets*. London: Routledge.

Hawkins, R. D., J. M. Williams, and Scottish Society for the Prevention of Cruelty to Animals (Scottish SPCA) (2017)."Childhood Attachment to Pets: Associations Between Pet Attachment, Attitudes to Animals, Compassion, and Humane Behaviour." *International Journal of Environmental Research and Public Health*, 14 (5): 490.

Khalid, A., and I. Naqvi (2016). "Relationship Between Pet Attachment and Empathy Among Young Adults." *Journal of Behavioural Sciences*, 26 (1): 66.

Paul, E. S. (2000). "Love of Pets and Love of People." In *Companion Animals and Us: Exploring the Relationships Between People and Pets*, 168–86. Cambridge: Cambridge University Press.

Paul, E. S., and J. A. Serpell (1993). "Childhood Pet Keeping and Humane Attitudes in Young Adulthood." *Animal Welfare*, 2 (4): 321–37.

EQUINES

The evolution of horses as companion animals has seen a significant transition from their traditional roles in agriculture and work to becoming cherished pets in contemporary situations. This change in perception is reflected in various scholarly works that delve into the relationship between humans and horses, particularly regarding welfare, companionship, and the increasing popularity of miniature equines.

Historically, horses have been utilized for agricultural tasks, transportation, and labor. However, the decline of agricultural reliance on horses has paved the way for their emergence in the leisure and companion animal sector. The growing engagement of adolescents with horses highlights their role as both recreational companions and therapeutic entities, suggesting that the bond between humans and horses can foster emotional well-being and social interaction, especially in therapeutic contexts such as equine-assisted activities for children with Autism Spectrum Disorder (Hauge et al., 2015; Petty et al., 2017). This indicates a broader shift in societal perceptions, whereby horses are increasingly seen as pets that provide companionship and emotional support.

Current studies illustrate the complexity of the human-horse relationship, particularly concerning horse welfare and owner perceptions. Horse owners often view their animals through the lens of companionship, particularly in recreational contexts (Dubois et al., 2018). This perspective aligns with findings where horses are conceptualized not just as livestock but as companions similar to traditional pets like dogs and cats (Keaveney, 2008). The emotional attachments that owners develop toward their horses may rival those found in other pet relationships, as demonstrated by studies exploring the psychological benefits of interacting with these animals (Petty et al., 2017).

Additionally, the phenomenon of miniature horses has become increasingly prominent, attracting specific interest as they are well-suited for urban environments and serve as therapy animals. The distinct neurological and behavioral responses seen in equines enhance their capacity for forming bonds with humans, thereby facilitating emotional interactions comparable to those observed in more common pet species (Rørvang, Nielsen, and McLean, 2020). This aligns with public sentiments addressing animal welfare; horses today are often maintained for recreational purposes, reflecting shifting societal values toward animal companionship and welfare.

Meanwhile, concerns surrounding horse welfare remain salient, with studies indicating that many owners struggle with accurately assessing their horse's body condition, which can lead to potential welfare issues (Ross et al., 2024; Golding et al., 2023). This underscores the need for improved education on proper equine care, emphasizing the necessity for horse owners to align their perceptions with best practices to ensure optimal welfare (Furtado et al., 2020). The notion of a horse as a pet parallels discussions in broader animal welfare discourse, where emotional considerations are progressively more integrated into the management practices of equines (Schuurman, 2015).

In conclusion, the literature indicates that horses are increasingly perceived and kept as pets, with miniature equines exemplifying a niche that caters to urban dwellers. The changing societal values toward companionship and welfare underscore the evolving role of horses within human lives, illustrating that they are not merely working animals but integral companions that contribute to mental health and emotional fulfillment.

See also: Bond, Human-Animal; Farm Animals as Pets; Therapy Pet

References and Further Reading

Dubois, C., L. Nakonechny, E. Derisoud, and K. Merkies (2018). "Examining Canadian Equine Industry Participants' Perceptions of Horses and their Welfare." *Animals*, 8 (11): 201.

Furtado, T., E. Perkins, G. Pinchbeck, C. McGowan, F. Watkins, and R. Christley (2020). "Exploring Horse Owners' Understanding of Obese Body Condition and Weight Management in UK Leisure Horses." *Equine Veterinary Journal*, 53 (4): 752–62.

Golding, E., A. Ansari, G. Sutton, N. Walshe, and V. Duggan (2023). "Rate of Obesity Within a Mixed-Breed Group of Horses in Ireland and their Owners' Perceptions of Body Condition and Useability of an Equine Body Condition Scoring Scale." *Irish Veterinary Journal*, 76 (1).

Hauge, H., I. Kvalem, M. Enders-Slegers, B. Berget, and B. Braastad (2015). "Persistence During Tasks with Horses in Relation to Social Support, General Self-Efficacy and Self-Esteem in Adolescents." *Anthrozoös*, 28 (2): 333–47.

Keaveney, S. (2008). "Equines and their Human Companions." *Journal of Business Research*, 61 (5): 444–54.

Petty, J., Z. Pan, B. Dechant, and R. Gabriels (2017). "Therapeutic Horseback Riding Crossover Effects of Attachment Behaviors With Family Pets in a Sample of Children with Autism Spectrum Disorder." *International Journal of Environmental Research and Public Health*, 14 (3): 256.

Pritchard, J., M. Upjohn, and T. Hirson (2018). "Improving Working Equine Welfare in 'Hard-Win' Situations, Where Gains are Difficult, Expensive or Marginal." *Plos One*, 13 (2): e0191950.

Rørvang, M., B. Nielsen, and A. McLean (2020). "Sensory Abilities of Horses and their Importance for Equitation Science." *Frontiers in Veterinary Science*, 7: 633.

Ross, M., K. Proudfoot, E. Nishimura, E. Morabito, K. Merkies, J. Mitchell, and C. Ritter (2024). "'It's More Emotionally Based': Prince Edward Island Horse Owner Perspectives of Horse Weight Management." *Animal Welfare*, 33: e14.

Schuurman, N. (2015). "Conceptions of Equine Welfare in Finnish Horse Magazines." *Society and Animals*, 23 (3): 250–68.

Seck, M., G. Carder, J. Wathan, M. Randau, K. Fletcher, and L. Proops (2024). "Working Horse Welfare in Senegal is Linked to Owner's Socioeconomic Status, their Attitudes and Belief in Horse Sentience." *Plos One*, 19 (10): e0309149.

ETHICS OF PET OWNERSHIP

The ethical implications of pet ownership have garnered increasing attention within philosophical discussions, particularly as they pertain to animal welfare and the moral responsibilities of pet owners. Recent literature reflects a growing unease about the ethics of keeping pets, highlighting various dimensions of this complex relationship between humans and animals.

Is the practice of keeping pets ethically justifiable (Kountaki, 2019)? Central to this question is the consideration of pets as sentient beings with their own needs and interests, which often conflict with human desires (Pierce, 2019).

Moreover, the structural and societal factors surrounding pet keeping also contribute to ethical concerns. For example, Mascitelli et al. highlight issues regarding housing policies that often impose restrictions on pet ownership, which can exacerbate the ethical implications of ownership, particularly for vulnerable populations who rely on the companionship of animals for emotional support (Mascitelli et al., 2024).

Socioeconomic factors further complicate the ethics of pet ownership. Research shows that economically disadvantaged owners may be less likely to provide appropriate care, leading to welfare concerns for the pets involved (Mee et al., 2022). Additionally, issues related to owners' socioeconomic status can influence their ability to care for pets properly and raise questions regarding how society can equitably ensure responsible ownership (Mota-Rojas et al., 2021). Correspondingly, another study indicates that individuals with more favorable attitudes toward animals are less likely to engage in neglectful behaviors, suggesting that fostering ethical consciousness can improve the welfare of pets (Busch et al., 2022).

The impact of pet ownership on ethical consumption behavior is significant as well. Liang et al. found that pet owners are typically more inclined to support animal welfare initiatives, indicating that ownership can foster a sense of responsibility toward the ethical treatment of animals more broadly (Liang et al., 2024). This suggests a potential for pet ownership to promote not only individual ethical considerations but also communal values regarding animal welfare (Pearce et al., 2023).

The welfare of exotic pets brings additional ethical challenges. Azevedo et al. emphasize the need for robust welfare standards for reptiles kept as pets, arguing that their particular needs often go unmet due to misconceptions surrounding their care (Azevedo et al., 2021). Furthermore, the rising trend of keeping exotic animals poses risks to both the species involved and the ecological balance, leading to calls for stricter regulations surrounding exotic pet ownership to prevent exploitation and ensure ethical treatment (McMillan et al., 2020).

In summary, the philosophical discourse on the ethics of pet keeping is highly nuanced, encapsulating a range of topics including individual moral responsibilities, societal inequalities, and the implications of consumer behavior. These discussions underline the necessity for ongoing critical evaluation of pet ownership practices, emphasizing a collective moral obligation to ensure the welfare and ethical treatment of companion animals.

See also: Affection and Control; Anthropomorphism; Empathy; Exotic Pets; Reptiles

References and Further Reading

Azevedo, A., L. Guimarães, J. Ferraz, M. Whiting, and M. Magalhães-Sant'Ana (2021). "Pet Reptiles—Are We Meeting Their Needs?" *Animals*, 11 (10): 2964.

Busch, G., A. Schütz, S. Hölker, and A. Spiller (2022). "Is Pet Ownership Associated With Values And Attitudes Towards Animals?" *Animal Welfare*, 31 (4): 447–54.

Finsen, S. (2020). "Keeping Companion Animals: Dilemmas of Domestication." *Journal of Animal Ethics*, 10 (1): 59–65.

Kountaki, D. (2019). "Jessica Pierce. Run, Spot, Run: The Ethics of Keeping Pets." Chicago: University of Chicago Press, 2016. *Conatus*, 4 (1): 137.

Liang, Y., M. Cheng-min, R. Chen, Y. Yang, and Y. Zeng (2024). "Pet Ownership and its Influence On Animal Welfare Attitudes and Consumption Intentions Among Chinese University Students." *Animals*, 14 (22): 3242.

Mascitelli, T., T. Graham, L. Loney, J. Applebaum, C. Murray, M. Binns-Calvey, and K. Morris (2024). "Barriers to Finding and Maintaining Pet-Inclusive Affordable Housing: Tenant Experiences in Houston, Texas." *Frontiers in Veterinary Science*, 11: 1465682.

McMillan, S., C. Dingle, J. Allcock, and T. Bonebrake (2020). "Exotic Animal Cafes are Increasingly Home to Threatened Biodiversity." *Conservation Letters*, 14 (1): 1–9.
Mee, G., E. Tipton, J. Oxley, and C. Westgarth (2022). "Owner Demographic Factors are Associated with Suitable Pet Rabbit Housing Provision in the United Kingdom." *Veterinary Record*, 190 (12): e1736.
Mota-Rojas, D., N. Calderón-Maldonado, K. Lezama-García, L. Sepiurka, and R. Garcia (2021). "Abandonment of Dogs in Latin America: Strategies and Ideas." *Veterinary World*, 14 (9): 2371–79.
Pearce, H., C. Neill, K. Royal, and M. Pairis-Garcia (2023). "Can Dogs Help Chickens? Pet Owners' Willingness to Pay for Animal Welfare-Friendly Pet Food in the United States." *Animal Welfare*, 32: e11.
Pierce, J. (2019). "The Ethics of Keeping Pets." In The Routledge Handbook of Animal Ethics, 316–25. New York: Routledge.

ETHNICITY, PETS AND

The relationship between pet ownership and ethnicity, race, and associated social dynamics reveals significant disparities influenced by historical and institutional factors.

Historically, pet adoption and ownership practices have varied widely across different racial and ethnic communities, reflecting broader societal norms and economic policies. For instance, African American communities have historically faced systemic barriers to pet ownership, including economic constraints and housing policies that often do not accommodate pet ownership, leading to lower rates of pet ownership compared to other ethnic groups, notably white families (Zhang, House, and Salois, 2024; Albright, Cui, and Allen, 2022; Bibbo et al., 2022). These institutional barriers underpin the difficulties that various demographic groups experience, notably due to financial strain and discriminatory housing practices that often stigmatize pet ownership among marginalized communities (Bibbo et al., 2022; Maris and Cameron, 2023).

Significant research indicates variations in how different ethnicities approach pet care and the economic burdens involved, contributing to disparities in veterinary care perceived within and among groups. For example, it has been shown that white families generally allocate more resources for veterinary care compared to their African American and Hispanic counterparts, pointing toward socioeconomic factors influencing pet care decisions and, by extension, pet ownership (Zhang, House, and Salois, 2024; Mueller et al., 2021). As noted by Girgenti (2020), there has been an observable shift with Hispanic households showing increased rates of pet ownership, potentially indicating changing cultural perceptions around pets (Zhang, House, and Salois, 2024).

Housing plays a critical role in pet ownership dynamics across different racial and ethnic groups. Renters, who are more likely to belong to minority groups, often experience significant challenges in finding pet-friendly housing, which forces them to compromise on location and living standards to retain their pets (Maris and Cameron, 2023). Additionally, this is compounded by the economic reality that many prospective pet owners face; financial barriers remain a pivotal issue inhibiting pet ownership, especially in lower-income communities where budget constraints are more pronounced (Bibbo et al., 2022; Meier and Maurer, 2022).

Furthermore, the implications of pet ownership extend to mental and physical health dimensions, suggesting that disparities in pet ownership could translate into broader implications for well-being in marginalized populations. Research indicates pets can have beneficial effects on mental health, but these benefits are less accessible to communities facing socioeconomic hardships (Saunders et al., 2017; Chin et al., 2023). The socioemotional benefits of pet ownership may be harder to realize for African Americans, who face unique challenges including historical trauma associated with animals and stigma surrounding pet care, which are less prevalent in white communities (Albright, Cui, and Allen, 2022; Irvine, Kahl, and Smith, 2012).

In addition, the problematic history of African Americans with dogs, particularly due to the association of dogs with state control over Black bodies during both slavery and the Civil Rights Movement, perpetuates a complex relationship between this group and pet ownership today. This cultural and historical context must be considered when examining current attitudes toward pets within African American and broader minority communities (Conway et al., 2022).

In conclusion, the intersection of race, ethnicity, and institutional barriers elucidates the complexity of pet ownership dynamics. Significant discrepancies in pet ownership across different ethnicities highlight not only historical and cultural factors but also socioeconomic constraints that inform access to pet-friendly housing and adequate veterinary care. Addressing these disparities is crucial for fostering inclusive practices that support pet ownership in diverse communities.

See also: Dogs; Social Capital, Pets And; Socioeconomic Class, Pets And

References and Further Reading

Albright, A., R. Cui, and R. Allen (2022). "Pet Ownership and Mental and Physical Health in Older White and Black Males and Females." *International Journal of Environmental Research and Public Health*, 19 (9): 5655.

Bibbo, J., J. Johnson, J. Drost, M. Sanders, and S. Nicolay (2022). "Pet Ownership Issues Encountered by Geriatric Professionals: Preliminary Findings from an Interdisciplinary Sample." *Frontiers in Psychology*, 13.

Chin, B., E. Casserly, S. Parsons, and B. Feeney (2023). "Pet Ownership and Mental Health in United States Adults During Covid-19." *Frontiers in Psychology*, 14: 1217059.

Conway, K., R. Jasuja, N. Hauser, and J. Foley (2022). "Benefits, Companion Animal Zoonotic Disease Prevalence and Public Perceptions of Pet Ownership Among People Experiencing Homelessness in Northern California." *Zoonoses and Public Health*, 69 (7): 806–15.

Girgenti, A. (2020). "North American Pet Ownership Demographics Shifting." Petfood Industry. https://www.petfoodindustry.com/articles/6938-north-american-pet-ownership-demographics

Irvine, L., K. Kahl, and J. Smith (2012). "Confrontations and Donations: Encounters Between Homeless Pet Owners and the Public." *Sociological Quarterly*, 53 (1): 25–43.

Maris, R., and M. Cameron (2023). "The Labour Market Trade-Offs of Pet Ownership." *Economics and Business Letters*, 12 (1): 75–84.

Meier, C., and J. Maurer (2022). "Buddy or Burden? Patterns, Perceptions, and Experiences of Pet Ownership Among Older Adults in Switzerland." *European Journal of Ageing*, 19 (4): 1201–12.

Mueller, M., E. King, K. Callina, S. Dowling-Guyer, and E. McCobb (2021). "Demographic and Contextual Factors as Moderators of The Relationship Between Pet Ownership and Health." *Health Psychology and Behavioral Medicine*, 9 (1): 701–23.

Saunders, J., L. Parast, S. Babey, and J. Miles (2017). "Exploring the Differences Between Pet and Non-Pet Owners: Implications for Human-Animal Interaction Research and Policy." *Plos One*, 12 (6): e0179494.

Zhang, X., L. House, and M. Salois (2024). "An Examination of US Pet Owners' Use of Veterinary Services, 2006–2018." *Veterinary Medicine and Science*, 10 (3): e1370.

EUROPE, PETS IN

Pets, in the modern sense of the word, are a European invention. While elites in particular kept pets throughout much of history, it was only in Europe in the eighteenth and nineteenth centuries that economic and cultural conditions allowed for the expansion of pet keeping into wider society; from there, this style of pet keeping spread around the world.

Initially, the notion of pets as potentially wasteful arose from the perspective that keeping animals primarily for companionship or as symbols of status detracted from more utilitarian agricultural or domestic purposes. This sentiment was evident in various discourses during the eighteenth and nineteenth centuries when economic historians critiqued indulgences in non-productive animal ownership, arguing that it diverted resources from essential societal needs (Strange, 2021).

As the Victorian era unfolded, marked by the Industrial Revolution, the rise of the middle class profoundly changed societal norms surrounding pet ownership. This period saw a shift where pets began to be integrated into the domestic sphere not only as symbols of affluence but also as integral family members. The bourgeois fantasy of pets—particularly the "family dog"—reflected broader changes in how social classes defined themselves, with pets serving as emotional and social facilitators in domestic spaces (Strange, 2021; Hamlett and Strange, 2021). Furthermore, the increasing affordability of certain pets and the emergence of pet care industries made pet keeping accessible to a wider demographic, ultimately embedding it deeper into middle-class identity (Strange, 2021).

The Victorian emphasis on educating children about kindness through pet care marked a pivotal development in family ideology. Parents viewed animals as crucial in teaching virtues such as empathy and responsibility among children. Scholarly discourse supports this, noting that domestic advisers of the period recommended pet ownership as a means to cultivate compassion, arguing that it especially helped boys develop emotional steadiness and moral character (Hamlett and Strange, 2021; Grier, 1999). The practice of engaging children with pets was thus not just an exercise in domesticity but an active strategy for societal moral training, reinforcing the idea that human–animal interactions foster moral values essential for both familial cohesion and community stability (Hamlett and Strange, 2021; Grier, 1999).

The historical trajectory of pet keeping in Europe underscores a significant interplay of social class, perceptions of wastefulness, and educational philosophies concerning kindness. The origin of pets as ostensibly frivolous companions transitioned into a more complex acknowledgment of their role in human emotional development and socialization, thus illustrating how social values around animals evolved alongside broader societal changes from the eighteenth to the nineteenth century.

See also: History of Pet Keeping: Middle Ages; History of Pet Keeping: Renaissance; History of Pet Keeping: Victorian Era; United States, Pets In

References and Further Reading

Grier, K. (1999). "Childhood Socialization and Companion Animals: United States, 1820–1870." *Society and Animals*, 7 (2): 95–120.

Hamlett, J., and J. Strange (2021). "Animals in the Family Mini-Special Issue Introduction and Historiographical Review." *The History of the Family*, 26 (2): 173–85.

Strange, J. (2021). "When John Met Benny: Class, Pets and Family Life in Late Victorian and Edwardian Britain." *The History of the Family*, 26 (2), 214–35.

EXOTIC PETS

The term "exotic pet" refers to the keeping of wild animals as pets, although the term has also been used to refer to the keeping of any "nontraditional" pet. The literature on exotic pets encompasses a broad range of interconnected issues, notably the wild pet trade, subsequent impacts on wildlife and local conservation efforts, and welfare concerns associated with keeping such animals. The growing interest in exotic pets, alongside their complex care needs, raises significant ethical and ecological questions.

The wild pet trade has escalated significantly over the years, driven largely by consumer demand for nontraditional pets, which typically involves the capture and transportation of animals from their natural habitats. This trend creates pressures on wild populations, contributing to declines in species and exacerbating the risk of extinction for many threatened species. For example, Mcmillan et al. highlight the adverse effects of exotic animal cafes on biodiversity, noting that increased demand for these animals puts pressure on their wild counterparts (McMillan et al., 2020). Similarly, the analysis of overseas wildlife trade mechanisms discussed by Bush et al. underscores a critical link between consumer habits in countries driving demand for exotic pets and the resultant ecological impacts on species populations and conservation efforts (Bush, Baker, and Macdonald, 2014).

From a conservation perspective, the implications are stark. The exotic pet trade not only affects biodiversity through the direct removal of species from their ecosystems but can also facilitate the introduction of invasive species, dramatically altering local ecosystems. Lockwood et al. emphasize that many non-native populations introduced through the exotic pet trade contribute significantly to vertebrate invasions (Lockwood et al., 2019). Furthermore, illegal wildlife trade remains a pressing issue correlated with the legal pet trade, as many legally sold exotic pets may originate from illegally sourced populations, hence complicating conservation efforts (Moorhouse, D'Cruze, and Macdonald, 2021).

Animal welfare in the context of exotic pet ownership is another critical aspect of this discussion. Exotic pets often have specialized needs that are largely unmet in domestic environments, leading to significant welfare issues. Elwin et al. detail the

This mother macaque monkey was once part of the illegal pet trade. She now lives at Folly Farm Zoo in Wales, the United Kingdom. (Gray Freeman/Dreamstime.com)

welfare challenges faced by exotic pets, which retain complex social and behavioral needs despite their presence in captivity (Elwin, Green, and D'Cruze, 2020). Animals in captivity may suffer from physical and psychological stress due to inappropriate care and environments, as noted by Lambert et al., who argue for the recognition of nonhuman sentience to improve husbandry practices in animal welfare contexts (Lambert, Carder, and D'Cruze, 2019). This issue is particularly acute for pets that are often abandoned or neglected once their care becomes too complex for the owner (Harrington, Macdonald, and D'Cruze, 2019).

Regulatory frameworks, such as the Convention on International Trade in Endangered Species (CITES), play a role in managing the trade in exotic pets. However, loopholes in enforcement enable continued issues with illegal trade (Aguiar et al., 2025). Public awareness and consumer behavior significantly influence the success of these regulations. Educational campaigns that elucidate the risks associated with zoonotic diseases—such as those discussed by Moorhouse et al.—can effectively reduce demand for exotic pets by informing potential owners about the inherent dangers (Moorhouse et al., 2021). Campaigns targeting the environmental and ethical challenges of the exotic pet trade have shown varying results in shifting public attitudes, as examined by Naito et al. (2024).

In summary, the exotic pet trade presents multifaceted challenges that touch on ecological preservation, animal welfare, and regulatory efficacy. The overlaps between these domains highlight the need for comprehensive strategies that address consumption behaviors, improve animal welfare standards, and enforce robust conservation initiatives to mitigate the negative impacts of this growing phenomenon.

See also: Birds; Pet Industry; Reptiles

References and Further Reading

Aguiar, R., R. Gray, E. Gallo-Cajiao, A. Rückert, C. Astbury, R. Labonté, P. Tsasis, A. M. Viens, and M. Wiktorowicz (2025). "Preventing Zoonotic Spillover Through Regulatory Frameworks Governing Wildlife Trade: A Scoping Review." *Plos One*, 20 (1): e0312012.

Bush, E., S. Baker, and D. Macdonald (2014). "Global Trade in Exotic Pets 2006–2012." *Conservation Biology*, 28 (3): 663–76.

Elwin, A., J. Green, and N. D'Cruze (2020). "On the Record: An Analysis of Exotic Pet Licences in the UK." *Animals*, 10 (12): 2373.

Harrington, L., D. Macdonald, and N. D'Cruze (2019). "Popularity of Pet Otters on Youtube: Evidence of an Emerging Trade Threat." *Nature Conservation*, 36: 17–45.

Lambert, H., G. Carder, and N. D'Cruze (2019). "Given the Cold Shoulder: A Review of the Scientific Literature for Evidence of Reptile Sentience." *Animals*, 9 (10): 821.

Lockwood, J., D. Welbourne, C. Romagosa, P. Cassey, N. Mandrak, A. Strecker, R. Keller, et al. (2019). "When Pets Become Pests: The Role of The Exotic Pet Trade in Producing Invasive Vertebrate Animals." *Frontiers in Ecology and the Environment*, 17 (6): 323–30.

McMillan, S. E., C. Dingle, J. A. Allcock, and T. C. Bonebrake (2021). "Exotic Animal Cafes are Increasingly Home to Threatened Biodiversity." *Conservation Letters*, 14 (1): e12760.

Moorhouse, T. P., N. C. D'Cruze, and D. W. Macdonald (2021). "Information about Zoonotic Disease Risks Reduces Desire to Own Exotic Pets Among Global Consumers." *Frontiers in Ecology and Evolution*, 9: 609547.

Naito, R., K. M. A. Chan, and J. Zhao (2024). "Combating the Exotic Pet Trade: Effects of Conservation Messaging on Attitudes, Demands, and Civic Intentions." *Conservation Science and Practice*, 6 (2): e13078.

F

FACILITY PETS

A facility pet is an animal used within institutional settings like schools, hospices, or hospitals to support the well-being of residents, patients, students, and staff.

While relatively new, the use of facility pets, especially in institutions with seniors or children—has grown considerably over the past few decades. Early initiatives in animal-assisted interventions can be traced to programs in correctional facilities and healthcare settings, where animals were intended to alleviate emotional distress and enhance rehabilitation through structured interactions (Allison and Ramaswamy, 2016; Scott and Kirnan, 2023). Over time, such practices expanded into a range of settings, including long-term care facilities, acute care hospitals, educational institutions, and substance abuse treatment centers, reflecting an increasing recognition of the benefits animals can offer in mitigating isolation, anxiety, and behavioral challenges (Allison and Ramaswamy, 2016; Coakley et al., 2021; Stanley et al., 2014).

Different types of facilities have adopted facility pet programs to address distinct needs. Prisons, for instance, have implemented animal-assisted therapy programs with structured sessions lasting several weeks, aimed at reducing stress, encouraging prosocial behavior, and supporting rehabilitative goals among inmates (Allison and Ramaswamy, 2016). In educational settings, particularly primary school classrooms, animals such as guinea pigs have been introduced as part of animal-assisted activities; these interventions are associated with improved social skills and increased psychological well-being (O'Haire et al., 2013; Brelsford et al., 2017). Similarly, nursing homes and other long-term care facilities have utilized facility pet programs to combat loneliness and encourage social interaction among residents, demonstrating measurable reductions in anxiety and improvements in overall mood (Stanley et al., 2014). Acute care settings and substance abuse treatment facilities have also experienced beneficial impacts, with animal-assisted activities yielding reductions in stress and providing a nonverbal channel for emotional expression to both patients and staff alike (Scott and Kirnan, 2023, Coakley et al., 2021).

The roles these facility animals play are multifaceted. In educational environments, they serve as a catalyst for improved social functioning, emotional stability, and even academic engagement by fostering a more relaxed classroom atmosphere (O'Haire et al., 2013; Brelsford et al., 2017). In residential and correctional facilities, animal-assisted interventions function therapeutically—helping to decrease depressive symptoms, reduce perceived loneliness, and promote an overall sense of comfort (Allison and Ramaswamy, 2016; Stanley et al., 2014). In clinical settings, the presence of facility pets has been linked to enhanced patient outcomes, including reduced anxiety and improved satisfaction with care (Scott and Kirnan, 2023; Coakley et al., 2021). The psychological and emotional benefits observed across these contexts underscore the role of facility pets in educational and therapeutic programming.

Regarding the origins and training of facility pets, contemporary practice involves the careful selection of animals based on temperament, adaptability, and ease of training. Animals often include dogs, guinea pigs, and other small mammals, typically sourced from specialized animal-assisted intervention organizations or accredited facilities that observe ethical treatment and strict welfare guidelines. Professional trainers use reward-based methods, such as positive reinforcement and gradual desensitization, to ensure reliability in environments that may be unfamiliar or stressful (Scott and Kirnan, 2023). These trained behaviors are essential not only for ensuring the safety and comfort of the facility's inhabitants but also for optimizing the therapeutic or educational outcomes of the interventions.

Initial deployments in prisons and healthcare facilities have expanded into educational and residential contexts, where facility pets yield tangible benefits for psychological, emotional, and social development. The adoption and training practices emphasize animal welfare and ethical treatment, reinforcing that when properly managed, facility pets are a valuable adjunct to interventions designed to reduce stress, foster empathy, and stimulate social and emotional growth among participants. Continued research is necessary to further elucidate long-term outcomes and best practices across diverse facility environments.

See also: Human Health, Pets And; Mental Health, Pets And; Therapy Pet

References and Further Reading

Allison, M., and M. Ramaswamy (2016). "Adapting Animal-Assisted Therapy Trials to Prison-Based Animal Programs." *Public Health Nursing*, 33 (5): 472–80.

Brelsford, V. L., K. Meints, N. R. Gee, and K. Pfeffer (2017). "Animal-Assisted Interventions in the Classroom—A Systematic Review." *International Journal of Environmental Research and Public Health*, 14 (7): 669.

Coakley, A. B., C. D. Annese, J. H. Empoliti, and J. M. Flanagan (2021). "The Experience of Animal-Assisted Therapy on Patients in an Acute Care Setting." *Clinical Nursing Research*, 30 (4): 401–5.

Murthy, R., G. Bearman, S. Brown, K. Bryant, R. Chinn, A. Hewlett, B. G. George, E. J. Goldstein, G. Holzmann-Pazgal, M. E. Rupp, T. Wiemken, J. S. Weese, D. J. Weber, et al. (2015). "Animals in Healthcare Facilities: Recommendations to Minimize Potential Risks." *Infection Control & Hospital Epidemiology*, 36 (5): 495–516.

O'Haire, M. E., S. J. McKenzie, A. M. Beck, and V. Slaughter (2013). "Social Behaviors Increase in Children with Autism in the Presence of Animals Compared to Toys." *PLOS One*, 8 (2): e57010.
Scott, T. M., and J. Kirnan (2023). "When Are the Dogs Coming Back? Animal-Assisted Activities with Men in Residential Substance Abuse Treatment." *The Humanistic Psychologist*, 51 (2): 173.
Stanley, I. H., Y. Conwell, C. Bowen, and K. A. Van Orden (2014). "Pet Ownership May Attenuate Loneliness Among Older Adult Primary Care Patients Who Live Alone." *Aging & Mental Health*, 18 (3): 394–99.
Stull, J. W., C. C. Hoffman, and T. Landers (2018). "Health Benefits and Risks of Pets in Nursing Homes: A Survey of Facilities in Ohio." *Journal of Gerontological Nursing*, 44 (5): 39–45.

FAMILY, PETS AS

Research on pets as family members has grown into a rich interdisciplinary field that draws on social work, sociology, and psychological research. Studies in these areas detail how nonhuman animals are increasingly incorporated into family roles, with pets now frequently regarded as kin and integral to household identity and emotional well-being.

Early sociological work established that pets are not merely property but function as members of the family, providing emotional support, companionship, and even serving as markers of identity within households Irvine and Cilia (2017). Qualitative research has emphasized that kinship with pets transgresses traditional household boundaries, giving rise to "more-than-human families" in which pets are central to communal rituals and interpersonal relationships Laurent-Simpson (2017). Scholars such as Laurent-Simpson (2017) have illustrated that the inclusion of pets in family narratives reinforces a sense of external support and companionship that mimics traditional human kinship, while studies by Ryan and Ziébland (2015) demonstrate how pets occupy significant roles in the lives of individuals dealing with chronic illness or social isolation.

Psychological research further supports these sociological findings by indicating that strong human–animal bonds can positively impact mental health. For example, research shows that children growing up in pet-owning families exhibit fewer peer problems and improved prosocial behavior, signifying that the presence of pets may contribute to healthier emotional and social development (Christian et al., 2020). In adults, the bond with companion animals, often characterized by attributes akin to familial closeness, has been linked to reduced loneliness, lower stress levels, and enhanced overall well-being. The emotional attachment formed with pets is sometimes described in ways analogous to the bond between a parent and child. Additionally, studies indicate that pet ownership can buffer the psychological impact of diminishing human social networks, particularly in demographic contexts of lowered birth rates and changing family structures Temesi et al. (2020).

Demographic shifts have also contributed to the changing role of pets in the family. Lower birth rates and evolving societal attitudes toward traditional family models have been associated with an increased tendency among individuals to form strong emotional bonds with their pets. Research by Temesi et al. (2020) suggests that women's cognitive representations of motherhood and family roles are increasingly mirrored in their attitudes toward pet keeping. This is complemented by findings that show households with fewer or no children are more likely to invest emotionally in pets, viewing them as surrogate family members (Westgarth et al., 2007). Such trends are further illuminated by studies on aging populations, where pets contribute significantly to maintaining social connectedness and emotional stability, effectively substituting for lost or diminished human kinship bonds (McGrane et al., 2024).

Finally, this convergence of social, psychological, and demographic factors underlines the profound impact of perceiving nonhuman animals as family. The conceptualization of pets as kin not only influences daily household dynamics and individual psychological outcomes but also reverberates through broader social practices, such as legal considerations in pet custody disputes and cultural representations of the "ideal family" (Rook, 2014). This multifaceted phenomenon underscores the importance of viewing pet ownership through a holistic lens, as it encompasses shifts in identity, social relations, and family dynamics in a modern context.

See also: Adults, and Pets; Children and Pets; Socialization, Pets And

References and Further Reading

Charles, N., and C. A. Davies (2008). "My Family and Other Animals: Pets as Kin." *Sociological Research Online*, 13 (5): 13–26.
Christian, H., F. Mitrou, R. Cunneen, and S. R. Zubrick (2020). "Pets Are Associated with Fewer Peer Problems and Emotional Symptoms, and Better Prosocial Behavior: Findings from the Longitudinal Study of Australian Children." *The Journal of Pediatrics*, 220: 200–6.
Irvine, L., and L. Cilia (2017). "More-Than-Human Families: Pets, People, and Practices in Multispecies Households." *Sociology Compass*, 11 (2): e12455.
Laurent-Simpson, A. (2017). "'They Make Me Not Wanna Have a Child': Effects of Companion Animals on Fertility Intentions of the Childfree." *Sociological Inquiry*, 87 (4): 586–607.
Lawton, L. E. (2025). "All In the Family: Pets and Family Structure." *Populations*, 1 (2): 8.
McGrane, C., L. Hjorth, and P. Murray (2024). "'We Make Them Our Kin': Australian Older Adults' Playful Kinship Practices Through Sharing Animal Images on Social Media." *Social Media + Society*, 10 (4): 20563051241306540.
Rook, D. (2014). "Who Gets Charlie? The Emergence of Pet Custody Disputes in Family Law: Adapting Theoretical Tools from Child Law." *International Journal of Law, Policy and The Family*, 28 (2): 177–93.
Ryan, S., and S. Ziebland (2015). "On Interviewing People with Pets: Reflections from Qualitative Research on People with Long-Term Conditions." *Sociology of Health & Illness*, 37 (1): 67–80.
Temesi, A., N. Bunford, and Á. Miklósi (2020). "Associations Among Attitudes Towards Motherhood, Pet-Keeping, and Postpartum Depression Symptoms." *Biologia Futura*, 71 (1): 153–64.
Westgarth, C., G. L. Pinchbeck, J. W. Bradshaw, S. Dawson, R. M. Gaskell, and R. M. Christley (2007). "Factors Associated with Dog Ownership and Contact with Dogs in a UK Community." *BMC Veterinary Research*, 3 (1): 5.

FANCY CLUBS

The history and current status of pet fancy groups, particularly those focused on dogs and cats, is rooted in the evolution of breed standards, the organization of competitive shows, and the interplay between aesthetic ideals and scientific theories of heredity that emerged in the nineteenth and early twentieth centuries.

Modern pet fancy movements emerged alongside organized shows, with the first modern cat show held in London in 1871, establishing breed characteristics and standards Plitman et al. (2019). Similar developments occurred in canine circles, as elite groups in Victorian England began forming clubs that organized dog shows and developed standardized criteria defining each breed's characteristics (Jung and Pörtl, 2019).

Initially, selective breeding practices were utilitarian, with dogs and cats serving practical roles such as hunting or vermin control; however, the rise of pet fancy shifted the focus toward breeding for specific aesthetic traits. By the late nineteenth century and into the Victorian era, the increasing emphasis on beauty led to the establishment of detailed breed standards (Morel et al., 2024). This new focus prompted the definition of ideal physical characteristics that often sidelined considerations of health and temperament in favor of appearance. The emergence of specialized veterinary roles, as seen in Edwardian London, further established the importance of maintaining these standards, as elite canine veterinarians were involved in addressing issues such as "faking" in dog shows and combating diseases like distemper that threatened pedigree breeding (Skipper, 2019).

A controversial aspect of the pet fancy movement is its historical association with eugenic ideas. Similar to the broader eugenics movement of the early twentieth century, proponents of selective breeding in the dog fancy often invoked notions of "purity" and "improvement" of bloodlines. The drive toward specific physical and behavioral standards was paralleled by a belief that controlled breeding could produce ideal organisms—an ideology that recent research has critiqued for its negative impact on animal welfare (Morel et al., 2024; Paul et al., 2023). Research on modern brachycephalic dog breeds indicates that prioritizing infant-like facial features—a trait that emerged from these aesthetic concerns—can lead to serious health issues, reflecting the unintended consequences of an overemphasis on appearance that has parallels with eugenic selection methods Paul et al., 2023).

In the current era, pet fancy groups continue to play a significant role in regulating and promoting purebred dogs and cats through organized shows and the maintenance of breed standards. Advances in genetics have allowed for a better understanding of breed variation and historical origins of modern breeds, as evidenced in studies indicating that over 85 percent of cat breeds have arisen in the past seventy-five years through selection on aesthetic traits (Kurushima et al., 2013). Despite their long history and ongoing influence, these groups face increasing scrutiny over welfare issues. Contemporary debates within the pet fancy community now emphasize balancing aesthetic standards with animal health and well-being, leading to calls for reform in breeding practices that historically prioritized appearance, sometimes at the expense of animal welfare (Morel et al., 2024; Paul et al., 2023).

In summary, pet fancy groups evolved from elite social and utilitarian origins into sophisticated organizations responsible for the establishment and enforcement of breed standards. Their historical link to selective breeding practices—once aligned with eugenic ideologies—has led to both remarkable developments in animal aesthetics and ongoing ethical concerns. Today, while dog and cat shows remain emblematic of the pet fancy tradition, they are also at the forefront of discussions surrounding animal welfare, reflecting an enduring tension between tradition and modern ethical standards.

See also: Cat Shows; Dog Shows; Selective Breeding

References and Further Reading

Comben, N. (1961). "The Cat Fancy." *Journal of Small Animal Practice*, 2 (1–4): 48–50.

Jung, C., and D. Pörtl (2019). "How Old Are (Pet) Dog Breeds?" *Pet Behaviour Science*, 7: 29–37.

Kurushima, J. D., M. J. Lipinski, B. Gandolfi, L. Froenicke, J. C. Grahn, R. A. Grahn, and L. A. Lyons (2013). "Variation of Cats Under Domestication: Genetic Assignment of Domestic Cats to Breeds and Worldwide Random-Bred Populations." *Animal Genetics*, 44 (3): 311–24.

Morel, E., L. Malineau, C. Venet, V. Gaillard, and F. Péron (2024). "Prioritization of Appearance over Health and Temperament Is Detrimental to the Welfare of Purebred Dogs and Cats." *Animals*, 14 (7): 1003.

Paul, E. S., E. Coombe, P. D. McGreevy, R. M. Packer, and V. Neville (2023). "Are Brachycephalic Dogs Really Cute? Evidence from Online Descriptions." *Anthrozoös*, 36 (4): 533–53.

Plitman, L., P. Černá, M. J. Farnworth, R. M. Packer, and D. A. Gunn-Moore (2019). "Motivation of Owners to Purchase Pedigree Cats, with Specific Focus on the Acquisition of Brachycephalic Cats." *Animals*, 9 (7): 394.

Ritvo, H. (1986). "Pride and Pedigree: The Evolution of the Victorian Dog Fancy." *Victorian Studies*, 29 (2): 227–53.

Skipper, A. (2019). "The Victorian Creation of the Pedigree Dog." *Journal of Victorian Culture*, 24 (3): 398–400.

FARM ANIMALS AS PETS

Research into keeping farm animals—or livestock—as pets has developed over time as societal attitudes toward animals have shifted from strictly utilitarian roles toward frameworks that recognize animals as members of the family. Early historical accounts, which primarily documented the use of these animals for food production, have been expanded by more recent studies that investigate the ethical, social, and practical implications of rearing farm animals in nontraditional, companion roles (Spencer et al., 2006). This literature synthesizes historical perspectives with contemporary concerns, noting that while the tradition of working animals has deep agrarian roots, the modern trend toward companion livestock is a relatively recent phenomenon spurred by urban farming and changing lifestyle preferences (Blouin, 2013).

A variety of farm animals such as goats, pigs, chickens, rabbits, and even cows have become popular as companion animals. Studies indicate that these species are chosen not only for their manageable size and ease of care in a domestic or semi-urban setting but also because they have come to represent desirable lifestyle choices aligned with organic and self-sufficient living practices (Blouin, 2013). However, the care requirements of these animals differ significantly from those of traditional pets like dogs and cats. Farm animals generally require more space, specialized diets, and access to outdoor areas that allow for natural behaviors; failure to meet these requirements can raise serious welfare concerns (Spencer et al., 2006).

Issues regarding the animals' care and protection have also been at the forefront of academic as well as public debates. Farm animals kept in domestic environments often present challenges associated with their husbandry, such as ensuring proper nutrition, appropriate environmental enrichment, and veterinary care that is geared to species-specific needs. Furthermore, the legal status of these animals typically differs from that of conventional companion animals. In many jurisdictions, state anti-cruelty laws offer extended protections to pets, but farm animals are often categorized differently as agricultural livestock. This legal distinction can lead to complications when a farm animal is kept primarily as a pet, with some states or regions providing limited protections compared to those afforded to companion animals. Veterinary practitioners, increasingly encountering farm animals on smallholdings or in urban settings, have highlighted the importance of implementing proper health-planning strategies and adhering to ethical breeding and husbandry practices.

In sum, the literature on keeping farm animals as pets documents a notable evolution in the human–animal relationship. Historically relegated to utilitarian roles, farm animals are now increasingly integrated as companions in domestic settings. This shift has fostered both enthusiasm and ethical debate regarding appropriate care standards, the adequacy of legal protections under anti-cruelty laws, and the overall well-being of animals traditionally raised for production purposes (Blouin, 2013). The current research underscores that while such practices can enrich human lives, they also necessitate a reexamination of husbandry practices, veterinary care protocols, and legal frameworks to ensure that the unique needs of these animals are adequately met.

See also: Equines; Trends in Pet Ownership

References and Further Reading

Blouin, D. D. (2013). "Are Dogs Children, Companions, or Just Animals? Understanding Variations in People's Orientations Toward Animals." *Anthrozoös*, 26 (2): 279–94.

Clement, G. (2011). "'Pets or Meat'? Ethics and Domestic Animals." *Journal of Animal Ethics*, 1 (1): 46–57.

Spencer, S., E. Decuypere, S. Aerts, and J. De Tavernier (2006). "History and Ethics of Keeping Pets: Comparison with Farm Animals." Journal of Agricultural and Environmental Ethics, 19 (1): 17–25.

FILM, PETS IN

Animals have been in film since the introduction of film as a new medium. One of the first moving images, in fact, is of an animal. Eadweard Muybridge's 1878 series of photos depicts an unnamed horse galloping; when viewed as a series, they appear to show the horse in motion. With the invention of the first true motion picture camera in 1895, animals fighting, walking, running, and performing were quickly recorded onto film.

With the rise of the movie industry in the early twentieth century and television beginning in the 1950s, not only did animals come to be not only been featured in movies and television shows, but many became stars in their own right. Rin Tin Tin, Lassie, Benji, Flipper, Toto, Morris the Cat, and the Taco Bell Chihuahua all became part of the American cultural vocabulary, thanks to their performances in dozens of movies, television shows, and TV commercials. (The characters listed here were portrayed by animal actors, sometimes several. Benji, e.g., was portrayed played by a dog named Higgins, while at least ten different dogs portrayed Lassie from 1943 until the last movie was filmed in 2005.) The history of television and film is filled with hundreds of memorable virtual animals as well—cartoon animals like Mickey Mouse, Bugs Bunny, Donald Duck, Dumbo, Bambi, Garfield, Roger Rabbit, or the animals from recent animated films such as *Zootopia, The Secret Life of Pets, Bolt*, *The Lion King, Madagascar, Kung Fu Panda, Ice Age*, and *Ratatouille*. It's difficult to imagine what Hollywood movies would be like without animals.

Horror films, in particular, have utilized animals. A popular sub-genre of horror film is the "eco-horror" film, which features nature run amok: monstrously big insects, rabbits, and other animals; killer dogs, rats, birds, and sharks; and human–animal hybrids like werewolves or the creature from *The Fly* (1958). Wild animals such as anacondas, sharks, and pythons are often treated in horror films as monsters, while domesticated animals like dogs, the most loyal animal of all, turn rogue and attack humans. In both cases, they are often difficult to kill or control. The 1950s and 1960s, during the height of the Cold War, saw the release of dozens of B movies featuring mutated animals wreaking havoc on humans, like—for example, *Earth vs. the Spider* (1958), *Beginning of the End* (1957), *Night of the Lepus* (1972), and the *Godzilla* films—and may have reflected the culture's fears about the nuclear age. More recent films involving animals that attack, such as *Jaws* (1975), *The Bees* (1978), *Cujo* (1983), or *Lake Placid* (1999), might have something to do with our fear that we have screwed up nature and that nature, and the animals that live in it, are "striking back." In either case, animals clearly are a useful symbol of "otherness." (For more on animals and horror, see Gregersdotter, Höglund, and Hållén et al. 2016; Gregersdotter and Hållén, 2015; Khapaeva, 2019; and Lennard, 2019).

Although animals are predominantly portrayed in cinema similarly to their representation in literature—as proxies for humans—recent years have witnessed some innovative portrayals of animals. Advances in technology, particularly those that enhance documentary filmmaking, have substantially simplified the process of capturing wildlife on film. As a result, audiences are afforded the chance to observe wild animals in ways previously unattainable. In recent years, several mainstream films have conveyed animal rights themes. A prime example is "Babe," a 1995 film centered on a pig who becomes aware of his impending slaughter and resolves to develop a unique ability—herding sheep—to render himself indispensable to the farmer, thereby securing his own safety. "Babe" resonated with global audiences, aided by the character's endearing traits portrayed through distinctive voice and behavior, as well as the film's moral lessons on overcoming adversity, confronting challenges, respecting others, and seizing opportunities. Animal advocacy groups praised "Babe," hoping that audiences would appreciate the individuality of pigs like the film's protagonist and reconsider the perception of such animals as mere food sources. However, much like "Charlotte's Web," the storyline concludes with the pig being saved, and while the farmer recognizes the pig's uniqueness, it fails to address the broader issue of the routine slaughter of farm animals for consumption. Neither "Babe" nor Wilbur dares to question the norm of consuming animals; their motivation centers purely on evading their grim fate. "Okja," a 2017 Korean film, tells the story of a genetically engineered "superpig" who shares a deep bond with Mija, a young girl involved in raising him. Unbeknownst to Mija, Okja has been bred for food production, and much of the narrative chronicles her efforts to save him, leading her on a journey from Seoul to New York. Following the film's release, numerous viewers reported feeling compelled to cease meat consumption. For a detailed examination of how this film confronts the legal perspective of animals as property, leading to Okja's moral significance, see Gunawan's analysis (2018). Art historian John Berger (1980) posits that the omnipresence of animal imagery in contemporary society compensates for a lost direct relationship with animals. This notion may elucidate the surging popularity of nature documentaries and television programs featuring wildlife. This raises an important question: what consequences arise from the pervasive animal imagery permeating our lives? Archaeologist and film historian Jonathan Burt (2002) argues that our perceptions and understanding of animals have evolved due to their depictions in film. It is evident that various films, television shows, and advertising campaigns featuring animals often precipitate spikes in both adoptions and purchasing of those animals—followed by subsequent abandonment. A notable case involves Dalmatians after the release of Disney's live-action "101 Dalmatians," as well as Chihuahuas promoted through Taco Bell commercials and "Beverly Hills Chihuahua," not to mention the surge in popularity of Jack Russell terriers spurred by the character Murray in "Frasier." Certain films may influence perceptions regarding contentious animal issues, as evidenced by increased interest in vegetarianism following the release of "Babe" or "Chicken Run," or heightened awareness concerning the captivity of whales post-"Free Willy." Ironically, even films that advocate for animal welfare, such as "Finding Nemo," can inadvertently harm animals; the film's release led to a significant surge in clownfish sales, resulting in many being confined to tanks and bowls, despite the film's message against such captivity. Additionally, animal rights films have gained traction, with organizations like People for the Ethical Treatment of Animals (PETA) and Mercy for Animals producing numerous films using undercover footage from fur farms, slaughterhouses, livestock yards, and animal research facilities. The graphic depictions of animal suffering in these films have significantly influenced societal views regarding animal exploitation. A poignant example is the 26-minute PETA film "Unnecessary Fuss" (1984), created from footage captured in a head injury lab at the University of Pennsylvania. This title was derived from a comment made by lead researcher Thomas Gennerelli, who expressed concern that public awareness of his work could provoke "unnecessary fuss" among animal advocates (Finsen and Finsen, 1994: 68). The film showcases disturbing scenes of lab workers inflicting injuries on baboons, at times lacking compassion for their suffering. Following the release of "Unnecessary Fuss," widespread public outrage led to the closure of the lab and the termination of Gennerelli's research. On the one hand, the increasing visibility of animals in film may foster greater awareness regarding their plights; however, it remains uncertain whether audiences will take meaningful action to mitigate the human impact on both wild and domestic animals. Furthermore, ethical considerations arise when viewing films or television shows with animal performers. The treatment and upbringing of animals cast in films is only loosely regulated in the United States, and productions shot outside the country have no obligation to adhere to US animal welfare laws; consequently, there is no accountability for incidents such as onset fatalities (Wilson, 2015). Although the American Humane Association often oversees the treatment of animals during filming in the United States, their monitoring does not extend beyond the production, and there is no statutory requirement for their presence. Additionally, as noted by Randy Malamud (2010), animals do not voluntarily agree to appear on screen. While some contemporary domestic animals in films might be cherished companions whose owners aspire to stardom for them, many others are kept by trainers in kennels between shooting schedules. Though some trainers provide exceptional care and primarily use animals sourced from shelters, the absence of comprehensive oversight for private trainers who work with both wild and domestic animals in film and television complicates the issue of humane treatment. Finally, critics Brett Mills (2010) and Randy Malamud raise important ethical questions about the privacy rights of animals. Malamud provocatively asks whether our fascination with wildlife films and documentaries reflects a genuine concern for the well-being of other animals or if it reduces these creatures to mere entertainment fodder (2010: 146).

See also: Breeds, Popularity Of; Literature, Pets In; Television, Pets On

References and Further Reading

Berger, J. (1980). *Why Look at Animals? About Looking.* New York: Pantheon.
Burnford, S. E. (1960). *The Incredible Journey*. London: Hodder & Stoughton.
Burt, J. (2002). *Animals in Film*. London: Reaktion Books.
Finsen, S., and L. Finsen (1994). *The Animal Rights Movement in America: From Compassion to Respect*. New York: Twayne Publishers.
Gregersdotter, K., J. Höglund, and N. Hållén (2016). *Animal Horror Cinema: Genre, History and Criticism*. Basingstoke: Springer; Palgrave Macmillan.
Gregersdotter, K., and N. Hållén (2015). "Anthropomorphism and the Representation of Animals as Adversaries." In *Animal Horror Cinema: Genre, History and Criticism*, 206–23. London: Palgrave Macmillan UK.
Khapaeva, D., ed. (2019). *Man-Eating Monsters: Anthropocentrism and Popular Culture*. Emerald Publishing Limited.
Lennard, D. (2019). *Brute Force: Animal Horror Movies*. Albany: State University of New York Press.
Malamud, R. (2010). "Animals on Film: The Ethics of the Human Gaze." *Spring*, 86: 1–26.
Mills, B. (2010). "Television Wildlife Documentaries and Animals' Right to Privacy." *Continuum*, 24 (2): 193–202.
Molloy, C. (2011). *Popular Media and Animals*. London: Springer; Palgrave Macmillan.
PETA (1984). *Unnecessary Fuss*. People for the Ethical Treatment of Animals.
Wilson, D. A. (2015). *The Welfare of Performing Animals: A Historical Perspective. Animal Welfare* 15. Berlin: Springer.

FISH

The literature on ornamental and pet fish encompasses studies of the motivations for ownership, welfare concerns, common species kept, and the scale of fish kept as pets. Research indicates that people are attracted to fish primarily for their aesthetic appeal, the calming effect of their presence, and the relatively low maintenance requirements compared to traditional companion animals. Qualitative studies have highlighted that fish offer a form of quiet companionship, and for some, particularly older adults, they serve as a means to alleviate social isolation and stress (Langfield and James, 2009; Takase et al., 2021). This psychological benefit, combined with the visual pleasure garnered from the vibrant colors and graceful movements of ornamental fish, explains why aquaria remain popular in domestic and even workplace settings.

The welfare of ornamental fish is a multifaceted topic that has received considerable attention in recent literature. Several studies have documented that husbandry practices, including handling techniques such as the use of specially designed nets to protect delicate scales and mucous layers, are critical for minimizing stress and the risk of injury in fish (Saengsitthisak et al., 2020). In parallel, reviews of stress indicators and mortality rates in the ornamental fish trade have stressed that outcomes in aquaculture can inform improvements in the aquarium industry, emphasizing the need for proper biosecurity, water quality maintenance, and overall management to reduce chronic health issues (Stevens et al., 2017). Moreover, the occurrence of diseases such as fish tuberculosis poses both welfare and zoonotic concerns, highlighting the necessity for improved fish management practices in retail settings (Kušar et al., 2017).

In terms of the species most commonly kept as pets, popular choices include goldfish, betta (Siamese fighting fish), guppies, tetras, and various cichlids. Betta fish, for instance, have been selectively bred for diverse phenotypes and are widely appreciated for their striking coloration and fin morphology (McLaughlin et al., 2022). The diversity in species selection reflects both regional preferences and the accessibility of different ornamental fish through the global pet trade. Multiple studies have estimated that the number of pet fish kept by households is typically high compared to other pets, with ornamental fish constituting one of the largest pet populations in many countries, although precise numbers vary across regions due to differences in survey methods and regulatory practices (Smith et al., 2008).

Overall, the literature paints a picture of ornamental fish as a unique pet category where aesthetics, accessibility, and low perceived responsibility synergies drive widespread ownership. However, despite their popularity, significant welfare challenges persist due to variations in husbandry practices and the highly commercialized nature of the ornamental fish industry. These findings call for ongoing research and regulatory efforts to ensure that the welfare of these animals is not compromised while maintaining the benefits they provide to their owners.

See also: Human Health, Pets And; Mental Health, Pets And

References and Further Reading

Kušar, D., U. Zajc, V. Jenčič, M. Ocepek, J. Higgins, M. Žolnir-Dovč, and M. Pate (2017). "Mycobacteria in Aquarium Fish: Results of a 3-Year Survey Indicate Caution Required in Handling Pet-Shop Fish." *Journal of Fish Diseases*, 40 (6): 773–84.

Lambert, H., A. Cornish, A. Elwin, and N. D'Cruze (2022). "A Kettle of Fish: A Review of the Scientific Literature for Evidence of Fish Sentience." *Animals*, 12 (9): 1182.

Langfield, J., and C. James (2009). "Fishy Tales: Experiences of the Occupation of Keeping Fish as Pets." *British Journal of Occupational Therapy*, 72 (8): 349–56.

McLaughlin, A., A. Emanuele, C. Hadfield, and K. Snekvik (2022). "Pigmentary Keratitis and Mixed Chromatophoroma in a Betta Fish (Betta splendens)." *Aquaculture, Fish and Fisheries*, 2 (6): 587–92.
Papavlasopoulou, I., L. Vardakas, C. Perdikaris, D. Kommatas, and I. Paschos (2014). "Ornamental Fish in Pet Stores in Greece: A Threat to Biodiversity?" *Mediterranean Marine Science*, 15 (1): 126–34.
Saengsitthisak, B., W. Chaisri, V. Punyapornwithaya, R. Mektrirat, S. Klayraung, J. K. Bernard, and S. Pikulkaew (2020). "Occurrence and Antimicrobial Susceptibility Profiles of Multidrug-Resistant Aeromonads Isolated from Freshwater Ornamental Fish in Chiang Mai Province." *Pathogens*, 9 (11): 973.
Smith, T. M., J. S. Hindell, G. P. Jenkins, and R. M. Connolly (2008). "Edge Effects on Fish Associated with Seagrass and Sand Patches." *Marine Ecology Progress Series*, 359: 203–13.
Stevens, C. H., D. P. Croft, G. C. Paull, and C. R. Tyler (2017). "Stress and Welfare in Ornamental Fishes: What Can Be Learned from Aquaculture?" *Journal of Fish Biology*, 91 (2): 409–28.
Walster, C., E. Rasidi, N. Saint-Erne, and R. Loh (2015). "The Welfare of Ornamental Fish in the Home Aquarium." *Companion Animal*, 20 (5): 302–6.

FREE-ROAMING CATS

Free-roaming cats, also known as stray cats or street cats, are cats that are not owned by a person and live outside of the control of people. Historically, free-roaming cats have inspired polarized responses among governments, private citizens, and animal welfare organizations. On the one hand, many governmental agencies and some private individuals have reacted negatively toward these cats—often resorting to violence or lethal control measures—in an attempt to protect public health and native wildlife populations (Scotney et al., 2023; Hadidian et al., 2012). Such extreme measures, however, frequently generate controversy, particularly among cat welfare advocates who argue that these approaches neglect the ethical considerations inherent in managing a sentient species.

In the West, and especially in the United States, Europe, Australia, and New Zealand, cats have been moving from outdoor animals to indoor pets. With this move comes a new focus on free-roaming cats, their welfare needs, and whether or not they pose a threat to biodiversity or even human health. This has resulted in a demonization of free-roaming cats.

In contrast, certain urban centers exemplify a more tolerant coexistence between humans and free-roaming cats. Istanbul, for instance, is well known for its thousands of street cats that live alongside city residents and benefit from the informal support of volunteer feeders and municipal policies that recognize their cultural value (Khor et al., 2018). This coexistence reflects a community-based ethic where free-roaming cats, rather than being viewed solely as pests or vectors for disease, are embraced as part of the urban ecosystem. Such examples illustrate that attitudes toward free-roaming cats are deeply context-dependent and culturally mediated (Khor et al., 2018).

Another significant aspect of the literature is the increasing use of trap-neuter-return (TNR) programs to manage free-roaming cat populations in a humane, non-lethal manner. TNR involves capturing cats, sterilizing them, vaccinating against diseases, and returning them to their original location, thereby stabilizing or reducing colony size over

A cat rooftop party in Istanbul, Turkey. The city is well known for its thousands of free-roaming cats, supported by municipal policies that recognize their cultural value. (Olgaturbina77/ Dreamstime.com)

time (Gonçalveset al., 2025; Levy et al., 2003). In the United States and various urban and suburban areas worldwide, TNR programs have been implemented by municipal authorities and community volunteers, whose involvement is critical for monitoring colony health and welfare (Spehar and Wolf, 2019; Wolf and Schaffner, 2019). Despite some debates regarding the short-term efficacy of these programs—especially in dense urban settings—the consensus in much of the literature is that TNR, coupled with public education and low-cost sterilization services, represents a promising strategy for balancing feline welfare with public and wildlife interests (Kennedy et al., 2020; Spehar and Wolf, 2019).

Amid these management strategies, a vigorous debate persists about the ecological impact of free-roaming cats. Numerous studies have estimated that these animals contribute significantly to the decline of small animal and bird populations through predation, with estimates of wildlife mortalities attributed to free-roaming cats ranging from millions to billions of bird fatalities annually (Peterson et al., 2012; Leong et al., 2020). This raises critical questions for conservation biologists, who argue that even well-intentioned TNR programs must be balanced with measures to protect vulnerable native species. Conversely, cat welfare proponents contend that free-roaming cats are part of urban biodiversity and that their welfare—and the humane management of their populations—should not be dismissed solely based on their predatory behavior (Hadidian et al., 2012).

In summary, the literature reveals a complex, multifaceted view of free-roaming cats that goes beyond simple categorizations as either pests or cherished urban dwellers. While negative responses—including violent interventions—persist in many regions, there is also a growing trend toward community-based, nonlethal management strategies such as TNR. These methods aim to stabilize or reduce cat populations while seeking to improve the welfare of the animals themselves, even as ongoing debates highlight the ecological costs associated with their predation on small animals and birds.

See also: Cats; Free-roaming Dogs

References and Further Reading

Gerhold, R. W., and D. A. Jessup (2013). "Zoonotic Diseases Associated with Free-Roaming Cats." *Zoonoses and Public Health*, 60 (3): 189–95.

Gonçalves, L. S., D. de Souza Machado, M. E. Caçador, G. A. Ferreira, C. R. Dickman, M. C. Ceballos, ... and A. C. Sant'Anna (2025). "The Wildcat That Lives in Me: A Review on Free-Roaming Cats (Felis catus) in Brazil, Focusing on Research Priorities, Management, and Their Impacts on Cat Welfare." *Animals*, 15 (2): 190.

Hadidian, J. (2021). "Cats and Wildlife: An Animal Welfare Perspective." WellBeing International Studies Repository. Potomac, MD: WellBeing International.

Hadidian, J., I. Gibson, S. Hagood, N. Peterson, B. Unti, B. McFarland, K. Lisnik, H. Bialy, I. Fricke, K. Schatzmann, J. Fearing, P. Runquist, and A. N. Rowan (2012). "Outdoor Cats: An Animal Welfare and Protection Perspective." In *Proceedings of the Vertebrate Pest Conference, 25* (25).

Kennedy, B. P., B. Cumming, and W. Y. Brown (2020). "Global Strategies for Population Management of Domestic Cats (Felis catus): A Systematic Review to Inform Best Practice Management for Remote Indigenous Communities in Australia." *Animals*, 10 (4): 663.

Khor, M. M., G. Davey, and X. Zhao (2018). "Why Do People Feed Free-Roaming Cats? The Role of Anticipated Regret in an Extended Theory of Planned Behavior in Malaysia." *Anthrozoös*, 31 (1): 101–16.

Leong, K. M., A. R. Gramza, and C. A. Lepczyk (2020). "Understanding Conflicting Cultural Models of Outdoor Cats to Overcome Conservation Impasse." *Conservation Biology*, 34 (5): 1190–9.

Levy, J. K., D. W. Gale, and L. A. Gale (2003). "Evaluation of the Effect of a Long-Term Trap-Neuter-Return and Adoption Program on a Free-Roaming Cat Population." *Journal of the American Veterinary Medical Association*, 222 (1): 42–6.

McLeod, L. J., D. W. Hine, and A. B. Driver (2019). "Change The Humans First: Principles for Improving the Management of Free-Roaming Cats." *Animals*, 9 (8): 555.

Peterson, M. N., B. Hartis, S. Rodriguez, M. Green, and C. A. Lepczyk (2012). "Opinions from the Front Lines of Cat Colony Management Conflict." *PLOS One*, 7 (9): e44616. https://doi.org/10.1371/journal.pone.0044616

Schmidt, P. M., R. R. Lopez, and B. A. Collier (2007). "Survival, Fecundity, and Movements of Free-Roaming Cats." *The Journal of Wildlife Management*, 71 (3): 915–19.

Spehar, D. D., and P. J. Wolf (2019). "Back to School: An Updated Evaluation of the Effectiveness of a Long-Term Trap-Neuter-Return Program on a University's Free-Roaming Cat Population." *Animals*, 9 (10): 768.

Wolf, P. J. and J. E. Schaffner (2019). "The Road to TNR: Examining Trap-Neuter-Return Through the Lens of Our Evolving Ethics." *Frontiers in Veterinary Science*, 5: 341.

FREE-ROAMING DOGS

Free-roaming dogs, also known as stray dogs or street dogs and sometimes village dogs or Rez dogs, are dogs that are not owned by an individual person and survive through their wits and often the assistance of humans, although they largely fall outside of human control. In Western countries, free-roaming dogs are demonized, but in other countries, they co-exist with humans.

In many urban settings, free-roaming dogs include both unowned individuals and owned dogs that are allowed to roam without restraint. Free-roaming dogs are implicated in various public health issues, notably dog bites and rabies transmission, as reported by Tayebwa et al. (2024). This contrasts with free-roaming cats, where concerns often focus on wildlife predation and ecological impacts (Scotney et al., 2023; Hadidian, 2021). However, both groups contribute to ongoing debates regarding animal welfare and human–animal interactions in public spaces.

Management strategies differ across countries and are influenced by cultural, economic, and legal factors. For free-roaming dogs, some governments rely on capture, neuter, vaccination, and return (CNVR) programs as a humane method to stabilize or reduce population growth, as illustrated in Bangkok, where high levels of community feeding coexist with CNVR efforts (Hiby et al., 2025). In other contexts, authorities may employ mandatory household restraint laws or capture and shelter strategies, as seen in various European and Asian cities (Smith et al., 2022). These efforts contrast with the widespread use of trap-neuter-return (TNR) programs for stray cats, which are often treated as community animals rather than public health hazards. Additionally, free-roaming dogs frequently present a higher risk to public safety through dog bites and rabies transmission, prompting more stringent regulatory responses compared to free-roaming cats (Tayebwa et al., 2024; Smith et al., 2022).

Welfare concerns for free-roaming dogs are significant. On the one hand, these animals face significant challenges, including malnutrition, injuries from cars, disease exposure, and potential abuse (Hiby et al., 2025; Belo et al., 2017). On the other hand, studies suggest that free-roaming dogs, in contrast to confined pet dogs, enjoy a degree of autonomy and natural behavior expression that can be associated with higher well-being in contexts where urban environments provide sufficient resources and lower levels of human aggression (de Santi et al., 2024; da Rocha et al., 2022). Thus, while increased autonomy may confer benefits such as enhanced natural social interactions and physical activity, it also exposes them to the unpredictable hazards of street life. In addition, free-roaming dogs are more frequently subject to enforcement actions and public health interventions because their mobility and behavior patterns are perceived as more threatening to humans and other animal populations (Tayebwa et al., 2024; Smith et al., 2022). Public opinion surveys suggest that attitudes toward free-roaming dogs are more polarized, with a higher likelihood of intervention through removal or compulsory sterilization, reflecting concerns about rabies, aggressive behavior, and dog bite incidents (Smith et al., 2022; Tenzin et al., 2015).

See also: Dogs; Free-roaming Cats

References and Further Reading

Belo, V. S., C. J. Struchiner, G. L. Werneck, R. G. Teixeira Neto, G. B. Tonelli, C. G. de Carvalho Júnior, … and E. S. da Silva (2017). "Abundance, Survival, Recruitment and Effectiveness of Sterilization of Free-Roaming Dogs: A Capture and Recapture Study in Brazil." PLOS One, 12 (11): e0187233.

da Rocha, Y. D. S. G., M. Ulanin, S. M. Barrero, A. J. Z. Lech, F. Argenton, M. Brugnerotto, G. Polo, M. Kopach, and R. D. C. M. Garcia (2022). "Physical Assessment of Free Roaming Dogs from Campo Magro–Paraná, Brazil." *Research, Society and Development*, 11 (14): e20111435839–e20111435839.

de Santi, C. E., W. A. Chiba de Castro, A. C. Sibim, R. D. Lopes, S. R. Galvão, G. M. Kurtz, L. M. Biondo, L. B. Kmetiuk, and A. W. Biondo (2024). "Spatial Distribution and Population Dynamics of Free-Roaming (Stray and Semi-Domiciled) Dogs in a Major Brazilian City." *Frontiers in Veterinary Science*, 11: 1417458.

Hadidian, J. H. (2021). "Cats and Wildlife: An Animal Welfare Perspective." WellBeing International Studies Repository. Potomac, MD: WellBeing International.

Hiby, E., T. Rungpatana, A. Izydorczyk, V. Benka, and C. Rooney (2025). "Population Demographics of Owned Dogs in Greater Bangkok and Implications for Free-Roaming Dog Population Management." *Animals*, 15 (9): 1263.

Scotney, R., J. Rand, V. Rohlf, A. Hayward, and P. Bennett (2023). "The Impact of Lethal, Enforcement-Centred Cat Management on Human Wellbeing: Exploring Lived Experiences of Cat Carers Affected by Cat Culling at the Port of Newcastle." Animals, 13 (2): 271.

Smith, L. M., R. J. Quinnell, C. Goold, A. M. Munteanu, S. Hartmann, and L. M. Collins (2022). "Assessing the Impact of Free-Roaming Dog Population Management Through Systems Modelling." Scientific Reports, 12 (1): 11452.

Tayebwa, D. S., D. Hyeroba, C. D. Dunn, E. Dunay, J. C. Richard, S. Biryomumaisho, J. O. Acai, and T. L. Goldberg (2024). "Viruses of Free-Roaming and Hunting Dogs in Uganda Show Elevated Prevalence, Richness and Abundance Across A Gradient of Contact with Wildlife." *Journal of General Virology*, 105 (7): 002011.

Tenzin, T., R. Ahmed, N. C. Debnath, G. Ahmed, and M. Yamage (2015). "Free-Roaming Dog Population Estimation and Status of the Dog Population Management and Rabies Control Program in Dhaka City, Bangladesh." PLOS Neglected Tropical Diseases, 9 (5): e0003784.

G

GENDER, PETS AND

Just like all animals, pets are gendered. Like humans, who are born as either biologically male or female, or occasionally, as intersex, most nonhuman animals are also born either male, female, or intersex, while there are a handful of species that do not sexually reproduce, producing themselves via cloning, which results in the species only having one sex: the female.

Gender, on the other hand, is a set of culturally constructed, historically contingent attributes that are then mapped onto biological sex. Saying that women are nurturing, or that men are aggressive, is a way of conflating sex (women or men) with gender (characteristics or behaviors). There is no question that gender is a critical aspect of all human societies, but very little is known about whether nonhuman animals have a conception of gender, and whether the concept itself can even be applied to nonhumans. One thing is very clear, though. Humans use our own understandings (and misunderstandings) about gender when thinking about animals, and, to complicate matters further, our own gender shapes this process as well.

For example, we have large amounts of literature on how (human) gender shapes attitudes toward animals (Herzog, Betchart, and Pittman, 1991; Eldridge and Gluck, 1996; Herzog, 2007; Allcorn and Ogletree, 2018; Randler et al., 2021; Allcorn and Ogletree, 2018), demonstrating once again how important gender is in shaping how humans see and perceive the world around them. There is also a fair amount of research from feminist scholars to suggest that on an institutional level, living in a society that is strictly gendered with a very clear preference for masculinity over femininity leads to policies and practices that are harmful to animals, as well as women, children, and the earth as a whole (Cudworth, 1998; Hunnicutt, 2019; Bossert et al., 2024).

When it comes to pets, we tend to use gender even more than we do with wildlife, livestock, or other types of animals. On one level, pets themselves can be thought of as generally female. Associated with women and the domestic space, cared for by women, and rescued by women, they still carry a shade of the ancient stigma that suggests that pets are trivial, and therefore women are trivial, unlike, for example, large working animals or charismatic megafauna. Even the veterinary profession is gendered (Irvine and Vermilya, 2010; Shaw et al., 2012, as is the animal rescue and rights movements (Gaarder, 2011; Markovits and Queen, 2009; Sinski, 2018; Taylor, Fraser, and Signal, 2022).

But even within that context, human gender norms shape preferences and behaviors among pet owners, influencing the type of pets chosen and the emotional bonding observed within these relationships (Ramirez, 2006; Mueller, Wingrove, and Murray, 2020). This is further supported by findings from Menchetti et al., which emphasize that women generally exhibit higher levels of care and empathy toward their pets, indicating different relational dynamics based on the owner's gender (Menchetti et al., 2020). Similarly, research by Abel and Krüger established patterns in naming practices for dogs that revealed gender-based distinctions; female dogs were often given more elaborate names compared to their male counterparts, suggesting societal norms regarding femininity and masculinity are projected even in how we address our pets (Abel and Krüger, 2007; Abel and Krüger, 2007).

Perhaps the influence of gender in terms of how we perceive and treat our pets is the most significant example of how we gender pets. For example, men might attribute more assertiveness or aggression to male dogs, reinforcing traditional stereotypes of masculinity, while women's perceptions tend to align with nurturing traits (Menchetti et al., 2018), or for female dogs that engage in behaviors that look aggressive, women may use terms like "sassy" to describe them instead.

But when we take a concept like gender—which is entirely socially constructed—and then apply those ideas, developed by humans in human societies, to nonhuman animals, it primarily results in a very odd form of anthropomorphism. It is one thing to dress our female small dogs in dresses and bows and ribbons, and to adorn our male pit bull-type dogs with hyper-aggressive collars and chains. But it is another thing to suggest that we *should* dress them this way, that small fluffy female dogs *enjoy* having bows in their hair, or that male dogs should get more playtime than females. Some companies are now even marketing gendered pet products that suggest that dogs and cats should get toys that are "appropriate" to their gender.

If animals did have gender, how do we see and interpret it? Would there just be two genders, or would there be more, given that we know that in human cultures, the number of genders can vary, with many Indigenous American tribes, for example, having "third genders." In people, most of us present ourselves as either male or female, and for most of us, this self-presentation mirrors our given sex. But for about half a percent of the human population, their sense of gender identity does not match their biological sex, resulting in dysphoria, where it feels like the person's sex is wrong. We can, for the most part, see sex in adults because our secondary sex characteristics—body and facial hair and increased size and strength are the most obvious in males, while for females, the development of breasts and greater body fat are obvious signs. Everything else—how we style our hair, what kinds of clothes we wear, whether we adorn our bodies with makeup or jewelry, and so on—is an indication of our gender presentation, whether or not it matches with our sex.

For animal species that are, like humans, at least somewhat sexually dimorphic (meaning that the male and the female of the species differ in size, coloration, strength, as well as other characteristics), it is relatively easy to determine their sex. Even without significant visible differences, we can often still tell whether an animal is male or female. Unneutered males of most mammal species, for example, are easily identified by their external genitalia.

Biological sex and reproductive status demonstratively shape some behaviors of nonhuman animals, just as they do to a limited extent in people. But beyond that, can we say that our male dogs or cats act differently from our females, especially when those animals are surgically desexed? One final complicating factor is that many scientists—especially those in the natural sciences—do not understand gender and use the term as if it were equivalent to sex. So while there are numerous articles that look at animal behaviors and attribute differences to gender, for the most part, based on this author's reading, they really should say sex instead.

I have located a single recent article that has taken on the question of whether animals do indeed have, or "do" gender. Meynell and Lopez (2021) suggest that at least some animals might have a concept of gender, in which certain behaviors are expected of individuals based on sex, and in which those behaviors are learned socially. This last part, that these gendered behaviors must be learned, is the most critical because gender *must be* learned. There is no real debate among most animal ethologists that many social species of animals have culture, which is broadly defined as a set of traditions, practices, and beliefs that hold a social group together and provide shared meaning within it. Using this as our starting point, I think it is not unreasonable to infer that because many social animals have complex social lives that are based on teaching and learning rather than instinct, one of the things that could be learned by the young is gendered behaviors.

As of this writing, this question, and the research that will allow us to answer it, is in the very preliminary stages. While there is hope that the question can be answered, especially as many young ethologists and ethnographers are committed to seeing animals in a less objectifying and anthropocentric way, how we use that knowledge is an open question. One thing that we can predict, however, is that pet product manufacturers will exploit the answer in order to sell more cute pet clothes.

See also: Animal Protection Movement; Children and Pets; Cruelty; History of Pet Keeping: Victorian Era; Link between Animal and Human Abuse

References and Further Reading

Abel, E., and M. Krüger (2007a). "Gender Related Naming Practices: Similarities and Differences Between People and Their Dogs." *Sex Roles*, 57 (1–2): 15–19.

Abel, E., and M. Krüger (2007b). "Stereotypic Gender Naming Practices for American and Australian Dogs and Cats." *Names*, 55 (1): 53–64.

Allcorn, A., and S. Ogletree (2018). "Linked Oppression: Connecting Animal and Gender Attitudes." *Feminism & Psychology*, 28 (4): 457–69.

Bossert, L., T. Crompton, A. Dutta, and J. Seager (2024). "Mapping the Patriarchy in Conservation." *npj Biodiversity*, 3 (1): 1–5.

Cudworth, E. (1998). *Gender, Nature and Dominance: An Analysis of Interconnections Between Patriarchy and Anthroparchy, Using Examples of Meat and Pornography*. Doctoral dissertation, University of Leeds.

Eldridge, J. J., and J. P. Gluck (1996). "Gender Differences in Attitudes Toward Animal Research." *Ethics & Behavior*, 6 (3): 239–56.

Gaarder, E. (2011). "Where the Boys Aren't: The Predominance of Women in Animal Rights Activism." *Feminist Formations*, 23 (2): 54–76.

Herzog Jr, H. A., N. S. Betchart, and R. B. Pittman (1991). "Gender, Sex Role Orientation, and Attitudes Toward Animals." *Anthrozoös*, 4 (3): 184–91.

Herzog, H. A. (2007). "Gender Differences in Human–Animal Interactions: A Review." *Anthrozoös*, 20 (1): 7–21.

Hunnicutt, G. (2019). *Gender Violence in Ecofeminist Perspective: Intersections of Animal Oppression, Patriarchy and Domination of the Earth*. Abingdon: Routledge.

Irvine, L., and J. R. Vermilya (2010). "Gender Work in a Feminized Profession: The Case of Veterinary Medicine." *Gender & Society*, 24 (1): 56–82.

Markovits, A., and R. Queen (2009). "Women and the World of Dog Rescue: A Case Study of the State of Michigan." *Society & Animals*, 17 (4): 325–42.

Menchetti, L., S. Calipari, G. Guelfi, A. Catanzaro, and S. Diverio (2018). "My Dog is not my Cat: Owner Perception of the Personalities of Dogs and Cats Living in the Same Household." *Animals*, 8 (6): 80. https://doi.org/10.3390/ani8060080.

Menchetti, L., S. Calipari, C. Mariti, A. Gazzano, and S. Diverio (2020). "Cats and Dogs: Best Friends or Deadly Enemies? What The Owners of Cats and Dogs Living in the Same Household Think About their Relationship with People and Other Pets." *Plos One*, 15 (8): e0237822.

Meynell, L., and A. Lopez (2021). "Gendering Animals." *Synthese*, 199: 4287–311.

Mueller, T., T. Wingrove, and S. Murray (2020). "The Human Response to Pet Rescue TV Commercials." *Pet Behaviour Science*, 9: 34–44.

Ramirez, M. (2006). "'My Dog's Just Like Me': Dog Ownership as a Gender Display." *Symbolic Interaction*, 29 (3): 373–91. https://doi.org/10.1525/si.2006.29.3.373.

Randler, C., A. Adan, M. M. Antofie, A. Arrona-Palacios, M. Candido, J. Boeve-de Pauw, C. Vollmer, et al. (2021). "Animal Welfare Attitudes: Effects of Gender and Diet in University Samples From 22 Countries." *Animals*, 11 (7): 1893.

Shaw, J. R., B. N. Bonnett, D. L. Roter, C. L. Adams, and S. Larson (2012). "Gender Differences in Veterinarian-Client-Patient Communication in Companion Animal Practice." *Journal of the American Veterinary Medical Association*, 241 (1): 81–8.

Sinski, J. B. (2018). *Gender and Leadership in Animal Sheltering Organizations*. PhD doctoral dissertation, University of Louisville.

Taylor, M., M. Rhodes, and S. Gelman (2009). "Boys will be Boys; Cows will be Cows: Children's Essentialist Reasoning About Gender Categories and Animal Species." *Child Development*, 80 (2): 461–81.

Taylor, N., H. Fraser, and T. Signal (2022). "'Rescued and Loved': Women, Animal Sanctuaries, and Feminism." In E. Cudworth, R. E. McKie, and D. Turgoose (eds), *Feminist Animal Studies*, 218–34. London: Routledge.

Wallace, J., and F. Kay (2021). "Supportive Relations in a Feminized Occupation: How Male and Female Veterinarians Compare." *Canadian Review of Sociology/Revue Canadienne De Sociologie*, 59 (1): 4–22.

GRIEF AND PET LOSS

One of the significant challenges of cohabitating with and forming bonds with companion animals arises from the reality that, in general, humans outlive their pets. In fact, the American Pet Products Association (APPA) reports that the sorrow felt when a pet passes away is the principal disadvantage noted by owners across all types of pets.

How do individuals manage this loss? Similar to other close relationships, the grief that accompanies the death of a pet can be profoundly intense. The stages of grief identified by Elizabeth Kübler-Ross in 1969 concerning human loss—namely denial, anger, bargaining, depression, and acceptance—are equally applicable to those grieving the loss of their animal companions. Recent sociological and psychological research has begun to examine human grief in relation to pet loss, employing instruments such as the Grief Experience Inventory and the Censhare Pet Attachment Survey to gain insights into the emotional depth of these experiences and the coping strategies employed by individuals. These investigations have found that the intensity of grief correlates with the level of attachment to the pet, the availability of supportive social networks, and the presence of additional stressors in one's life (Wrobel and Dye, 2003).

Contrary to some assumptions, research indicates that grief intensity does not significantly differ between individuals without children and those with children, nor does it necessarily increase for individuals who own a single pet as opposed to multiple pets. Moreover, while it appears that women may express greater grief over pets than men, this discrepancy could stem from societal expectations that discourage men from overtly showing their emotions in public.

Philosopher Yi-Fu Tuan (2017) asserts that the loss of a pet differs from that of a human, as the death of a loved one is often final, whereas the space left by a deceased pet might be filled with another companion animal. This notion of replaceability is echoed in sociologist David Redmalm's 2006 examination of condolence cards designed for pet loss.

Nevertheless, it is undeniable that we form deep attachments to our pets and often experience significant sorrow following their passing. The APPA notes that over half of

Products and services related to pet memorialization have become more available and popular, providing comfort and acknowledging the emotional significance of a pet's life. (Olha Vorotyntseva/Dreamstime.com)

dog, bird, and small animal owners (though not cat owners) would consider memorializing their pets after death. A small but expanding movement advocates for "whole-family cemeteries" in the United States and other locations, allowing the burial of pets alongside their human counterparts—an enduring practice that dates back to the late Paleolithic. At present, individuals have access to a multitude of options for addressing the death of a companion animal, including burial, cremation, taxidermy, and the incorporation of the animal's remains (such as ashes, fur, or other parts) into jewelry, tattoos, or artwork.

Professionals such as social workers and counselors now cater specifically to those mourning their companion animals, while classes for veterinarians focus on assisting clients through the grieving process. Tips are also available to support surviving animals who may be grieving the loss of their companions. Various services, such as pet psychics who claim to communicate with deceased animals for grieving owners, memorial websites providing support for individuals lacking a personal network, and support groups offered by universities and veterinary practices, further illustrate the available resources for those suffering from the loss of a pet. Interestingly, an Indiana funeral home has introduced a unique practice of offering a support dog to assist families mourning human loss. Fortunately, the acknowledgment of the significance of companion animals has grown, enabling many to find solace in shared experiences of grief, rather than navigating their sorrow in isolation.

Research on pet bereavement suggests that the loss of a companion animal can elicit grief comparable to that experienced following the death of a human loved one. Studies emphasize that the strength of the emotional bond formed with pets significantly influences the experience of profound sorrow, guilt, and anger, shaped by factors such as the number of pets owned and the depth of the relationship (Silva et al., 2025; Uccheddu et al., 2019). In many cases, pet loss remains socially unrecognized, leading to a phenomenon termed "disenfranchised grief," wherein individuals feel isolated in their sorrow due to the absence of formal rituals or empathetic recognition typically accorded to human deaths (Packman et al., 2014). To support individuals coping with pet loss, various mechanisms have been established. These include traditional avenues such as support groups, counseling sessions often conducted by veterinarians, and peer discussions (Edwards-Callaway et al., 2022).

The role of veterinarians is particularly crucial, as they frequently guide clients toward appropriate professional and community-based support resources. Additionally, the burgeoning pet death industry serves as a significant cultural response, establishing normative practices to facilitate bereavement and legitimize the grief associated with pet loss. Increasingly popular offerings such as pet condolence cards, obituaries, online memorials, and dedicated pet cemeteries provide tangible means for memorialization, affirming the emotional import of a pet's life (Kogan et al., 2022). These memorial practices not only enhance individual comfort but also create public spaces for communal mourning and remembrance, fostering connections among pet owners. The incorporation of memorial practices into societal norms is further supported by the rise of online communities and social media platforms, where pet obituaries and digital tributes help to address the traditional invisibility surrounding pet loss.

Such platforms enable collective grieving and encourage emotional expression, facilitating psychological healing (Packman et al., 2014). Furthermore, literature posits that children, who often develop strong emotional bonds with their pets, benefit from structured support, such as school-based counseling and family discussions, ensuring that they can integrate their experience of loss into their developmental narratives in a healthy way (Field et al., 2009; Wrobel and Dye, 2003).

Beyond personal grieving processes, the relationship between individual attachment and cultural practices concerning pet memorialization underscores a broader societal transformation. As pets increasingly come to be viewed as family members, the evolving landscape surrounding pet loss and memorialization seeks to ensure that grief is recognized and addressed within both personal and communal frameworks. Research evidence continues to support that structured interventions—encompassing both professional counseling and peer support mechanisms—are instrumental in alleviating the severity of grief responses, thereby promoting healthier adjustment following the passing of a pet (Silva et al., 2025a; Edwards-Callaway et al., 2022; Field et al., 2009).

See also: Bond, Human-Animal; Pet Death Industry

References and Further Reading

Barnard-Nguyen, S., M. Breit, K. A. Anderson, and J. Nielsen (2016). "Pet Loss and Grief: Identifying at-Risk Pet Owners During The Euthanasia Process." *Anthrozoös*, 29 (3): 421–30.

Cordaro, M. (2012). "Pet Loss and Disenfranchised Grief: Implications for Mental Health Counseling Practice." *Journal of Mental Health Counseling*, 34 (4): 283–94.

Eckerd, L. M., J. E. Barnett, and L. Jett-Dias (2016). "Grief Following Pet and Human Loss: Closeness is Key." *Death Studies*, 40 (5): 275–82.

Edwards-Callaway, L., H. Simpson, N. Román-Muñiz, C. Cramer, S. Mijares, L. Stallones, J. R. Gonzalez, and J. Aberle (2022). "Preliminary Exploration of Weekly Peer Group Discussions as a Strategy for Coping with Feelings Associated with Euthanasia in Dairy Caretakers." *International Journal of Environmental Research and Public Health*, 19 (4): 2177.

Field, N. P., L. Orsini, R. Gavish, and W. Packman (2009). "Role of Attachment in Response to Pet Loss." *Death Studies*, 33 (4): 334–55.

Kogan, L. R., C. Bussolari, J. Currin-McCulloch, W. Packman, and P. Erdman (2022). "Disenfranchised Guilt—Pet Owners' Burden." Animals, 12 (13): 1690.

Packman, W., B. J. Carmack, R. Katz, F. Carlos, N. P. Field, and C. Landers (2014). "Online Survey as Empathic Bridging for the Disenfranchised Grief of Pet Loss." *OMEGA-Journal of Death and Dying*, 69 (4): 333–56.

Pierce, J. (2014). *The Last Walk: Reflections on Our Pets at the End of their Lives*. Chicago: University of Chicago Press.

Pierce, J., and R. Taylor (2019). "Last Moments: Witnessing and Representing the Death of Pets." In T. K. Shackelford and V. Zeigler-Hill (eds), *Evolutionary Perspectives on Death*, 161–76. Cham: Springer.

Redmalm, D. (2015). "Pet Grief: When is Non-Human Life Grievable?" *The Sociological Review*, 63 (1): 19–35.

Silva, M. V., R. R. Santos, and M. Barbosa (2025a). "Euthanasia and Prolonged Grief: A Cross-Sectional Study with Bereaved Pet Owners." *Journal of Veterinary Behavior*, 79: 60–7.

Silva, S., M. M. Martins, and M. Tavares (2025b). "The Grieving Process and Reorganization: Experiences and Needs of Post-Caregivers After the Death of the Person Cared For." Psychology International, 7 (3): 70.

Tuan, Y. F. (2017). *Dominance and Affection*. New Haven: Yale University Press.
Uccheddu, S., L. De Cataldo, M. Albertini, S. Coren, G. Da Graça Pereira, A. Haverbeke, ... and F. Pirrone (2019). "Pet Humanisation and Related Grief: Development and Validation of a Structured Questionnaire Instrument to Evaluate Grief in People Who Have Lost a Companion Dog." Animals, 9 (11): 933.
Wrobel, T. A., and A. L. Dye (2003). "Grieving Pet Death: Normative, Gender, and Attachment Issues." *OMEGA-Journal of Death and Dying*, 47 (4): 385–93.

H

HEALTH RISKS OF PET OWNERSHIP

While it is clear that pets provide a variety of health-improving functions for people, there are also health risks associated with pets, particularly those risks that arise from bites, scratches, and zoonotic transmission.

Studies have documented that close contact with household pets creates favorable conditions for the transmission of bacterial pathogens such as Bartonella spp. (cat scratch disease), Leptospira spp. (leptospirosis), Campylobacter, and Salmonella, all of which can cause systemic infections in humans (Damborg et al., 2016; Guptill, 2012). Such zoonotic infections may occur through direct interactions such as bites and scratches, via contact with contaminated saliva, urine, or feces, or as a result of inhaling aerosols that contain pathogens (Damborg et al., 2016; Chitty and Hendricks, 2007).

Household practices, including the sharing of beds and the close presence of pets in living spaces, may further heighten the risk of zoonotic transmission. Research indicates that a significant proportion of pet owners allow their companion animals into their bedrooms, inadvertently creating opportunities for close contact and increased exposure to pathogens (Chomel and Sun, 2011). Additionally, improper handling of pets or inadequate hand hygiene can serve as key risk factors for infection, particularly among high-risk populations such as young children and immunocompromised individuals (Whitfield and Smith, 2014). Veterinary literature underscores that while pets offer numerous health benefits, they also require proper management and preventive measures—such as regular veterinary care, vaccination, and parasite control—to mitigate these risks (Basit et al., 2024).

The transmission dynamics of zoonoses are further complicated by the changing landscape of pet ownership, including the increasing popularity of exotic pets. These animals may harbor less familiar pathogens and potentially contribute to the emergence of novel zoonotic diseases if proper contact safety and veterinary oversight are lacking (Arnecke et al., 2024). In this context, effective risk communication and interdisciplinary collaboration among veterinary and medical professionals are essential for informing pet owners and managing public health threats (Nyokabi et al., 2024; Giessen et al., 2022).

In summary, the literature indicates that the risks posed by pet-associated zoonoses are significant and multifactorial. Transmission through bites, scratches, and other forms of close contact is well-documented, and these risks are exacerbated by factors such as intimate household contact, inadequate hygiene practices, and the growing popularity of nontraditional pets. Public health strategies that include effective risk communication, interdisciplinary collaboration, and improved veterinary oversight are essential for minimizing these risks and ensuring that the benefits of pet ownership are not outweighed by potential health hazards.

See also: Healthcare Spending on Pets; Human Health, Pets And

References and Further Reading

Arnecke, A. L., S. Schwarz, A. Lübke-Becker, K. C. Jensen, C. Herre, and M. Bahramsoltani (2024). "Risk Communication on Zoonoses and Antimicrobial Resistance—How Do Exotic Pet Owners Perceive the Communication of Their Veterinarians?" Animals, 14 (14): 2035.

Basit, A., U. Yasin, H. Hashmi, A. Kiran, H. Ali, B. Keerio, M. A. Rana, M. T. Ahmad, K. Zeb, and H. Riaz (2024). "Pets Diseases and Public Health: Zoonosis, Transmission and Treatment: A Review." *IJBR*, 2 (2): 1059–71.

Chan, E., Y. Gao, L. Li, and P. Lee (2017). "Injuries Caused by Pets in Asian Urban Households: A Cross-Sectional Telephone Survey." *BMJ Open*, 7 (1): e012813. https://doi.org/10.1136/bmjopen-2016–012813.

Cherniack, E., and A. Cherniack (2014). "The Benefit of Pets and Animal-assisted Therapy to the Health of Older Individuals." *Current Gerontology and Geriatrics Research*, 2014: 1–9.

Chitty, J., and A. Hendricks (2007). "Zoonotic Skin Disease in Small Animals." In Practice, 29 (2): 92–7.

Chomel, B. B., and B. Sun (2011). "Zoonoses in the Bedroom." Emerging Infectious Diseases, 17 (2): 167.

Damborg, P., E. M. Broens, B. B. Chomel, S. Guenther, F. Pasmans, J. A. Wagenaar, ... and L. Guardabassi (2016). "Bacterial Zoonoses Transmitted by Household Pets: State-of-the-Art and Future Perspectives for Targeted Research and Policy Actions." Journal of Comparative Pathology, 155 (1): S27–40.

Guptill, L. (2012). "Bartonella Infections in Cats: What Is the Significance?" In Practice, 34 (8): 434–45.

Johnson, N., and A. Fooks (2014). "Jet Set Pets: Examining the Zoonosis Risk in Animal Import and Travel Across the European Union." *Veterinary Medicine Research and Reports*, 6: 17–25.

Kilduff, S., B. Steinman, Y. Xie, T. Kiss-Farengo, M. Foca, and N. Hayde (2023). "Pet Safety Guidelines for Pediatric Transplant Recipients." *Pediatric Transplantation*, 28 (1): e14527.

Lee, E., and J. Devlin (2021). "Knowledge of Pet-Related Zoonotic Diseases and Pet Care in Hong Kong, A Heavily Crowded Urban Setting." *Veterinary Medicine and Science*, 8 (1): 130–8.

Ma, J., B. Almanza, L. Ge, E. Her, Y. Liu, A. Lando, F. Wu, and L. Verrill (2020). "Pet Ownership and Pet Type Influence Food Safety in the Home: Evidence from a National Survey." *Journal of Food Protection*, 83 (9): 1553–60.

Nyokabi, N. S., L. Phelan, J. F. Lindahl, S. Berg, E. Muunda, A. Mihret, ... and H. L. Moore (2024). "Exploring Veterinary Students' Awareness and Perception of Zoonoses Risks, Infection Control Practices, and Biosecurity Measures in Ethiopia." Frontiers in Veterinary Science, 11: 1385849.

Platero, L., P. Sánchez, T. Sainz, C. Calvo, I. Iglesias, F. Esperón, R. de la Fuente, E. Frauca, A. Perez-Martinez, and A. Méndez-Echevarría (2022). "Pets for Pediatric Transplant Recipients: To Have or Not to Have." *Frontiers in Veterinary Science*, 9.

Stull, J., A. Peregrine, J. Sargeant, and J. Weese (2012). "Household Knowledge, Attitudes and Practices Related to Pet Contact and Associated Zoonoses in Ontario, Canada." *BMC Public Health*, 12 (1).

Tamiru, Y., D. Abdeta, and M. Amante (2022). "Knowledge, Attitude, and Practice Toward Pet Contact Associated Zoonosis in Western Ethiopia." *Veterinary Medicine Research and Reports*, 13: 47–58.

Van Der Giessen, J., F. Vlaanderen, T. Kortbeek, C. Swaan, H. van den Kerkhof, E. Broens, ... and C. Maassen (2022). "Signalling and Responding to Zoonotic Threats Using a One Health Approach: A Decade of the Zoonoses Structure in the Netherlands, 2011 to 2021." Eurosurveillance, 27 (31): 2200039.

Whitfield, Y. and A. Smith (2014). "Household Pets and Zoonoses." *Environmental Health Review*, 57 (2): 41–9.

HEALTHCARE SPENDING ON PETS

Over the past few decades, Americans have increasingly devoted substantial financial resources to their pets, signifying a deepening of the human–pet bond that treats companion animals as family members. While it is difficult to find statistics on consumer spending regarding pets outside of the United States, we can assume that spending increases in the United States might parallel similar increases in other parts of the world.

In the United States, significant expenditures are directed toward pet healthcare, including routine veterinary services, emergency interventions, diagnostic procedures, and specialized treatments. One prominent indication of this trend is that annual spending on veterinary care has surpassed 11 billion dollars, reflecting, in part, the evolution of pets from working animals or property into cherished family members. This financial commitment parallels investments in other pet-related expenses such as pet food, which has evolved into a multi-billion dollar industry focused on the health and quality of life for animals (Pearce et al., 2023).

As pet owners increasingly view their animals not merely as property but as integral members of the family, there is a corresponding willingness to bear rising healthcare costs to maintain the health, longevity, and quality of life of their pets. Additionally, the development of markets for premium pet healthcare services, advanced diagnostic techniques, and specialized treatments has driven innovation within the veterinary sector, mirroring trends seen in human healthcare. This is accompanied by the growing availability of pet insurance and the expanding role of veterinarians as key figures in both clinical care and public education regarding pet health.

Furthermore, consumer spending on pet healthcare reflects broader cultural shifts. The prioritization of pet health intersects with trends in pet humanization and heightened concerns about animal welfare. Spending patterns not only reveal our desire to care for pets with the same attention given to human family members but also serve as a barometer for societal attitudes toward animals. High financial outlays in pet healthcare indicate both the diversification of pet-related services and a strengthened commitment to ensuring pets lead healthy, enriched lives in an era where they occupy roles that extend far beyond traditional companionship (Pearce et al., 2023).

These trends are indicative of broader societal changes, where increasing financial and emotional investments in companion animals are reshaping industries, influencing product development, and reinforcing the notion that pets are essential members of modern households.

On the other hand, as of this writing, the world may be headed into a recession, which will provide greater financial pressures on people living with pets. The result of any economic downturn is often a decline in spending for pets and an increase in pet abandonment. Even as we become more deeply invested in having relationships with pets, those relationships are still subject to economic issues, and thus complicate matters further.

See also: Attachment; Family, Pets As; Pet Food; United States, Pets In; Veterinary Care

References and Further Reading

Hannay, C., and E. Alini (2024). "Pet Price Shock: Why Owning A Dog or Cat has Become so Expensive; From Private Equity Overtaking Veterinary Clinics to a Longstanding Shortage of Veterinarians, A Complex Web of Factors is Making it Ever More Expensive for Canadians Who Own a Pet." *Globe & Mail (Toronto, Canada)*, B10–B10.

Lloyd, J. (2023). "Pet Healthcare in the US: Another Look at the Veterinarian Workforce." *Mars Veterinary Health*.

Pearce, H., C. L. Neill, K. Royal, and M. Pairis-Garcia (2023). "Can Dogs Help Chickens? Pet Owners' Willingness to Pay for Animal Welfare-friendly Pet Food in the United States." *Animal Welfare*, 32: e11.

Williams, A., B. Williams, C. R. Hansen, and K. H. Coble (2020). "The Impact of Pet Health Insurance on Dog Owners' Spending for Veterinary Services." *Animals*, 10 (7): 1162.

Wolf, C. A., J. W. Lloyd, and J. R. Black (2008). "An Examination of US Consumer Pet-related and Veterinary Service Expenditures, 1980–2005." *Journal of the American Veterinary Medical Association*, 233 (3): 404–13.

HISTORY OF PET KEEPING: ANCIENT CLASSICAL WORLD

Research on pet keeping in the ancient world demonstrates that domestic animals were more than mere utilitarian assets; they were also important emblematic and emotional companions whose influence is seen in literature, visual arts, and funerary monuments.

It is almost certain that dogs, the first animal to be domesticated and one of the only animals not domesticated for food, were the first pets. As they were domesticated at least 15,000 years ago as hunting partners, it is logical to assume that hunters developed close relationships with some of their canine partners. The first pets, then, most likely served two primary purposes—hunting *and* companionship. But it probably was not until after the rise of the ancient civilizations that some dogs stopped working and turned into full-time pets for those with the means to support them.

Wild-caught birds were another early pet, kept in cultures as diverse as ancient Rome and the Aztec empire (Tella, 2011; MacKinnon, 2014; Bodson, 2000). In fact, ancient rulers kept wild animals of all sorts in menageries that were intended to demonstrate their mastery over nature; but birds, because of their size, could easily be kept indoors, in cages. Birds were beautiful and exotic, and could sing—a benefit in the era before recorded music became available. Ornamental fish, another wild but easily kept and controlled wild animal, were also kept as pets, most prominently in Japan and China, where the keeping of pet fish probably originated as early as the seventh century (Schafer, 1968).

Cats were another early companion (Pennisi, 2004; Driscoll et al., 2009). Like dogs, cats were not domesticated for food—they were enticed by early farmers to live near human fields and granaries to keep down the rodent populations—and thus were good candidates to live indoors with humans. They were famously kept as pets (and sacred animals) by the Egyptians. Both the Greeks and Romans kept dogs and parrots as pets, and the Romans buried their dead dogs with marble grave markers. Because Romans believed in an afterlife that would be shared with animals, many Romans had their companion animals killed after their own deaths, in order that their fates might be shared.

In ancient Greece, evidence from vase paintings, literary texts, and inscriptions indicates that a range of animals—including dogs, birds, and small exotic creatures—were kept as companions and helpers. Greek writers often referenced animals in their descriptions of

A lead-glazed earthenware dog sculpture from China, Eastern Han Dynasty (25–220 CE). Dogs were the first animals to be domesticated, likely becoming full-time pets with the rise of ancient civilizations. (The Cleveland Museum of Art, Purchase from the J. H. Wade Fund 1983.1)

household life, and artistic depictions reveal that pets were part of daily social and ritual activities, reflecting both status and intimate family bonds.

In Rome, pet keeping became a well-established institution that transcended social classes. Roman literature, from satires to personal letters, frequently mentions the role of dogs, birds, and other companion animals as loyal friends and symbols of virtue. Moreover, archaeological finds such as tomb inscriptions and monuments sometimes depict pets, underlining the emotional bonds between owners and their animals. Such evidence suggests that for many Romans, the relationship with pets was deeply integrated into expressions of loyalty, mourning, and remembrance—so much so that pets were occasionally honored with funerary rites analogous to those for human family members.

Ancient Egypt offers perhaps the most striking example of pet keeping because of its profound religious and cultural significance. The Egyptians revered certain animals, most notably cats, which were believed to be sacred and connected to deities such as Bastet. Artistic representations, mummification practices, and funerary monuments attest to the status of cats not only as household companions but also as symbols of divine protection and afterlife continuity. In addition to cats, evidence attests to the companionship provided by dogs, birds, and other domesticated species, which were integral to household and ritual life.

Together, these sources illustrate that pet keeping in the ancient classical world was a multifaceted practice. It was documented through various media—literature that celebrated the virtues of companion animals, visual art that depicted them in everyday and ceremonial contexts, and monuments that memorialized them after death—indicating that both the elite and ordinary citizens valued the companionship, loyalty, and symbolic power of their pets.

See also: Grief and Pet Loss; History of Pet Keeping: Middle Ages; History of Pet Keeping: Prehistory; History of Pet Keeping: Renaissance; History of Pet Keeping: Victorian Era

References and Further Reading

Bodson, L. (2000). "Motivations for Pet-Keeping in Ancient Greece and Rome: A Preliminary Survey." In A. L. Podberscek, E. S. Paul, and. J. A. Serpell (eds), Companion Animals and Us: Exploring the Relationships Between People and Pets, 27–41. Cambridge: Cambridge University Press.

Bradley, K. (1998). "The Sentimental Education of the Roman Child: the Role of Pet-keeping." *Latomus*, 57 (Fasc. 3): 523–57.

Calder, L. (2017). "Pet and Image in the Greek World: the Use of Domesticated Animals in Human Interaction." In T. Fögen and E. V. Thomas (Eds.), *Interactions between Animals and Humans in Graeco-Roman Antiquity*, 61–88. Berlin: Walter de Gruyter GmbH & Co KG.

Driscoll, C. A., J. Clutton-Brock, A. C. Kitchener, and S. J. O'Brien (2009). "The Taming of the Cat." *Scientific American*, 300 (6): 68.

MacKinnon, M. (2010). "'Sick as a Dog': Zooarchaeological Evidence For Pet Dog Health and Welfare in the Roman World." *World Archaeology*, 42 (2): 290–309.

MacKinnon, M. (2014). "Fauna of the Ancient Mediterranean World." In G. L. Campbell (ed.), *The Oxford Handbook of Animals in Classical thought and Life*, 156–79. Oxford: Oxford University Press

Pennisi, E. (2004). "Burials in Cyprus Suggest Cats were Ancient Pets." *Science*, 304 (5668): 189.
Schafer, E. H. (1968). "Hunting Parks and Animal Enclosures in Ancient China." Journal of the Economic and Social History of the Orient, 11: 318–43.
Tella, J. L. (2011). "The Unknown Extent of Ancient Bird Introductions." *Ardeola*, 58 (2): 399–404.

HISTORY OF PET KEEPING: MIDDLE AGES

The history of pet keeping in Medieval Europe and Medieval Asia reveals diverse cultural practices and beliefs regarding animals. In Medieval Europe, pet keeping was found among royalty and other elites, as well as among members of the clergy.

Archaeological evidence from Europe—including iconographic material, grave inscriptions, and written texts—demonstrates that dogs, cats, and birds were sometimes maintained as companions, symbols of loyalty, and even occasionally as objects of mourning. This was not a widespread practice, however, and for most, owning or caring for a pet animal was not just economically unfeasible, but forming a close bond with a nonhuman animal could easily subject someone to accusations of witchcraft.

During the Middle Ages, the Catholic Church viewed the practice of pet keeping as a form of heresy, in that it challenged the human/animal boundary, even though many nuns and monks kept pets (Maehle, 2002). Another reason why pet keeping had been seen as trivial—and still is today, though to a much lesser extent—seen as trivial is the fact that historically, as today, women are the primary caretakers of companion animals. Because of women's association with pets, and in particular, with lap dogs, pets were seen as feminizing and symbolized women's inferiority. Even worse, a woman in Medieval Europe could be accused of being a witch if she had too close a relationship with an animal.

European elites were often satirized in the press for their keeping of pets, although noblemen who kept hunting dogs were exempt from this stigma. The poor, on the other hand, were prohibited from hunting in England (as lands were owned either by either the Church, the Crown, or the nobility); consequently, they were prohibited from keeping hunting dogs. Although people from all socioeconomic statuses keep pets today, high income remains correlated with pet keeping—the higher one's income, the greater the likelihood of living with a pet. In addition, purebred pets remain associated with elites, whereas the poor kept animals that the wealthy saw as dirty and diseased. Even today, purebred animals are still linked to elites.

Turning to Medieval Asia, particularly in China and Japan, evidence—though less voluminous in the archaeological record—points to sophisticated systems of pet keeping intertwined with courtly culture, art, and ritual practices; in other words, it was most common, and certainly most well documented, among the elites. In Medieval China, pet keeping was often associated with refined leisure activities and garden culture, where companion animals such as small dogs and birds were appreciated for both their aesthetic qualities and the harmony they symbolized within Confucian and Taoist frameworks.

Similarly, in Medieval Japan, the adoption of pet keeping practices is attested by surviving literary texts, paintings, and ceramics that depict animals in intimate, domesticated settings.

Scholars like Keirstead have argued that during the Medieval period, pet keeping became part of elite practices, reflecting ideals of refined living and the symbolic integration of nature with human society.

These practices highlight that, while the social and theological contexts differed markedly from their European counterparts, the mutual human–animal bond was a small but growing element of Medieval life across both regions.

See also: China, Pets In; Elite Pet Keeping; Europe, Pets In; History of Pet Keeping: Ancient Classical World; History of Pet Keeping: Renaissance; Japan, Pets In

References and Further Reading

Figg, K. M. (2015). "Pets in the Middle Ages: Evidence from Encyclopedias and Dictionaries." *Enarratio*, 18.

Gerhart, K. M. (2016). "13 Medieval Scavengers: Dogs in Japanese Handscrolls." In L. D. Gelfand (ed.), *Our Dogs, Our Selves*, 303–24. Leiden: Brill.

Knapp, K. N. (2019). "The Use and Understanding of Domestic Animals in Early Medieval Northern China." *Early Medieval China*, 25 (1): 85–99.

Maehle, A. H. (2002). "Cruelty and Kindness to the 'Brute Creation': Stability and Change in the Ethics of the Man-Animal Relationship, 1600–1850." In A. Manning and J. Serpell (eds), Animals and Human Society, 113–37. London: Routledge.

O'Connor, T. (2017). "Animals in Urban Life in Medieval to Early Modern England." In U. Albarella, M. Rizzetto, H. Russ, K. Vickers, and S. Viner-Daniels (eds), *The Oxford Handbook of Zooarchaeology*, 214. New York: Oxford University Press.

Schäfer, D., M. Siebert, and R. Sterckx (2018). "Knowing Animals in China's History." *Rough Chinese History*, 1.

Thomas, R. (2005). "Perceptions Versus Reality: Changing Attitudes Towards Pets in Medieval and Post-Medieval England." *Bar International Series*, 1410, 95.

Walker-Meikle, K. F. (2008). *Late Medieval Pet Keeping: Gender, Status and Emotions*. London: University of London, University College London.

HISTORY OF PET KEEPING: PREHISTORY

What we know about human relationships with nonhuman animals from prehistory—the time before writing emerged—is scant and comes primarily from the archaeological record. In addition, anthropologists have long taken the practices and beliefs of contemporary indigenous people to act as a kind of glimpse into the past, with the understanding that contemporary tribal peoples can help us to imagine the lives of our own ancestors. Even this is a problematic way to look at the past, as contemporary peoples are not a mirror of the past; they, like all human societies today, have evolved in concert with broader planetary changes.

One thing that we know for sure is that only one species entered fully into human culture in the period before writing, and thus history, emerged. The domestication of the wolf was an event that not only transformed a wild canine into the modern domestic dog but also inaugurated humanity's longstanding relationship with animals as partners and companions. Even the process of domestication, which has been largely reconceived in recent years from a one-way process led by humans, to a two-way process initiated probably by wolves, demonstrates that the first dogs were partners, not tools or resources. This early process of domestication appears to have involved a commensal relationship, wherein wolves that exhibited reduced fear of humans were more likely to be nurtured and incorporated into human groups—a hypothesis known as cross-species adoption (Serpell, 2021). Over time, such behaviors laid the foundation for a symbiotic relationship, as humans benefited from the wolves' roles in hunting assistance and protection, while the wolves received food and shelter.

In addition to the domestication of wolves, our ancestors likely engaged in taming and cohabitation with other wild animals, initiating a broader tradition of keeping animals as companions. However, the wolf-to-dog transition is particularly pivotal because it not only set the stage for an enduring partnership between humans and canids but also influenced broader patterns of pet keeping across cultures in subsequent millennia.

Together, these lines of evidence illustrate that the prehistoric process of domesticating the wolf was not a singular event but a gradual and multifaceted transformation embedded in the daily lives of early humans. This transformation—fueled by both natural selection for socially compatible traits and later intensification through human-directed breeding—illustrates how our ancestors began shaping a relationship with animals that has evolved into the diverse and emotionally rich practice of pet keeping seen in modern societies.

See also: Dogs; Domestication; History of Pet Keeping: Ancient Classical World; Selective Breeding

References and Further Reading

Herzog, H. (2014). "Biology, Culture, and the Origins of Pet-Keeping." Animal Behavior and Cognition, 1 (3): 296–308.

Hill, E. (2013). "Archaeology and Animal Persons: Toward a Prehistory of Human-Animal Relations." *Environment and Society*, 4 (1): 117–36.

Russell, N. (2011). *Social zooarchaeology: Humans and Animals in Prehistory*. New York: Cambridge University Press.

Serpell, J. (2014). "Pet-keeping and Animal Domestication: A Reappraisal." In J. Clutton-Brock (ed.), *The Walking Larder*, 10–21. Routledge.

Serpell, J. A. (2021). "Commensalism Or Cross-Species Adoption? A Critical Review of Theories of Wolf Domestication." *Frontiers in Veterinary Science*, 8: 662370.

HISTORY OF PET KEEPING: RENAISSANCE

During the Renaissance, the period in European history from about the fourteenth to the sixteenth centuries, pet keeping underwent a significant transformation as animals increasingly came to symbolize personal refinement, social status, and an emerging humanist interest in nature. Members of the aristocracy and emerging middle classes—toward the end of the period—adopted a wide range of companion animals—not only for utilitarian purposes such as hunting and pest control but also as expressions of taste, leisure, and identity. Contemporary literary accounts, personal diaries, and visual art from the period document the presence of dogs, cats, birds, and even horses in noble households, where these animals were carefully bred, trained, and celebrated.

Dogs, in particular, were esteemed for their loyalty and versatility. They served in both functional roles—such as assisting in hunts—and as everyday companions whose gentle presence was emblematic of refined domestic life. Portraits and engravings from the period often include dogs as symbols of fidelity and virtue, mirroring classical traditions revived during the Renaissance. A similar shift is noted in the changing perception of cats. Once regarded largely as wild or even pestiferous creatures, cats began to be seen as refined companions whose independent yet affectionate nature meshed well with the period's evolving aesthetic sensibilities and ideas about controlled domesticity. This evolving attitude toward felines laid early groundwork for the modern conception of the pet, one based on mutual companionship rather than strict utilitarian function (Schuurman and Dirke, 2020).

Horses also played an exceptional role during this era. As the Renaissance saw the blossoming of organized equestrian sports and elaborate displays of horsemanship, horses became both a practical necessity and an object of leisure and refinement. The culture of horse racing and performance—which involved intricate breeding programs and selective training practices—demonstrates an early form of the "pet fancy" mindset applied to equines. Specialists and noble patrons engaged in aesthetic manipulation of animal form and function, selecting for traits that enhanced performance in races while also signifying beauty and noble lineage. Such practices contributed not only to the enjoyment of the animal but also to a broader reimagining of human–animal relationships, intertwining art, science, and social status (Cooley, 2023).

In summary, Renaissance Europe witnessed a redefinition of pet keeping—from a primarily utilitarian activity to an expression of personal and cultural values. The period's renewed interest in classical antiquity, combined with innovations in art, literature, and natural philosophy, provided fertile ground for developing sophisticated practices of animal care, breeding, and companionship. By the end of the Renaissance, keeping animals as pets had assumed roles that were both intimately personal and overtly symbolic, reflecting a society that increasingly viewed its animal companions as extensions of its own refined identity.

See also: Elite Pet Keeping; Europe, Pets In; History of Pet Keeping: Middle Ages; History of Pet Keeping: Victorian Era

References and Further Reading

Cooley, M. (2023). *The Perfection of Nature: Animals, Breeding, and Race in the Renaissance.* Chicago: University of Chicago Press.

Schuurman, N., and K. Dirke (2020). "From Pest to Pet. Liminality, Domestication and Animal Agency in the Killing of Rats and Cats." *TRACE∴ Journal for Human-Animal Studies*, 6: 2–25.

HISTORY OF PET KEEPING: TWENTIETH AND TWENTY-FIRST CENTURIES

The evolution of pet keeping over the twentieth and twenty-first centuries demonstrates a shift from utilitarian or functional roles toward complex, emotional, and culturally embedded relationships in both Western societies and beyond. In the West—specifically the United States and the United Kingdom—pet keeping has transitioned from a function primarily associated with work (such as hunting and pest control) to an expression of family bonds and social identity.

Research indicates that by the twentieth century, emotional investment in pets had become a public display of acceptable sentiment, intertwining notions of class, gender, and modern family life (Hamlett, Hoskins, and Preston, 2021). In this context, pets were viewed increasingly as valued family members, a transition further influenced by cultural forces and trends in preferred breeds (Kubinyi, 2025; Herzog, 2014). These shifts not only reflect changes in the economic and social fabric of post-industrial societies but also mirror broader patterns of consumer behavior, where pet keeping is part of a wider inheritance of socially transmitted norms (Herzog, 2016).

In the United States, pet ownership patterns can vary significantly by ethnicity and socioeconomic status. Studies show white families tend to have pets more frequently than children, while some Asian families exhibit the opposite trend (Kubinyi, 2025). These dynamics highlight how pet keeping in the United States is deeply embedded within identity formation and lifestyle choices, contributing to a market that adapts to changing tastes and the symbolism of specific breeds. Such trends underscore the role of pets in reflecting and reinforcing changing societal ideals regarding family structure and companionship (Herzog, 2014).

In East Asia, particularly Japan, pet keeping has become increasingly prevalent in urban and suburban life throughout the late twentieth and early twenty-first centuries. The rapid urbanization and demographic changes in Japan have fostered a growing market for companion animals, as families become smaller and the functions of pets shift from utility to companionship and emotional support (Rock and Degeling, 2013). Comparative studies reveal that early exposure to pets can shape positive attitudes toward animal welfare and establish a foundation for adult interactions with pets (Miura, Bradshaw, and Tanida, 2002). In Japan, pets are often regarded not only as companions but also as a means to mitigate modern urban isolation, reflecting broader societal changes.

Overall, the history of pet keeping across these cultural contexts illustrates both convergence and divergence. In Western societies, the progression toward the "humanization" of pets is well-documented, showing how pet keeping intertwines with identity and emotional life (Hamlett et al., 2021; Kubinyi, 2025). In contrast, in Japan, rising pet ownership is closely linked to modernization and evolving familial structures, where pets serve roles as both traditional companions and facilitators of social interaction (Rock and Degeling, 2013; Miura, Bradshaw, and Tanida, 2002). As the twentieth century transitioned into the twenty-first, these trends have become increasingly pronounced, revealing how pet keeping functions as an arena for expressing cultural values and responding to societal evolution (Herzog, 2016, 2014).

See also: Family, Pets As; Healthcare Spending on Pets

References and Further Reading

Hamlett, J., L. Hoskins, and R. Preston (2021). "Pets and Family Relationships in Twentieth-Century British Diaries." *The History of the Family*, 26 (2): 266–87.

Herzog, H. (2014). "Biology, Culture, and the Origins of Pet-Keeping." Animal Behavior and Cognition, 1 (3): 296–308.

Herzog, H. (2016). "Does Cultural Group Selection Explain the Evolution of Pet-keeping?" *Behavioral and Brain Sciences*, 39: e41.

Kubinyi, E. (2025). "The Link Between Companion Dogs, Human Fertility Rates, and Social Networks." *Current Directions in Psychological Science*, 34 (4): 232–9.

Miura, A. J. W. S., J. W. S. Bradshaw, and H. Tanida (2002). "Childhood Experiences and Attitudes Towards Animal Issues: A Comparison of Young Adults in Japan and the UK." *Animal Welfare*, 11 (4): 437–48.

Rock, M., and C. Degeling (2013). "Public Health Ethics and a Status for Pets as Person-Things: Revisiting the Place of Animals in Urbanized Societies." *Journal of Bioethical Inquiry*, 10: 485–95.

HISTORY OF PET KEEPING: VICTORIAN ERA

Pet keeping as we know it today did not really emerge until the nineteenth century, when enough people had the disposable resources to keep animals solely for companionship. This period, known in England as the Victorian era as it largely coincides with the reign of Queen Victoria (1819–1901), also marked the rise of the commercial pet industry—an industry that began with companies selling food, medicine, and cages for companion animals but which grew to include breeders, dealers, and all of the associated businesses, like grooming salons and pet spas. In 1860, the first commercial dog food was available for purchase in England (but such commercially produced food didn't reach America until the early twentieth century), and the first commercial cat litter was produced in 1947 (Kete, 1994; Grier, 2010; Ritvo, 1987).

Queen Victoria is important in the history of pet keeping not just because she reigned during the period when the pet industry developed. Her marriage to Prince Albert and their large family of nine children and multiple pets became the model for a new type of family and a new type of childrearing which emphasized domesticity, humane child raising, and creating a home environment that is pleasing to the husband. This type of family structure became the norm for middle-class families throughout England and beyond.

Historian Harriet Ritvo (1987) points out that while the modern pet industry arose during the nineteenth century, it emerged during a time of incredible cruelty toward animals. Most Europeans and Americans were indifferent to animal pain and suffering. But slowly, during this century, an attitude that we might call the beginnings of the humane impulse began to emerge. Ritvo suggests that one reason for that change was that Europeans

Saturday Night: Returning from Labour, print by E. B. and E. C. Kellogg. Romanticized scenes of Victorian domestic life often emphasized devoted husbands, dutiful and contented wives, and obedient children and pets. (National Museum of American History, Harry T. Peters "America on Stone" Lithography Collection)

and Americans had by then felt that they had "conquered" nature and vanquished much of the threat that it posed to human culture. Only with the conquest of human cities over the natural world could humans begin to incorporate animals into their homes.

And with industrialism, and the changes in the agriculture industry which removed farm animals from most communities, animals largely disappeared from many people's lives, leaving a gap (compounded by the fact that middle-class families were having fewer children as large families make the most economic sense in an agricultural economy) to be filled by the development of the modern pet industry.

In addition, widespread pet keeping was also enabled in part by the rise of a middle class, developed through industrialism, with incomes to support what had been an elite, somewhat frivolous, hobby. We see not only pet keeping rise during this period, but also the fancy dog, cat, and rabbit breeding clubs.

Historian Katherine Grier (2010) also points out that the pet movement was fueled by a desire on the part of Victorian families to use animals to teach children middle-class, bourgeois virtues, such as kindness and self-control. Moralists of the period saw having a relationship with a pet as a way to instill positive virtues in a child, which Grier calls the "domestic ethic of kindness," which might have played a role in reducing some of the casual violence toward animals prevalent that was at that time. (Cruelty to animals was seen during this era as a sign of "inward moral collapse.")

It was expected that children would learn kindness toward all those who were dependent on others for their care, including pets, the elderly, and even (in the United States) slaves. This domestic ethic was a major part of the expansion of pet keeping in American society; it helped lead to the rise of the animal welfare movement in this country at the end of the nineteenth century, and is still very much with us today. But as Grier points out, while it changed the way that people thought about and dealt with animals within the home, it did not transform the ways in which people thought about animals outside of the home, or animals in general.

Even though the nineteenth century marked the rise of the commercial pet industry and the modern pet keeping movement, it really wasn't until the mid-twentieth century that many of the practices that we associate with modern pet keeping emerged. For instance, when a companion animal became sick, it either died on its own or the owners often killed it, as the practice of bringing animals to a veterinarian did not become common until the twentieth century.

See also: Europe, Pets In

References and Further Reading

Flegel, M. (2015). *Pets and Domesticity in Victorian Literature and Culture: Animality, Queer Relations, and the Victorian Family*. New York: Routledge.
Grier, K. C. (2010). *Pets in America: A History*. Chapel Hill, NC: UNC Press Books.
Howell, P. (2002). "A Place for the Animal Dead: Pets, Pet Cemeteries and Animal Ethics in Late Victorian Britain." *Ethics, Place & Environment*, 5 (1): 5–22.
Kete, K. (1994). The Beast in the Boudoir: Petkeeping in Nineteenth-Century Paris. Berkeley: University of California Press.
Ritvo, H. (1986). "Pride and Pedigree: The Evolution of the Victorian Dog Fancy." *Victorian Studies*, 29 (2): 227–53.
Ritvo, H. (1987). "The Emergence of Modern Pet-Keeping." *Anthrozoös*, 1 (3): 158–65.
Roussillon-Constanty, L., and S. Thornton (2018). "Comforting Creatures: Changing Visions of Animal Otherness in the Victorian Period." *Cahiers victoriens et édouardiens*, (88 Automne).

HOARDING

Animal hoarding is a psychological disorder in which a person collects a large number of animals, without the resources or ability to care for them properly.

Researchers have increasingly characterized animal hoarding as a specific manifestation of hoarding disorder—a psychiatric condition often linked with obsessive–compulsive spectrum disorders, dysfunctional early attachment, and impaired cognitive performance (Nguyen et al., 2022; Ferreira et al., 2020; Paloski et al., 2020). Individuals who hoard animals frequently display concurrent object hoarding behaviors, and clinical studies have highlighted high levels of comorbidity with depression, anxiety, and social isolation (Nguyen et al., 2022; Ferreira et al., 2020).

Demographically, a large number of studies converge on a profile in which animal hoarders are predominantly female, middle-aged or older, and often economically disadvantaged; however, cases have been documented across diverse socioeconomic and gender groups (Joffe et al., 2014; Cunha et al., 2022). The animals hoarded tend to be primarily cats and dogs, though other species are often involved. The problematic conditions found in hoarding environments typically include severe neglect, overcrowding, and unsanitary conditions, resulting in both animal suffering and heightened public health risks (Frost et al., 2000; Sacchettino et al., 2023). Mortality among the animals is common, and instances of dead animals are reported frequently as indicators of the extreme neglect observed in many cases (Sacchettino et al., 2023).

Intervention strategies have evolved from punitive approaches to more collaborative, long-term monitoring and support measures involving both human and animal welfare agencies. For example, Hill et al. (2019) reported that early identification of animal hoarding cases and integrated interventions, which include veterinary assessments and ancillary human services, can help speed case resolution, although recidivism remains high. Similarly, Tamimi et al. (2024) demonstrated that voluntary spay/neuter and relinquishment programs in New York City can yield favorable outcomes by helping to stabilize the animal population and improve overall welfare. Nonetheless, a lack of centralized reporting and statutory coordination complicates efforts to track and systematically address hoarding behavior and its associated recurrences.

Legal interventions vary considerably between jurisdictions. In places such as Australia and the United Kingdom, legal actions against animal hoarders tend to focus on extreme cases, with prosecutions often representing only the most severe manifestations of the problem (Joffe et al., 2014; Elliott et al., 2019). In many regions, however, law enforcement faces challenges in applying standard animal welfare laws to hoarding cases because of the diffuse nature of responsibility among animal, health, and social services Wilkinson et al., 2022). When authorities intervene, animals are frequently removed from homes and transferred to shelters or specialized care facilities, although the long-term outcomes for these animals depend on the resources available for rehabilitation and rehoming, and sometimes animals face euthanasia when conditions are deemed irreparable (Sacchettino et al., 2023).

Overall, the literature underscores that animal hoarding is not only a public and animal health issue but also a deep-seated psychiatric disorder with significant recidivism rates. Effective management and intervention require multidisciplinary approaches that address both the mental health of the hoarders and the welfare of the animals, alongside improved legal and administrative frameworks to ensure early detection and sustained support (Hill et al., 2019; Wilkinson et al., 2022; Cunha et al., 2022).

See also: Attachment; Cruelty; Neglect

References and Further Reading

Cunha, G. R. D., M. Pellizzaro, C. M. Martins, S. M. Rocha, A. C. Yamakawa, E. C. Da Silva, … and A. W. Biondo (2022). "Serological Survey of Anti-Leptospira spp. Antibodies in Individuals With Animal Hoarding Disorder and Their Dogs in a Major City of Southern Brazil." Veterinary Medicine and Science, 8 (2): 530–6.

Dozier, M. E., C. Bratiotis, D. Broadnax, J. Le, and C. R. Ayers (2019). "A Description of 17 Animal Hoarding Case Files from Animal Control and a Humane Society." *Psychiatry Research*, 272: 365–8.

Elliott, R., J. Snowdon, G. Halliday, G. E. Hunt, and S. Coleman (2019). "Characteristics of Animal Hoarding Cases Referred to the RSPCA in New South Wales, Australia." Australian Veterinary Journal, 97 (5): 149–56.

Ferreira, L., L. H. Paloski, D. B. Costa, C. M. Tatay, and T. Q. Irigaray (2020). "Psychopathological Comorbid Symptoms in Animal Hoarding Disorder." *Psychiatric Quarterly*.

Frost, R. O., G. Steketee, and L. Williams (2000). "Hoarding: A Community Health Problem." Health & Social Care in the Community, 8 (4): 229–34.

Hill, K., D. Yates, R. Dean, and J. Stavisky (2019). "A Novel Approach to Welfare Interventions in Problem Multi-Cat Households." BMC Veterinary Research, 15 (1): 434.

Joffe, A., D. O'Shannessy, N. K. Dhand, M. Westman, and A. Fawcett (2014). "Characteristics of Persons Convicted for Offences Relating to Animal Hoarding in New South Wales." *Australian Veterinary Journal*.

Lockwood, B. (2018). "Animal Hoarding: The Challenge for Mental Health, Law Enforcement, and Animal Welfare Professionals." *Behavioral Sciences & the Law*.

Nadal, Q., M. Ferrari, J. Lora, A. Revollo, F. Nicolás, S. Astegiano, and M. D. Videla (2022). "Noah's Syndrome: Systematic Review of Animal Hoarding Disorder." *Human–Animal Interaction Bulletin*.

Nguyen, B. K., J. J. Zakrzewski, L. Sordo Vieira, and C. A. Mathews (2022). "Impact of Hoarding and Obsessive–Compulsive Disorder Symptomatology on Quality of Life and Their Interaction With Depression Symptomatology." Frontiers in Psychology, 13: 926048.

Paloski, L. H., E. A. Ferreira, D. B. Costa, C. R. de Oliveira, C. Moret-Tatay, and T. Q. Irigaray (2020). "Cognitive Performance of Individuals With Animal Hoarding." Health and Quality of Life Outcomes, 18 (1): 40.

Paloski, W., E. A. Ferreira, D. B. Costa, M. L. Huerto, C. R. Oliveira, Irani I. L. Argimon, and T. Q. Irigaray (2017). "Animal Hoarding Disorder: A Systematic Review." *Psico*.

Sacchettino, L., C. Gatta, V. O. Giuliano, F. Bellini, A. Liverini, F. Ciani, … and F. Napolitano (2023). "Description of Twenty-Nine Animal Hoarding Cases in Italy: The Impact on Animal Welfare." Animals, 13 (18): 2968.

Steketee, G., A. Gibson, R. O. Frost, J. Alabiso, A. Arluke, and G. Patronek (2011). "Characteristics and Antecedents of People Who Hoard Animals: An Exploratory Comparative Interview Study." *Review of General Psychology*.

Tamimi, A., L. Kisiel, E. Dolan, and E. Berliner (2024). "A Retrospective Study of Cat Hoarding Cases and their Management Through Voluntary Spay/neuter And Relinquishment in New York City." *Journal of Shelter Medicine and Community Animal Health*.

Ung, D., M. E. Dozier, C. Bratiotis, and C. R. Ayers (2016). "An Exploratory Investigation of Animal Hoarding Symptoms in a Sample of Adults Diagnosed with Hoarding Disorder." *Journal of Clinical Psychology*.

Wilkinson, J., M. Schoultz, H. M. King, N. Neave, and C. Bailey (2022). "Animal Hoarding Cases in England: Implications for Public Health Services." *Frontiers in Public Health*.

HOLOCAUST, PETS DURING THE

The Holocaust refers to the genocide of European Jews during the Second World War.

Although there is some documentation about animal companionship under Nazi rule, existing literature provides indirect evidence that clarifies two interrelated dimensions. On the one hand, several Nazi leaders cultivated images of natural affinity and domestic normalcy through the ownership of pets, while on the other, discriminatory state policies contributed to the broader exclusion of Jewish citizens from aspects of everyday life that might otherwise include pet keeping.

Nazi leaders were known for their carefully managed public images, and pet keeping played a role in that construction. Historical accounts document that prominent figures, notably Adolf Hitler with his German Shepherd Blondi, maintained close relationships with their animals. This practice was not merely a private matter but was used propagandistically to signify loyalty, purity, and a connection to a mythologized, Norse-based, German nature.

Cole (2014) underscores how the regime's fascination with nature was a recurring theme in Nazi ideology. This idealization extended into personal behavior and was strategically deployed to contrast the "Aryan" way of life with the supposedly inferior practices of those they persecuted. Schafft (2004) further details how pseudo-scientific ideas and cultural images were manipulated to legitimize Nazi racial policies, suggesting that even domestic practices such as pet keeping were not immune from ideological interpretation.

In stark contrast, the policies imposed on Jewish citizens during the Nazi era reflect a broader system of exclusion and dehumanization that likely extended to everyday domestic practices, including pet keeping. Cole (2022) documents state-driven measures of exclusion that denied Jews access not only to public and economic life but arguably to the normative cultural practices of a racially defined "ideal domestic sphere." Wiebe (2008) emphasizes that the regime's progression from marginalization to systematic extermination was accompanied by the stripping away of cultural privileges. Although specific studies on pet keeping restrictions for Jewish households are sparse, the overall framework of exclusion and differential treatment—as seen in housing, travel, and public behavior policies—suggests that pet ownership, when permitted in the "Aryan" context, would have been discouraged or restricted among Jews as part of a wider process of symbolic and material disenfranchisement.

Together, these strands indicate that pet keeping in Nazi Germany cannot be regarded solely as a matter of personal preference. The selective privileging of animal companionship by Nazi leaders served as an emblem of racial purity and natural order, while the broader discriminatory practices that targeted Jewish citizens effectively excluded them from such symbols of an idealized domestic life. Although the literature directly addressing these issues remains limited, the synthesis of available studies highlights significant gaps and opportunities for further research in understanding how everyday practices were mobilized to reinforce exclusionary ideologies during the Holocaust.

See also: Ethnicity, Pets And; Europe, Pets In

References and Further Reading

Cole, Phillip (2022). "PATHOLOGIES OF EXCLUSION." In *Global Displacement in the Twenty-First Century: Towards an Ethical Framework*, 26–39. Edinburgh: Edinburgh University Press.
Cole, T. (2014). "'Nature Was Helping Us': Forests, Trees, and Environmental Histories of the Holocaust." *Environmental History*, 19 (4): 665–86.
Schafft, G. E. (2004). *From Racism to Genocide: Anthropology in the Third Reich*. Urbana: University of Illinois Press.
Wiebe, M. (2008). "Extermination and Euthanasia: Animal Symbolism in Nazi Germany." *The Mirror-Undergraduate History Journal*, 28 (1): 21–38.

HOMELESSNESS, PET OWNERSHIP AND

The literature on pets and people without homes reveals a multifaceted interplay between the human–animal bond and the challenges of homelessness. Studies consistently document that companion animals offer substantial psychosocial benefits to individuals experiencing homelessness, including alleviation of loneliness and isolation, emotional stability, and even motivation to pursue avenues toward rehousing or self-care (Cleary et al., 2020; Rhoades, Winetrobe, and Rice, 2014; Maujean et al., 2024).

For many people living on the streets, pets are often regarded as "lifesavers," providing not only a sense of responsibility and attachment but also an assertion of identity and moral worth in otherwise marginalized circumstances (Irvine, 2013). In many cases, the bond between homeless individuals and their companion animals is so critical that many would rather remain on the streets than give up their pets when faced with shelter or housing that restricts animals (Rhoades, Winetrobe, and Rice, 2014; Maujean et al., 2024; Singer, Hart, and Zasloff, 1995).

However, the benefits of pet companionship are counterbalanced by significant practical and systemic challenges. Homeless pet owners frequently encounter barriers when attempting to access pet-friendly shelter or rental accommodations—with some studies reporting that a considerable percentage (up to 80 percent) of homeless people with pets struggle to find stable housing that accepts animals (Cleary et al., 2020; Scanlon, McBride, and Stavisky, 2020). This difficulty is compounded by the stigma and exclusion imposed by some housing agencies and social services, leading to situations where pet owners are forced into either relinquishing their animals or remaining in unsustainable living conditions (Scanlon, McBride, and Stavisky, 2020).

Research now illustrates how pets mediate social connections and even serve as conduits for accessing broader support networks among homeless populations. For example, studies document that interactions involving pets can lead to increased social integration and act as a bridge to human services (Bower et al., 2023; Garland-Lewis et al., 2024). In parallel, service-oriented interventions—ranging from veterinary outreach programs to integrated human care services—are increasingly viewed as vital components in a humane, One Welfare approach that recognizes the interdependence of human and animal well-being (Kerman et al., 2020; Covner et al., 2022). Nonetheless, many service providers continue to struggle with balancing public health mandates (such as the management of flea infestations) and the imperative to preserve the unique bond between homeless individuals and their pets (Carpenter et al., 2025).

The literature also highlights demographic variations, noting that among homeless youth, pet ownership can significantly influence mental health and social engagement, even as some young people must ultimately rehome their pets to access critical services (Schmitz, Carlisle, and Tabler, 2021). Similarly, while most companion animals in homeless populations tend to be dogs, other species are present, and there is growing research into the physical and behavioral health of these animals (King et al., 2024; Scanlon et al., 2021). In some cases, pet food banks and emergency boarding programs demonstrate a positive return on investment by mitigating the psychosocial and logistical challenges associated with pet ownership among the unhoused (Conway et al., 2022).

See also: Capitalism, Pets And; Socioeconomic Class, Pets And

References and Further Reading

Bower, M., M. Carvalheiro, K. Gournay, J. Perz, and E. Conroy (2023). "When More Satisfying and Supportive Relationships Increase Loneliness: The Social Worlds of People With Lived Experience of Homelessness." Health & Social Care in the Community, 2023(1): 9836967.

Carpenter, A., S. N. Rich, B. Dell, S. Adams, N. Bestul, R. Henderson, ... and D. W. McCormick (2025). "Veterinary Care and Flea Preventatives Are Limited in Homeless Shelters and Outreach Organizations Serving People Experiencing Homelessness." Journal of the American Veterinary Medical Association, 263 (3): 330–34.

Cleary, M., S. West, D. Visentin, M. Phipps, M. Westman, K. Vesk, and R. Kornhaber (2020). "The Unbreakable Bond: the Mental Health Benefits and Challenges of Pet Ownership for People Experiencing Homelessness." *Issues in Mental Health Nursing*, 42 (8): 741–6.

Conway, K., R. Jasuja, N. Hauser, and J. Foley (2022). "Benefits, Companion Animal Zoonotic Disease Prevalence and Public Perceptions of Pet Ownership Among People Experiencing Homelessness in Northern California." *Zoonoses and Public Health*, 69 (7): 806–15.

Covner, A. L., H. Ekstrom, N. Kalaf, and B. Sherman (2022). *Creating Evidence-Driven Practices to Enhance Human Care Services for Unhoused and Low-Income Pet Owners*. https://doi.org/10.21203/rs.3.rs-1761858/v1

Garland-Lewis, G., E. Scott, V. Ramirez, and P. Rabinowitz (2024). "Shifting the Focus: A Photovoice Exploration of the Benefits and Barriers of Having a Pet while Experiencing Homelessness." *Plos One*, 19 (3): e0295588.

Irvine, L. (2013). "Animals as Lifechangers and Lifesavers: Pets in the Redemption Narratives of Homeless People." *Journal of Contemporary Ethnography*, 42 (1): 3–30.

Irvine, L. (2023). "Homelessness and Pet Ownership." In H. Aubrey, Megan K. Mueller, Zenithson Y. Ng, Alan M. Beck, and Jose M. Peralta (eds), *The Routledge International Handbook of Human-Animal Interactions and Anthrozoology*, 272–85. London: Routledge.

Kerman, N., M. Lem, M. Witte, C. Kim, and H. Rhoades (2020). "A Multilevel Intervention Framework for Supporting People Experiencing Homelessness with Pets." *Animals*, 10 (10): 1869.

Kidd, A., and R. Kidd (1994). "Benefits and Liabilities of Pets for the Homeless." *Psychological Reports*, 74 (3): 715–22.

King, C., T. J. Smith, K. Kabrick, A. Dzur, and T. Grandin (2024). "Physical and Behavioural Health of Dogs Belonging to Homeless People." Animal Welfare, 33: e11.

Maujean, A., M. Downes, E. Sofija, J. Byrnes, and M. Downes (2024). "'Caring For Their Pet More Than Themselves': Attitudes Toward Homeless People With a Companion Animal." Society & Animals, 1(aop): 1–19.
Rhoades, H., H. Winetrobe, and E. Rice (2014). "Pet Ownership Among Homeless Youth: Associations with Mental Health, Service Utilization and Housing Status." *Child Psychiatry & Human Development*, 46 (2): 237–44.
Scanlon, L., A. McBride, and J. Stavisky (2020). "Prevalence of Pet Provision and Reasons for Including or Excluding Animals by Homelessness Accommodation Services." *Journal of Social Distress and the Homeless*, 30 (1): 77–83.
Scanlon, L., P. Hobson-West, K. Cobb, A. McBride, and J. Stavisky (2021). "Homeless People and their Dogs: Exploring the Nature and Impact of the Human–companion Animal Bond." *Anthrozoös*, 34 (1): 77–92.
Schmitz, R., Z. Carlisle, and J. Tabler (2021). "'Companion, Friend, Four-Legged Fluff Ball': The Power of Pets in the Lives of LGBTQ+ Young People Experiencing Homelessness." *Sexualities*, 25 (5–6): 694–716.
Singer, R., L. Hart, and R. Zasloff (1995). "Dilemmas Associated with Rehousing Homeless People who have Companion Animals." *Psychological Reports*, 77 (3): 851–7.

HOUSING RESTRICTIONS AND PET OWNERSHIP

The literature on housing restrictions and pet keeping reveals that policies and landlord practices—ranging from prohibitive pet fees to blanket no-pet rules—can significantly disrupt the human–animal bond and contribute to pet relinquishment.

Several studies have documented that difficulties in finding pet-inclusive affordable housing are among the significant reasons why people are forced to give up their pets, especially during moves or when seeking new rental accommodations (Hoffman et al., 2021; Mascitelli et al., 2024). Housing policies that exclude pets tend to disproportionately affect marginalized groups, including older adults, low-income tenants, and renters who are more likely to be forced to relocate or face eviction because of pet-related fees and restrictions (Applebaum et al., 2021; Toohey and Krahn, 2018).

Research indicates that systemic inequities and ageism in affordable housing systems have led to barriers that prevent people from aging in place with their companion animals; studies focusing on older adults have found that rules implicit in eligibility for social assistance and affordable housing programs can force a difficult choice between personal well-being and maintaining a cherished pet relationship (McLennan et al., 2022; Branson et al., 2019). Similarly, landlord practices that impose additional pet fees or strict pet restrictions have been critiqued as discriminatory and may drive residential relocation to "less desirable" neighborhoods (Applebaum et al., 2021; Power, 2016). In the United States, for example, the scarcity of pet-friendly rental properties drives many pet owners—especially younger adults who predominantly rent—to move frequently or give up their pets when faced with inflexible housing markets (Hoffman et al., 2021; Mascitelli et al., 2024).

Moreover, the literature suggests that inclusive policies, such as those that incorporate pet-friendly design in social and affordable housing, can potentially alleviate the cycle of pet relinquishment and housing instability. Researchers have proposed preventative approaches—like offering pet deposits, temporary boarding support, or low-cost veterinary and training assistance—that could help retain the human–animal bond even during housing transitions (Eagan et al., 2022; Ly et al., 2021; Applebaum et al., 2021). There is also a growing call for policies that recognize the psychosocial benefits of pet ownership and integrate pet-inclusive housing in public services, which support both animal welfare and contribute to improved mental health and stability among vulnerable populations (Irvine, 2013; McDonald et al., 2021).

The interrelationship between pet ownership and housing instability is also evident in studies of vulnerable populations. Some research (Rhoades, Winetrobe, and Rice, 2014; Garland-Lewis et al., 2024) points to the fact that homeless individuals and youth with pets often face additional barriers in accessing traditional housing services, with many reporting reluctance to enter shelters that do not accommodate pets. Such studies underscore the importance of integrated support measures, including affordable emergency pet care services and pet-inclusive shelter policies, to mitigate the adverse impacts on these populations. These findings reinforce arguments for a holistic review of housing policies that account for the needs of pet owners, especially among marginalized groups Garland-Lewis et al., 2024; Rhoades, Winetrobe, and Rice, 2014).

Conceptually, a broader urban policy perspective is emerging that considers the public health and social equity implications of pet-friendly housing. For some scholars, a healthy city or town is a city or town that integrates pet ownership within urban planning and health promotion contexts. Such frameworks emphasize the importance of inclusive policy design that navigates the trade-offs between high-density urban living, pet-related care, and regulatory oversight. By situating pet-friendly housing within the larger discourse of housing affordability and urban health equity, researchers highlight the potential benefits of policy reforms that can bridge the gap between tenant needs and market practices (Rock et al., 2014).

See also: Homelessness, Pet Ownership And; Laws and Regulations; Socioeconomic Class, Pets And

References and Further Reading

Applebaum, J., K. Horecka, L. Loney, and T. Graham (2021). "Pet-Friendly for Whom? An Analysis of Pet Fees in Texas Rental Housing." *Frontiers in Veterinary Science*, 8: 767149.

Branson, S. M., L. Boss, N. S. Padhye, N. R. Gee, and T. T. Trötscher (2019). "Biopsychosocial Factors and Cognitive Function in Cat Ownership and Attachment in Community-Dwelling Older Adults." Anthrozoös, 32 (2): 267–82.

Eagan, B. H., E. Gordon, and A. Protopopova (2022). "Reasons for Guardian-Relinquishment of Dogs to Shelters: Animal and Regional Predictors in British Columbia, Canada." Frontiers in Veterinary Science, 9: 857634.

Garland-Lewis, G., E. Scott, V. Ramirez, and P. Rabinowitz (2024). "Shifting the Focus: A Photovoice Exploration of the Benefits and Barriers of Having a Pet While Experiencing Homelessness." *Plos One*, 19 (3): e0295588.

Graham, T., K. Milaney, C. Adams, and M. Rock (2018). "'Pets Negotiable': How Do The Perspectives of Landlords and Property Managers Compare with Those of Younger Tenants with Dogs?" *Animals*, 8 (3): 32.

Hoffman, C. L., M. Thibault, and J. Hong (2021). "Characterizing Pet Acquisition and Retention During the COVID-19 Pandemic." Frontiers in Veterinary Science, 8: 781403.
Irvine, L. (2013). "Animals as Lifechangers and Lifesavers: Pets in the Redemption Narratives of Homeless People." *Journal of Contemporary Ethnography*, 42 (1): 3–30.
Ly, L. H., E. Gordon, and A. Protopopova (2021). "Inequitable Flow of Animals in and Out of Shelters:Comparison of Community-Level Vulnerability for Owner-Surrendered and Subsequently Adopted Animals." Frontiers in Veterinary Science, 8: 784389.
Mascitelli, T., T. Graham, L. Loney, J. Applebaum, C. Murray, M. Binns-Calvey, S. M. Hawes, and K. Morris (2024). "Barriers to Finding and Maintaining Pet-inclusive Affordable Housing: Tenant Experiences in Houston, Texas." *Frontiers in Veterinary Science*, 11: 1465682.
McDonald, S. E., K. E. O'Connor, A. Matijczak, C. A. Tomlinson, J. W. Applebaum, J. L. Murphy, and B. A. Zsembik (2021). "Attachment to Pets Moderates Transitions in Latent Patterns of Mental Health Following the Onset of the COVID-19 Pandemic: Results of a Survey of US Adults." *Animals*, 11 (3): 895.
McLennan, K., M. J. Rock, E. Mattos, and A. M. Toohey (2022). "Leashes, Litterboxes, and Lifelines: Exploring Volunteer-Based Pet Care Assistance Programs for Older Adults." Frontiers in Psychology, 13: 873372.
Power, E. (2016). "Renting with Pets: A Pathway to Housing Insecurity?" *Housing Studies*, 32 (3): 336–60.
Rhoades, H., H. Winetrobe, and E. Rice (2014). "Pet Ownership Among Homeless Youth: Associations with Mental Health, Service Utilization and Housing Status." *Child Psychiatry & Human Development*, 46 (2): 237–44.
Rock, M., C. Adams, C. Degeling, A. Massolo, and G. McCormack (2014). "Policies on Pets for Healthy Cities: A Conceptual Framework." *Health Promotion International*, 30 (4): 976–86.
Rose, D., C. McMillian, and O. Carter (2020). "Pet-Friendly Rental Housing: Racial and Spatial Inequalities." *Space and Culture*, 26 (1): 116–29.
Toohey, A. M., and T. M. Krahn (2018). "'Simply to Be Let In': Opening the Doors to Lower-Income Older Adults and Their Companion Animals." Journal of Public Health, 40 (3): 661–5.

HUMAN HEALTH, PETS AND

While the history of working animals around the world is primarily a utilitarian one—animals recruited, generally without their input and often against their wills, to provide services to humans—the more recent examples of the ways in which animals work for humans are a better indication of the human–animal bond. In particular, the rise of animal-assisted therapy could not have occurred if it were not clear that humans benefit from animals in ways that are not strictly utilitarian. Assistance animals, for example, do not merely push wheelchairs or help guide the blind across streets; they make people feel less lonely and more connected. Military dogs, too, do not just detect bombs or guard installations; they also give soldiers comfort during extremely stressful times. The wealth of research on the power of the human–animal bond tells us that the benefits that animals can provide to humans go far deeper than we ever thought before.

Over the past several decades, a substantial body of research has surfaced that illustrates the numerous ways in which animals contribute to human health. For instance, animals promote increased physical activity, leading to improved physical well-being. Research indicates that elderly individuals who have dogs visit the doctor less frequently than those without (Siegel, 1990) and show a better capacity to handle stressful life events without resorting to healthcare services. Studies reveal that pet owners exhibit lower blood pressure, triglycerides, and cholesterol levels compared to non-owners (Anderson, Reid, and Jennings, 1992), experience fewer minor health concerns (Friedmann, Son, and Saleem, 2015), have diminished rates of heart attack mortality (Friedmann et al., 1980), and demonstrate better one-year survival rates following incidents of coronary heart disease (Friedmann et al., 1980). Additionally, children who are exposed to pets during their first year of life tend to have a lower incidence of allergic rhinitis and asthma. Moreover, there are observable emotional advantages associated with living with companion animals.

The presence of animals alleviates feelings of loneliness (Sable, 1995), assists children in adjusting to significant illnesses and bereavement within the family, and enhances psychological well-being and self-esteem. Pets provide emotional solace and support for children coping with parental divorce and for seniors mourning the loss of loved ones (Mallia, 2006). Furthermore, pet owners experience reduced fear of becoming victims of crime when they are accompanied by a dog or when sharing their residence with one. The presence of a dog during a child's medical appointment or in legal situations, such as a jury trial, can significantly reduce the child's anxiety and provide comfort (Hansen et al., 1999). In interactions involving children, animals can capture attention, foster cognitive development, and exert a calming influence, which may enhance school attendance, adherence to rules, and the retention of educational material.

Some studies suggest that interaction with pets nurtures empathetic behavior in children, potentially leading to more nurturing traits in adulthood. Additionally, individuals with HIV/AIDS who own pets report experiencing less depression and lower stress levels (Hansen et al., 1999).

The health and emotional advantages of cohabiting with companion animals have become so pronounced that, over the last two decades, various organizations have emerged to assist individuals with physical disabilities or serious health challenges in maintaining relationships with pets. Known as human–animal support services, these entities offer either financial or practical assistance to help people retain their pets. One such organization, San Francisco's Pets Are Wonderful Support (PAWS), provides support specifically for people with HIV/AIDS by offering services such as veterinary care, pet food provisions, and transportation to veterinary appointments. Volunteers from PAWS also visit HIV/AIDS patients in their homes to ensure their companion animals are adequately cared for, promoting the patients' overall well-being and health. Research has demonstrated that pet ownership, particularly of dogs, often promotes increased physical activity through regular exercise routines, such as daily walks, which can lead to improved cardiovascular health and reduced blood pressure (Parslow and Jorm, 2003; Arhant-Sudhir, Arhant-Sudhir, and Sudhir, 2011). The physical activity associated with pet care routines appears to alleviate some of the risks linked to sedentary lifestyles and positively influences recovery processes following acute medical events (Wells, 2009). Furthermore, findings suggest that pet ownership may indirectly encourage healthier behaviors by motivating individuals to maintain active lifestyles (Hodgson et al., 2015). In addition to health benefits, living with pets confers significant social advantages. Animals are often described as "social lubricants," as they enhance our capacity to connect with others by facilitating conversation and alleviating anxiety. Pets stimulate social interactions, discussions, and rapport across diverse settings.

For example, the presence of animals in nursing homes has been shown to increase social and verbal interactions among residents (Bernstein, Friedmann, and Malaspina, 2000), while children with autism who have pets display more prosocial behaviors and fewer self-absorbed tendencies. Furthermore, children residing with pets tend to engage more deeply in sports, hobbies, clubs, and chores, and they score significantly higher on empathy and prosocial orientation assessments than their peers without pets. The companionship provided by animals is particularly beneficial for older adults, as it can mitigate feelings of loneliness and provide both cognitive and emotional support (McDonough et al., 2022). Additionally, pet-facilitated interventions and therapeutic programs have been effectively implemented in various contexts, ranging from clinical care to educational settings, where the presence of animals contributes to stress reduction and overall well-being (Brodie and Biley, 1999). These findings underscore the multifaceted role pets play in enhancing individual health and contributing to the well-being of society as a whole.

This research has gained such prominence that the National Institutes of Health has advocated for the inclusion of pet-related queries in national health surveys for more than three decades to better understand the relationship between human health and the presence or absence of animals. Regardless, it is evident that the presence of animals in human lives, which may have initially been viewed as a utilitarian concern, has evolved into a central aspect of our functioning and emotional health. Why do animals exert such a profound influence on human well-being? Scholars have proposed several theories to elucidate the substantial benefits humans derive from their relationships with companion animals.

The biophilia hypothesis, introduced by biologist Edward Wilson (1986), posits that humans and other animals possess an inherent attraction to one another, resulting in mutually beneficial relationships. This concept not only accounts for the bond between humans and companion animals but also elucidates the prominent presence of animals in children's literature, art, and games. Another perspective suggests that humans are neurologically predisposed to attend to animals, as throughout much of human evolution, animals served as a crucial source of sustenance. Thus, the inclination toward animal companionship may only follow our innate interest in animals primarily as food sources.

A separate explanation, known as the social support theory, argues that any form of social support—whether derived from marriage, religious affiliation, or participation in social clubs—contributes positively to human health due to our innate need for social contact (Meehan, Massavelli, and Pachana, 2017). The nature of the social contact—whether it involves animals or other humans—is deemed inconsequential; what matters is the interaction with another being. Furthermore, some researchers posit that since men typically have broader social support networks than women, the latter may demonstrate a greater need for companion animals.

In conclusion, current literature substantiates that pet ownership generally yields positive contributions to human health through enhanced emotional support, increased physical activity, and improved social interactions. However, while numerous studies highlight these beneficial effects, the overall impact is intricate and contingent upon variables such as the type of pet, the characteristics of the owner, and the larger context. Future research aimed at clarifying these moderating factors will be crucial for developing more focused interventions that leverage the health-promoting potential of human–animal interactions.

See also: Allergies; Health Risks of Pet Ownership; Mental Health, Pets And

References and Further Reading

Amiot, C., C. Gagné, and B. Bastian (2022). "Pet Ownership and Psychological Well-being During the Covid-19 Pandemic." *Scientific Reports*, 12 (1): 6091.

Anderson, W. P., C. M. Reid, and G. L. Jennings (1992). "Pet Ownership and Risk Factors for Cardiovascular Disease." *Medical Journal of Australia*, 157 (5): 298–301.

Arhant-Sudhir, K., R. Arhant-Sudhir, and K. Sudhir (2011). "Pet Ownership and Cardiovascular Risk Reduction: Supporting Evidence, Conflicting Data and Underlying Mechanisms." *Clinical and Experimental Pharmacology and Physiology*, 38 (11): 734–8.

Bennetts, S., T. Howell, S. Crawford, F. Burgemeister, K. Burke, and J. Nicholson (2023). "Family Bonds with Pets and Mental Health During Covid-19 in Australia: A Complex Picture." *International Journal of Environmental Research and Public Health*, 20 (7): 5245.

Bernstein, P. L., E. Friedmann, and A. Malaspina (2000). "Animal-Assisted Therapy Enhances Resident Social Interaction and Initiation in Long-Term Care Facilities." *Anthrozoös*, 13 (4): 213–24.

Brodie, S., and F. Biley (1999). "An Exploration of the Potential Benefits of Pet-facilitated Therapy." *Journal of Clinical Nursing*, 8 (4): 329–37.

Friedmann, E., A. H. Katcher, J. J. Lynch, and S. A. Thomas (1980). "Animal Companions and One-Year Survival of Patients After Discharge from a Coronary Care Unit." *Public Health Reports*, 95 (4): 307.

Friedmann, E., H. Son, and M. Saleem (2015). "The Animal–Human Bond: Health and Wellness." In *Handbook on Animal-Assisted Therapy*, 73–88. San Diego, CA: Academic Press.

Gee, N., M. Mueller, and A. Curl (2017). "Human–animal Interaction and Older Adults: An Overview." *Frontiers in Psychology*, 8: 1416.

Hansen, K. M., C. J. Messinger, M. M. Baun, and M. Megel (1999). "Companion Animals Alleviating Distress in Children." *Anthrozoös*, 12 (3): 142–8.

Herzog, H. (2011). "The Impact of Pets on Human Health and Psychological Well-being." *Current Directions in Psychological Science*, 20 (4): 236–9.

Hodgson, K., L. Barton, M. Darling, V. Antao, F. Kim, and A. Monavvari (2015). "Pets' Impact on Your Patients' Health: Leveraging Benefits and Mitigating Risk." *The Journal of the American Board of Family Medicine*, 28 (4): 526–34.

Mallia, M. E. (2006). *The Role of Companion Animals in Senior Well-Being*. Rensselaer Polytechnic Institute.

Martins, C. F., J. P. Soares, A. Cortinhas, L. Silva, L. Cardoso, M. A. Pires, and M. P. Mota (2023). "Pet's Influence on Humans' Daily Physical Activity and Mental Health: A Meta-Analysis." *Frontiers in Public Health*, 11: 1196199.

McConnell, A., C. Brown, T. Shoda, L. Stayton, and C. Martin (2011). "Friends with Benefits: on the Positive Consequences of Pet Ownership." *Journal of Personality and Social Psychology*, 101 (6): 1239–52.

McDonough, I. M., H. B. Erwin, N. L. Sin, and R. S. Allen (2022). "Pet Ownership is Associated with Greater Cognitive and Brain Health in a Cross-Sectional Sample Across the Adult Lifespan." *Frontiers in Aging Neuroscience*, 14: 953889.

Meehan, M., B. Massavelli, and N. Pachana (2017). "Using Attachment Theory and Social Support Theory to Examine and Measure Pets as Sources of Social Support and Attachment Figures." *Anthrozoös*, 30 (2): 273–89.

Parslow, R., and A. Jorm (2003). "Pet Ownership and Risk Factors for Cardiovascular Disease: Another Look." *The Medical Journal of Australia*, 179 (9): 466–8.
Sable, P. (1995). "Pets, Attachment, and Well-Being Across the Life Cycle." *Social Work*, 40 (3): 334–41.
Siegel, J. M. (1990). "Stressful Life Events and Use of Physician Services Among the Elderly: The Moderating Role of Pet Ownership." *Journal of Personality and Social Psychology*, 58 (6): 1081.
Wells, D. (2009). "The Effects of Animals on Human Health and Well-being." *Journal of Social Issues*, 65 (3): 523–43.
Wilson, E. O. (1986). *Biophilia*. Cambridge, MA: Harvard University Press.

HUMANE SOCIETIES

A humane society is a nonprofit organization, supported primarily by private donations (rather than public funds), which takes in homeless pets and adopts them out to new homes.

The history of humane societies in the United States and around the world has its roots in nineteenth-century England, where industrialization, urban growth, and evolving attitudes toward moral and social reform created a fertile environment for establishing organizations focused on animal welfare and humanitarian efforts (Woodger, 2018). Early humane societies in England, following the founding of the Royal Society for the Prevention of Cruelty to Animals (RSPCA), set a precedent by advocating for the treatment of animals as a measure of a society's moral character. These organizations were not isolated; they represented a wider impulse among reformers who viewed progress as linked to ethical considerations toward both humans and nonhuman animals (Woodger, 2018).

This reformist spirit soon spread beyond the British Isles, influencing the formation of humane societies in North America and other regions. For example, the Toronto Humane Society, which operated between 1887 and 1891, explicitly drew from the ideas and structures of its British and American counterparts. It prioritized the moral reformation of the working classes, connecting animal welfare to a broader spectrum of social and ethical improvements (Woodger, 2018). These patterns illustrate that humane societies worldwide were intertwined with contemporary movements aimed at enhancing social standards and addressing the challenges presented by rapid urbanization and industrial capitalism.

The evolution of humane societies in the United States is well documented in the literature as part of a broader reform movement that emerged in the mid-nineteenth century. Early nonprofit initiatives, most notably the American Society for the Prevention of Cruelty to Animals (ASPCA), the first animal protection organization in the United States, pioneered the concept of organized animal welfare, championing legal reforms and public education to combat abuse and neglect. These privately run organizations have traditionally relied on philanthropic donations, volunteer labor, and fundraising to support their programs of rescue, rehabilitation, and adoption. Their mission has been largely proactive, focusing not only on care but also on legislative advocacy and public outreach that redefines the relationship between people and animals.

A Royal Society for the Prevention of Cruelty to Animals ambulance in Sydney, Australia, 1935. The organization's fleet of ambulances were used to rescue nearly 2,000 stray, diseased, unwanted, and disabled dogs and cats every month in Sydney. (State Library of New South Wales)

In contrast, municipal animal shelters in the United States are government-operated facilities designed to serve a primarily public health and safety function. While nonprofit humane societies often have more latitude to innovate and implement programs that emphasize rehoming and compassionate care, municipal shelters must adhere to local ordinances and often operate under limited budgets. Studies have shown that municipal shelters tend to follow stringent regulatory protocols aimed at controlling animal populations and minimizing risks of disease outbreaks. Moreover, differences in how these two sectors operate are evident in outcomes; municipal shelters have sometimes been associated with higher euthanasia rates in part due to the reactive nature of their operations and resource constraints, while privately run charities are more apt to develop specialized adoption programs and community-based support systems.

In addition, these divergent organizational structures reflect key ideological differences: privately run humane societies are often viewed as advocates for animals' intrinsic value as sentient beings, whereas municipal shelters, as instruments of local government, are primarily mandated to enforce animal control and public safety laws. This has led to a dual system in which the humane treatment of animals is pursued along parallel, albeit interdependent, tracks. The nonprofit model, which emphasizes advocacy, fundraising, and grassroots engagement, frequently serves as a counterbalance to the more bureaucratic and public health-driven approach of municipal shelters.

In addition, humane societies are closed access, which means that they take in animals only as space allows. For that reason, humane societies often do not euthanize healthy animals; instead, they keep them as long as they need to in order to find them a permanent home. Municipal shelters, on the other hand, are open access, meaning that they must, by law, take in every stray or unwanted animal that is presented to them. Because of this pressure, many municipal shelters euthanize animals when they do not have enough space to keep them all. This has resulted in the "kill/no kill" distinction in the animal sheltering world and unfairly stigmatizes the work of municipal shelters who, thanks to government mandates, cannot close their doors to the public when they are full. Today, most shelters use the terms "open access" vs. "closed access" to better describe the distinction.

The history of humane societies in the United States reveals an ongoing transformation. On the one hand, privately run nonprofit charities have been at the forefront of animal welfare, employing innovative strategies and public outreach campaigns to rescue and rehome animals. On the other hand, municipal animal shelters, while critical to maintaining public health and safety, often face challenges stemming from regulatory and fiscal constraints. Together, these complementary systems illustrate a complex landscape in which the goals of compassionate care, public health, and animal protection are pursued through different operational and ideological frameworks.

The evolution of humane societies can thus be viewed as part of a global wave of social reform that originated in nineteenth-century England and progressively influenced policies and practices in the United States and other areas. These societies not only tackled local issues of cruelty and neglect but also became integral to the larger discourse on morality, reform, and the evolving understanding of social responsibility during significant socioeconomic transformations (Woodger, 2018).

See also: Rescue Groups; Sanctuaries; Shelters

References and Further Reading

Beers, D. L. (2006). *For the Prevention of Cruelty: The History and Legacy of Animal Rights Activism in the United States*. Athens: Ohio University Press.

Garner, R. (2016). *Political Animals: Animal Protection Politics in Britain and the United States*. Athens: Springer.

Horecka, K., and S. Neal (2022). "Critical Problems for Research in Animal Sheltering, A Conceptual Analysis." *Frontiers in Veterinary Acience*, 9: 804154.

Sternbach, G. L., J. Varon, R. Fromm Jr, and P. J. Baskett (2000). "The Humane Societies." *Resuscitation*, 45 (2): 71–5.

Woodger, K. (2018). "We Speak for Those who Cannot Speak for Themselves." *Ontario History*, 105 (2): 143–59.

I

IMMIGRATION, PETS AND

The literature on immigration and pets is relatively limited compared to broader studies on international migration policy; however, emerging research underscores several critical challenges faced by migrants—including refugees—when attempting to relocate with their companion animals. Much of the academic focus on migration policy emphasizes institutional, legal, and economic mechanisms for regulating human migration, but this has increasingly drawn attention to how these frameworks also affect pet ownership during migration.

Migrants and refugees confront a complex regulatory landscape when moving with pets. Many destination countries require extensive health certifications, mandatory vaccinations, and quarantine periods before admitting animals—a process that can be cost prohibitive and logistically overwhelming. Such requirements not only delay the migration process but may also force potential migrants to choose between relocating with their pets or abandoning them due to practical constraints. This difficulty is compounded by variations in international policies: for instance, while some regions (such as the European Union) have streamlined procedures like the pet passport, these systems may not be available or accessible to all migrants, especially those fleeing conflict or extreme poverty.

Additionally, the issue of pets is rarely incorporated into the legal and humanitarian frameworks that govern refugee resettlement. Refugee camps are generally designed to meet basic human needs and typically do not allow for pets due to health, security, and sanitation concerns. Although anecdotal evidence suggests that some refugees might covertly keep small animals as emotional supports, the lack of official policies or designated pet-friendly accommodations means that migrants are often forced to forfeit their relationships with companion animals. Such separations can have profound psychological and emotional consequences, given the well-documented benefits of pet companionship for mental health and social stability.

Overall, while international migration policies continue to evolve to address human mobility, there is a pressing need for policymakers to develop integrated frameworks that acknowledge the role of pets in migrants' lives. A more holistic approach would consider the psychosocial benefits of pet ownership by creating pet-inclusive migration policies, clarifying quarantine and health certification procedures, and exploring the feasibility of pet-friendly spaces in refugee camps. Research in this area remains at an early stage, underscoring the need for interdisciplinary studies that bring together insights from migration policy, animal welfare, and public health to better support the needs of both migrants and their animal companions.

See also: Natural Disasters and Pets; War, Pets And

References and Further Reading

Fox, R., and K. Walsh (2011). "Furry Belongings: Pets, Migration and Home." In J. Bull (ed.), *Animal Movements, Moving Animals: Essays on Direction, Velocity and Agency in Humanimal Encounters*, University Printers, Uppsala University, Uppsala, 97–117.

Lopez, H. (2023). "Companion Pets in Displacement." *Displaced Voices: A Journal of Archives, Migration and Cultural Heritage*, 3 (2): 77–95.

Sandvik, K. B. (2023). "The Ukrainian Refugee Crisis: Unpacking the Politics of Pet Exceptionalism." *International Migration*, 61 (4): 292–304.

INDIA, PETS IN

Pet keeping in India has been influenced by historical and cultural practices within the subcontinent, as well as by newer practices and ideas that emerged through colonialism.

Early investigations and historical analyses suggest that pet keeping in India was initially a practice reserved for the elite, with cultural practices possibly influenced by indigenous traditions and later modified by colonial interaction. Although explicit empirical detail on the colonial period in India is scarce, broader works on the historical and cultural aspects of pet keeping provide an interpretive framework whereby the practice was linked to status, household hierarchy, and the symbolic appropriation of nonhuman companions (Serpell, 2014). These sources imply that during the colonial era, pet keeping was interwoven with the dynamics of power and social stratification, much like similar patterns observed in other parts of the world, where the introduction of European norms modulated local practices (Serpell, 2014).

In the post-colonial and modern era, the review of contemporary scholarship indicates a marked shift in both the scale and nature of pet keeping in India. The transition from extended family configurations, typical of Indian society, to nuclear households has intensified the need for companionship, leading to an increase in the adoption of pets, particularly dogs. Brill et al. (2024) notes that during the last two to three decades, there has been a noticeable rise in keeping dogs as pets in Indian homes. This trend is driven not only by the practical benefits of having a companion for security and entertainment but also by broader social transformations that emphasize individualized lifestyle choices. Volsche et al. (2019) further reinforce this argument by providing empirical evidence from urban college settings in Bangalore, where changing attitudes emphasize the role of pet dogs in shaping social interactions and personal identity in increasingly urbanized contexts.

This rise in pet keeping has taken place alongside the much more ancient practice of living alongside free-roaming dogs, both in the cities as well as in the countryside.

In summary, the literature on pet keeping in India demonstrates a historical evolution from a practice deeply embedded in sociocultural hierarchies during the colonial period to a modern phenomenon that is increasingly valued for its contribution to individual well-being and lifestyle aesthetics. The interplay of historical legacies, cultural group dynamics, and socioeconomic transformations has thus given rise to a distinctive pet keeping culture in India that aligns with both local traditions and global trends.

See also: Dogs; Free-Roaming Dogs

References and Further Reading

Brill, G., A. Chaudhari, K. Polak, S. Rawat, D. Pandey, P. Bhatt, ... and A. Murali (2024). "Owned-Dog Demographics, Ownership Dynamics, and Attitudes Across Three States of India." *Animals*, 14 (10): 1464.

Serpell, J. (2014). "Pet-keeping and Animal Domestication: A Reappraisal." In J. Clutton-Brock (ed.), *The Walking Larder*, 10–21. London: Routledge.

Volsche, S., M. Mohan, P. Gray, and M. Rangaswamy (2019). "An Exploration of Attitudes Toward Dogs Among College Students in Bangalore, India." *Animals*, 9 (8): 514.

INDIGENOUS PET KEEPING

While pet keeping has been practiced in societies around the world for millennia, it has only been in the last hundred or so years that pet keeping has exploded, creating multi-billion dollar industries focused on producing, feeding, caring for, medically treating, and even disposing of millions of animals per year.

People have long kept animals as companions; even hunter-gatherers with no domesticated animals kept pets. The animals kept by such groups were tamed wild animals, usually caught as babies, used for companionship. Aboriginal Australians, for example, once kept dingoes and wallabies as pets, Polynesians kept dogs and parrots, South American tribes kept monkeys, parrots, and even wild cats, and Native Americans kept deer, dogs, and crows (Serpell, 1987; Cormier, 2010; Gray and Young, 2011). Even though these same animals were hunted and consumed in these cultures, generally once an animal was adopted as a pet, it was exempt from consumption.

Indigenous pet keeping has been widely documented as a culturally embedded practice among various indigenous communities, where the care of captured animals is intertwined with reciprocity and kinship-building. Anthropological studies indicate that in certain groups, the act of hunting and subsequently killing an animal with a baby is not solely driven by subsistence needs but also initiates a social contract, whereby the infant—frequently a primate—is retained and reared as a pet. For instance, fieldwork with the Guajá people reveals that primate infants, whose mothers are killed for meat, are incorporated into the kinship system as quasi-human entities and receive care paralleling that given to human children (Parathian and Maldonado, 2010; Roncal, Bowler, and Gilmore, 2018). This duality in use—both for nutritional purposes and social integration—demonstrates a nuanced form of reciprocity that extends beyond simple utilitarianism, encapsulating elements of moral economy and cultural symbolism (Estrada et al., 2022; Anca, Shanee, and Svensson, 2023). Moreover, the retention of animals as pets following hunting practices aligns with indigenous perspectives that view such animals as possessing a unique inherent essence, further blurring the lines between the human and nonhuman (Estrada et al., 2022).

In addition to the social dimensions of reciprocity, certain indigenous cultures extend their nurturing practices to include direct caregiving behaviors such as breastfeeding (Erikson, 2000; Parathian and Maldonado, 2010). In fact, in cultures around the world, women (and children) generally have had the closest relationships with pet animals, and that remains the case today. In these contexts, the act of breastfeeding an animal, particularly infant primates, symbolizes the enactment of a profound kinship bond and the integration of the animal into the family structure. This practice, though striking from an outsider's perspective, is deeply rooted in indigenous worldviews that see nurturing as a

A member of the Indigenous Ticuna tribe in the Amazon rainforest of Colombia feeding pet monkeys. (Laszlo Mates/Dreamstime.com)

communal responsibility transcending conventional species boundaries and emphasizing a holistic interspecies relationship (Cormier, 2010; Roncal, Bowler, and Gilmore, 2018). By breastfeeding and otherwise caring for the animal, indigenous caregivers not only provide physical sustenance but also impart social and cultural values, thereby reinforcing the animal's status as an active participant in the community's social fabric (Roncal, Bowler, and Gilmore, 2018; Anca, Shanee, and Svensson, 2023). Such practices underline the importance of understanding indigenous pet keeping within broader sociocultural and ecological frameworks, where care for animals is both a reflection of social reciprocity post-hunting and a manifestation of deeply held cultural beliefs regarding kinship and the natural world.

See also: Birds; Dogs; Pigs

References and Further Reading

Anca, E., S. Shanee, and M. Svensson (2023). "Ethnoprimatology of the Shipibo of the Upper Ucayali River, Perú." *Journal of Ethnobiology and Ethnomedicine*, 19 (1): 1–19.

Cormier, L. (2010). "The Historical Ecology of Human and Wild Primate Malarias in the New World." *Diversity*, 2 (2): 256–80.

Erikson, P. (2000). "The Social Significance of Pet-Keeping Among Amazonian Indians." In A. L. Podberscek, E. S. Paul, and. J. A. Serpell (eds), *Companion Animals and Us: Exploring the Relationships Between People and Pets*, 7–26. Cambridge: Cambridge University Press.

Estrada, A., P. Garber, S. Gouveia, Á. Fernández-Llamazares, F. Ascensão, A. Fuentes, S. Volampeno, et al. (2022). "Global Importance of Indigenous Peoples, Their Lands, and Knowledge Systems for Saving the World's Primates from Extinction." *Science Advances*, 8 (32): 1–19.

Gray, P. B., and S. M. Young (2011). "Human–Pet Dynamics in Cross-Cultural Perspective." *Anthrozoös*, 24 (1): 17–30.

Parathian, H., and A. Maldonado (2010). "Human–nonhuman Primate Interactions Amongst Tikuna People: Perceptions and Local Initiatives for Resource Management in Amacayacu in the Colombian Amazon." *American Journal of Primatology*, 72 (10): 855–65.

Roncal, C., M. Bowler, and M. Gilmore (2018). "The Ethnoprimatology of the Maijuna of the Peruvian Amazon and Implications for Primate Conservation." *Journal of Ethnobiology and Ethnomedicine*, 14 (1): 1–16.

Serpell, J. A. (1987). "Pet-Keeping in Non-Western Societies: Some Popular Misconceptions." *Anthrozoös*, 1 (3): 166–74.

INFLUENCERS, PET

A pet influencer is an animal, most commonly a dog or a cat, that has a large social media presence and following and uses that presence to promote products, services, or even lifestyles. Even though the term is pet influencer, the bulk of the work of "influencing" occurs through a human whose presence may not be visible to the public but who creates the content, works with brands, arranges videos and photos, and more.

The literature on pet influencers highlights a dynamic marketing strategy that leverages both traditional pet mascots, such as Morris the Cat from cat food advertising, and Internet celebrities like Lil Bub and Grumpy Cat to engage consumers and influence purchasing behavior. Empirical research indicates that pet influencers generate higher engagement and increase consumers' willingness to buy through heightened perceived credibility, with cuteness and whimsicality being central to their appeal (Cioccio, Pozharliev, and Angelis, 2024). This elevated credibility is contrasted with that of human influencers, positioning pet influencers as particularly effective endorsers for pet-related products.

Moreover, studies suggest that pet influencers occupy a unique niche in digital marketing by embodying distinct personalities and narratives that resonate with audiences on an emotional level. Influencers such as Doug the Pug create content that highlights their individual charm, which drives brand awareness while serving as brand ambassadors independent of their owners (Haigh and Wigley, 2024). This narrative-driven approach enhances brand relatability and strengthens the emotional connection between consumers and the product.

Furthermore, research has shown that pet mascots and influencers contribute to relational branding strategies that humanize companies and create affective exchanges with consumers. These strategies have historical antecedents, where brands utilize pet images to activate positive emotions and secure a competitive advantage in marketing communications (Hidalgo and Ortega, 2023; Keller and Gierl, 2020). Such affective responses are critical in transforming consumer attitudes and driving purchase decisions. Additionally, case studies illustrate that integrating pet mascots as central elements of brand identity can support product promotions and overall business growth, as demonstrated through the implementation of pet-in-residence strategies in service-oriented industries (Larrea and Smith, 2024).

Cumulatively, the literature establishes that pet influencers, whether emerging as digital stars on social media or as established mascots in traditional media, wield significant power in marketing by appealing to consumer emotions, enhancing product credibility, and driving engagement. This multifaceted approach underscores the importance of pet influencers in reinforcing brand narratives and directly impacting consumer behavior in the pet industry.

See also: Celebrities, Pet; Internet, Pets on The

References and Further Reading

Cioccio, M., R. Pozharliev, and M. Angelis (2024). "Pawsitively Powerful: Why and When Pet Influencers Boost Social Media Effectiveness." *Psychology and Marketing*, 41 (7): 1614–28.

Haigh, M. M., and S. Wigley (2024). "Doug the Pug, Eggnog, and Boo: The Impact of Pet Influencers on Stakeholders' Purchase Intentions and Attitudes Toward the Brand." In N. C. Bi and R. Zhang (eds), *Global Perspectives on Social Media Influencers and Strategic Business Communication*, 162–80. IGI Global Scientific Publishing.

Hidalgo, M., and H. Ortega (2023). "Brands and Their Pets for A Relational Branding That Humanizes." *Scientific Journal of Applied Social and Clinical Science*, 3 (23): 2–5.

Keller, B., and H. Gierl (2020). "Effectiveness of Animal Images In Advertising." *Marketing ZFP*, 42 (1): 3–32.

Larrea, G., and S. Smith (2024). "Bear E. Drooley: Achieving Business Growth with the Help of Pets-in-residence and Resort Mascots." *Journal of Hospitality & Tourism Cases*, 13 (1): 3–11.

INTERNET, PETS ON THE

Animals enjoy comparable—if not greater—popularity on the Internet than the humans who utilize it for various activities such as shopping, global exploration, and building relationships. While animals have increasingly captured the public's attention in television and film in recent years, the Internet has witnessed a veritable explosion of animal content. Images and videos of cute and amusing animals are ubiquitous online, attracting considerable traffic to websites dedicated to such content.

For instance, The Dodo, a site that shares stories and videos centered around animals, garners approximately one hundred million views each month. A notable Instagram account called Chonky Animals, which features various videos and images of overweight ("thicc," "chonky") animals, exemplifies this trend. Unlike many other aspects of human culture—encompassing mythology, literature, or cinema—where animals often symbolize deeper meanings, the representations of animals on the Internet tend to portray them as their authentic selves.

The widespread availability of cell phone cameras has enabled billions of individuals worldwide to capture and upload footage of their pets or wildlife to platforms such as YouTube, TikTok, and Instagram, where certain clips achieve viral status. The evolution of animal-focused websites began with early creations like Hamster Dance (1997) and hoaxes such as Bonsai Kitten (2003), culminating in successful endeavors like Lolcats (2006). These forms of animal observation on the Internet have evolved into popular distractions during work hours.

Across the past decade, animal imagery and videos have become central to many of the most prevalent memes—a concept characterized by its viral dissemination and mutation across the web. A prime example is the "Dramatic Chipmunk," a five-second clip featuring a prairie dog that turns its head, scored with music from the film Young Frankenstein. Despite its simplicity, this video from 2007 has achieved millions of views and has sparked numerous adaptations, parodies, and even served as the backdrop for advertisements, such as a CarMax commercial.

Similarly, Keyboard Cat, an orange tabby named Fatso, famously appears to play the keyboard while wearing a t-shirt. This video not only amassed millions of views but also became emblematic of a meme genre where Keyboard Cat is often appended to videos showcasing human bloopers, essentially "playing off" the mishaps of a performer, akin to the act of "gonging" a subpar contestant in a television competition. Additional animal-centric Internet memes encompass "Harambe," "Socially Awkward Penguin," "Animals Sitting Like Humans," "I Have No Idea What I'm Doing," "Lawyer Dog," "Advice Animal," "Pedo Bear," "Surprised Kitty," "Spaghetti Cat," and Maru, a live cat.

Unlike the majority of prominent animal figures online, Maru—a Scottish fold who resides in Japan and frequently entertains himself with cardboard boxes—has achieved star status in his own right. His content has been viewed hundreds of millions of times, and in 2013, Guinness World Records recognized Maru as the most-watched animal on YouTube. Numerous commentators have highlighted cats' overwhelming dominance as the Internet's favorite animals.

Joel Veitch, the mastermind behind the website RatherGood.com, created a video titled "The Internet Is Made of Cats" that posits that without cats such as Maru and Keyboard Cat, the Internet as we know it would cease to exist. Cats indeed appear to be an ideal fit for online popularity. Their innate difficulty in being trained means that when they perform clever acts, it is particularly noteworthy. Additionally, their aloofness compared to dogs renders them less predictable, adding to the amusement of attempting to discern their thoughts. Cats' prevalence on the Internet mirrors their ubiquity in reality—they rank as the most popular pets worldwide, making it unsurprising that they are prominently featured in our photos and videos. Moreover, their often-serious demeanor juxtaposed with their absurd antics captures a unique humor, contributing to the widespread appeal of memes such as Lolcats—images of cats paired with humorously incorrect captions in a pseudo-language referred to as "lolspeak" or "catois."

In most Lolcat instances, captions aim to convey the thoughts of the feline subjects, frequently depicting them engaging in behavior reminiscent of human actions. The pioneering Lolcat emerged in early 2007, created by programmer Eric Nakagawa, showcasing an overweight gray cat exhibiting a pleading expression, captioned "I Can Has Cheezburger?" The appeal of Lolcats lies not only in their feline infatuation—many of which derive humor from typical cat traits, such as their disregard for humans—but also in their reflection of the humorous, often darker thoughts of the creators. Therefore, Lolcats serve as unique vessels for human sentiment, as viewers recognize their own feelings echoed in the captions.

Recent literature examining the role of pets in Internet culture reveals a nuanced landscape characterized by both communicative practices and market-driven representations. Researchers have noted that pet profiles on social media serve not just as platforms for personal expression but have also transformed into commodified spaces that reshape cultural identities. Studies indicate that Instagram accounts dedicated to pets function as curated representations in the "cute economy," where selected images are intentionally crafted for aesthetic enjoyment and market engagement (Maddox, 2020).

This commodification of pet narratives not only fosters personal branding but also enhances broader social media marketing strategies that exploit the intrinsic appeal of animal companionship. Social media platforms like Facebook increasingly serve as communal environments where pet owners exchange health information, advice, and emotional support through shared images of their animals (Kogan, Little, and Oxley, 2021). These online communities underline a dual function of social media: while they promote practical knowledge about pet care, they also cultivate a sense of belonging among pet enthusiasts, underscoring the position of pets as pivotal family members in the evolving landscape of digital culture (Kogan, Little, and Oxley, 2021). This merging of practical and emotional discourse emerges as a recurring motif in contemporary studies related to Internet-mediated pet communication. In addition, investigations into the representation of exotic species in online media illuminate how pet-related content encompasses a range from familiar domestic animals to more unusual, exotic creatures. Although many studies focus on the dominant narrative surrounding common companion animals—particularly cats and dogs—the integration of exotic pet content adds complexity by normalizing, and occasionally commodifying, less conventional animal species for online audiences

(Moloney et al., 2021). This inclusion broadens the discourse surrounding pet imagery, revealing that the cultural significance of online pet content intersects with larger social issues, including animal welfare and consumer behavior concerning exotic animals (Moloney et al., 2021).

Through curated social media profiles that reinforce a commodified aesthetic, interactive online forums that provide both practical advice and emotional backing, and comical videos and memes that reshape our engagement with pet imagery (Marwick, 2013), the burgeoning presence of animals on the Internet provides profound insights into the dynamic significance that pet content holds in modern Internet culture (Moloney et al., 2021).

See also: Celebrities, Pet; Film, Pets In; Influencers, Pet; Television, Pets On

References and Further Reading

Kogan, L., S. Little, and J. Oxley (2021). "Dog and Cat Owners' Use of Online Facebook Groups for Pet Health Information." *Health Information & Libraries Journal*, 38 (3): 203–23.

Maddox, J. (2020). "The Secret Life of Pet Instagram Accounts: Joy, Resistance, and Commodification in the Internet's Cute Economy." *New Media & Society*, 23 (11): 3332–48.

Marwick, A. (2013). "Memes." *Contexts*, 12 (4): 12–3.

Moloney, G., J. TUKe, E. Grande, T. Nielsen, and A. Chaber (2021). "Is Youtube Promoting the Exotic Pet Trade? Analysis of the Global Public Perception of Popular Youtube Videos Featuring Threatened Exotic Animals." *Plos One*, 16 (4): e0235451.

J

JAPAN, PETS IN

Pets are incredibly popular in Japan, but for all their popularity, the explosion of pets is a very recent phenomenon. The transformation began during Japan's rapid modernization period following its opening to the West in the Meiji era (1868–1912), when Western ideas and practices—including the keeping of companion animals—were introduced and adapted within a uniquely Japanese context (Toprak 2006). This period set the stage for the evolution of pet keeping from a utilitarian past to a more emotionally charged practice, particularly as postwar affluence enabled more households to embrace the idea of animals as family members (Herzog, 2014).

Demographic changes in Japan have also played a crucial role in shaping pet keeping trends. Japan's declining and aging population, compounded by the practice of living in small apartments in urban centers, has led to a new focus on pets, and small pets in particular. With fewer children and a weakening traditional extended family structure, pets have increasingly come to fill roles of companionship and emotional support (Herzog, 2014; Hanihara, 1993). In these contexts, domestic animals, especially smaller pets like certain dog breeds and cats, have become particularly appealing owing to their suitability for limited living spaces (Herzog, 2014; Kubinyi, 2025).

In parallel with demographic and spatial constraints, Japanese popular culture has significantly influenced pet keeping practices. The emergence and globalization of the kawaii (cute) aesthetic—a cultural phenomenon that manifests in various media including anime, manga, and fashion—has deeply affected the way pets are perceived and celebrated (Pellitteri, 2018). The kawaii trend not only enhances the visual and emotional appeal of pets but also permeates the broader social values related to care, gentleness, and novelty. This aesthetic has encouraged trends such as specialized pet accessories and boutique grooming, framing pets as "cute companions" in alignment with broader cultural narratives (Pellitteri, 2018).

The advent of virtual and robotic pets further illustrates how modern pet keeping in Japan adapts to contemporary socioeconomic challenges. With a growing elderly population and many individuals living alone, technological innovations—exemplified by robotic pets such as AIBO—have emerged to provide companionship and mitigate feelings of isolation (Galanos and Reisel, 2020). These developments indicate how technological advances, combined with cultural imperatives rooted in both affective and aesthetic values, continue to redefine the boundaries of pet keeping in Japan (Galanos and Reisel, 2020).

Modern Japanese pet keeping is characterized not only by traditional domestic animals but also by innovative virtual and robotic companions that cater to the evolving needs of an increasingly urban and aging society (Galanos and Reisel, 2020). This multifaceted evolution is a testament to the dynamic interplay between historical legacy, socioeconomic realities, and cultural creativity.

See also: China, Pets In; History of Pet Keeping: Middle Ages

References and Further Reading

Galanos, V., and M. Reisel (2020). "Assessing the Japanese Turn in AI and Robot Ethics: Extracting Meaningful Principles Between Exoticism and Empiricism in the Case of AIBO." In *IFIP International Conference on Human Choice and Computers*, 141–57. Cham: Springer International Publishing.

Hanihara, K. (1993). "The Population History of the Japanese." *Nippon Ronen Igakkai Zasshi Japanese Journal of Geriatrics*, 30 (11): 923–31.

Herzog, H. (2014). "Biology, Culture, and the Origins of Pet-Keeping." *Animal Behavior and Cognition*, 1 (3): 296–308.

Kubinyi, E. (2025). "The Link Between Companion Dogs, Human Fertility Rates, and Social Networks." *Current Directions in Psychological Science*, 34 (4): 232–9.

Pellitteri, M. (2018). "Kawaii Aesthetics From Japan To Europe: Theory of the Japanese 'Cute' and Transcultural Adoption of its Styles in Italian and French Comics Production and Commodified Culture Goods." *Arts*, 7 (3): 24.

Toprak, B. (2006). "Economic Development Versus Cultural Transformation: Projects of Modernity in Japan and Turkey." *New Perspectives on Turkey*, 35: 85–127.

L

LANGUAGE, PETS AND

Language is more than a system of communication. Language both reflects and shapes how we see the world. On one level, language is just a system of categories. We create words to categorize the world around us—words like "furniture" refer to a class of objects such as tables and chairs that are used in a particular way. Without such linguistic categories, it would be difficult to teach our children about the world in which we live.

Words about animals shape our understanding of them. We know that terms like "pet" and "livestock" reflect a particular understanding of animals and then shape our treatment of them. Once an animal has been classified as a pet, it would, in our culture, be difficult to turn that animal into meat. Similarly, we don't think twice about consuming animals that have been classified as livestock.

So, if language shapes how we see the world and teaches us cultural values and norms, what does it say about how we see the world when we use words like "dog," "cow," or "fox" to describe people, or when we call someone bitchy, catty, or chicken? In addition to using specific language to talk about nonhuman animals, we also use animal terms to refer to humans, and most commonly, those terms are slurs, and are used to insult a person or a group of people. A number of related issues are at play here. First is the way in which humans so easily draw on animals to make sense of human realities. This is the case both when we call people by animal terms like "bitch" or "sow" and when we use phrases and idioms such as "flog a dead horse" or "skin a dead cat."

Second, we need to look at the implications of using animals for these purposes. What are the implications of calling people by animal names, especially when those names are used as pejoratives? Animal pejoratives reinforce attitudes toward marginalized humans—most commonly, people of color and women—by comparing them with another marginalized group: animals.

Words like "bitch," "cow," and "sow" almost exclusively refer to women and girls and imply a number of negative traits, such as stupidity, passivity, moodiness, and obesity.

Calling a woman a dog or a pig not only implies that the woman is ugly or fat, but it says something about all women as well: that they are to be judged primarily on the basis of their physical appearance, and that if they fail to achieve the cultural standard of beauty, they deserve to be called an animal. In other words, the huge number of expressions linking negative traits in women with animals demonstrates disdain not just for the animal but for women as well.

Third, there are implications for animals as well. How does using animal idioms and animal pejoratives affect the real-world treatment of animals? Language is never neutral—it shapes behavior. According to Joan Dunayer (2001), animal pejoratives denigrate certain categories of people—women and minorities, primarily—but they denigrate animals as well. Anthropologist Edmund Leach (1964) noted that by creating pejoratives from the names of animals, humans establish distance between themselves and the animals they regularly abuse. The fact that dog, cow, pig, sow, bitch, and chick are thought of as negative terms means that those animals are thought of negatively. A woman who is called a cow is thought to be fat and dull; likewise, the cow is thought of as dull. A dull animal is an animal who ultimately does not deserve to live (Dunayer 2001).

An example of the way in which animal slurs are not just gendered but feminized is the English language word "cunt"—which, for American speakers of English, is one of the worst things you can call a woman. This term derives from the old English term for rabbit (coney, cunny, which comes from the Latin *cuniculus*, which means burrowing; rabbits are thus "burrowing hares" even though, of course, they aren't hares and hares don't burrow) and refers not just to a woman but to her genitalia. That's why it's the worst thing we can call a woman, because it associates a woman with two things that many societies find distasteful: an animal (and a degraded, feminized animal at that) and female genitalia. This kind of linguistic violence is another way of affirming the "killability" of rabbits (Davis and DeMello 2003).

In her analysis of the history and mythology of the turkey, Karen Davis (2001) makes the same claim: when we're determined to do violence to an animal, we must first turn the victim into a despicable thing that deserves such treatment.

Haslam, Loughnan, and Sun (2011) conducted two surveys about animal slurs and found that one of the reasons these are so offensive to people is that the terms (and the animals behind them) convey stupidity, depravity, and disagreeableness and that, ultimately, by using an animal term to refer to people, they dehumanize the subject. (On the other hand, the authors showed that the offensiveness of these terms is mitigated when the person using the slur and the subject of the slur share the same status, such as women.)

In addition, idioms such as "skin a dead cat" contribute to a permissive social attitude toward the abuse of animals. Language influences the ways in which animals are socially constructed and therefore treated in human society. Negative animal idioms normalize or trivialize violence toward animals. When sayings like "flog a dead horse" become a normal part of our vocabulary, we can no longer see the implications of human violence against animals. These expressions mask the real violence within them and demonstrate human power over animals. Further, what does it mean when we use phrases such as "slaughter," "butcher," or "hunt" to discuss the killing of humans? It makes killing humans vivid and awful, yet we don't consider that when we use those terms to discuss animals; in those cases, it's normal and happens every day.

Another example of the use of language to reinforce attitudes surrounding the exploitation of animals is the different terms for describing the killing of humans and nonhumans. We routinely use words such as "murder" or "execution" when talking about the killing of a person, but we never use those words when referring to animals. Instead, we use "slaughter" or "euthanize," both of which have specific meanings and neither of which includes the sense of culpability (or illegality) that "murder" conveys. On the other hand, we often use words intended for animals, like slaughter, to refer to humans. Using slaughter in this way implies a particularly brutal murder, yet when we use it to refer to animals, most of us never think about the brutality of the killings.

Most negative animal idioms about companion animals are about cats. This reflects a long history of mistreatment of cats. At one time, cats were stuffed into bags and drowned, skinned, and swung around. Although we no longer skin cats, they and other pets are routinely abused, neglected, and abandoned. But when we use an expression such as "swing a dead cat," it seems funny and unreal. Rabbits are another animal that is routinely trivialized yet subjected to horrific abuse. Terms like "dumb bunny," "bunny slope," "snow bunnies," and "ski bunnies," all of which are aimed at women, paint both women and rabbits as being dumb and childish. Most terms associated with rabbits are derogatory, including "harebrained" (meaning frivolous or stupid), "rabbity" (meaning small, cowardly, or rabbit-like), and "rabbit shouldered" (meaning slumped in the shoulders).

The treatment of animals is shaped by another linguistic practice as well. A linguistic convention in popular media is to refer to animals as "it" unless the writer or speaker knows that the animal in question is male or female. Even then, animals still might be referred to as "it." Likewise, addressing an animal directly in print is almost never done. The introduction to Dave Aftandilian's *What Are the Animals to Us?* (2007) stated that the book was devoted to animals: "it is *you* that we have written this book about and for" (xvii; emphasis mine).

In recent years, we have seen a change in terminology referring to both pets and the people who live with them. Increasingly, animal welfare advocates and the general public have begun using the term "companion animal" rather than "pet" when describing the animals with whom we share our lives. But another, perhaps more important change involves switching from the term "owner" to the term "guardian" or "caretaker" when referring to a person who lives with a companion animal.

In Defense of Animals (IDA), an international animal rights organization that has spearheaded the campaign to end the use of the term "owner," argues that this term is linked to the treatment of animals. In particular, using the term owner means that companion animals are considered to be mere commodities or property, not individual beings, and IDA argues that this classification underlies their exploitation. The organization and other advocates argue that it was not so long ago that women, children, and others were seen, in legal terms, as merely property. IDA created their Guardian Campaign in 1999 as a nationwide platform to redefine society's standards for relating to and treating animals. Since that time, dozens of North American cities and counties, as well as the state of Rhode Island, have changed the language used in their legislation to replace "owner" with "guardian." A few years later, Carlisle-Frank and Frank conducted a study (Carlisle-Frank and Frank 2006) to test whether using the term "owner" or "guardian" is correlated with attitudes toward pets. The researchers found that, indeed, it was; those who self-identified as animal guardians engaged in more responsible behavior toward their companion animals than did those who called themselves owners, and the guardians appeared to have more intimate relationships with their pets as well. The study did not demonstrate causation, however, so it's not yet known whether differences in attitude and treatment are shaped by the use of the different terms, or whether different attitudes toward animals shape which term a person chooses to use.

See also: Communication, Human-Pet

References and Further Reading

Aftandilian, D., ed. (2007). *What are the Animals to Us?: Approaches From Science, Religion, Folklore, Literature, and Art*. Knoxville: University of Tennessee Press.
Carlisle-Frank, P., and J. M. Frank (2006). "Owners, Guardians, and Owner-guardians: Differing Relationships with Pets." *Anthrozoös*, 19 (3): 225–42.
Davis, K. (2001). *More Than a Meal: The Turkey in History, Myth, Ritual, and Reality*. New York: Lantern Books.
Davis, S., and M. DeMello (2003). *Stories Rabbits Tell: A Natural and Cultural History of a Misunderstood Creature*. New York: Lantern Books.
Dunayer, J. (1995). "Sexist Words, Speciesist Roots." In *Animals and Women: Feminist Theoretical Explorations*, ed C. J. Adams and J. Donovan, 11–31. Durham, NC: Duke University Press.
Dunayer, J. (2001). *Animal Equality: Language and Liberation*. Derwood, MD: Ryce Publishing.
Haslam, N., S. Loughnan, and P. Sun. (2011). "Beastly: What Makes Animal Metaphors Offensive?" *Journal of Language and Social Psychology* 30 (3): 311–25.
Leach, E. (1964). "Anthropological Aspects of Language: Animal Categories and Verbal Abuse." In *New Directions in the Study of Language*, ed. E. Lenneberg. Cambridge, MA: MIT Press.

LATIN AMERICA, PETS IN

The literature on pet keeping in Latin America reveals a complex panorama characterized by a high prevalence of companion animals, significant species diversity, and notable challenges regarding animal welfare.

Research in Costa Rica indicates that household pet ownership is highly common, with approximately 71 percent of households keeping animals, predominantly dogs and cats; specifically, around 53 percent of households own dogs compared to 15 percent owning cats. Additionally, about 24 percent maintain wild species as pets (Drews, 2001). Similar trends are evident throughout the region, where pet ownership extends to exotic and wild species, necessitating a closer examination of both the sociocultural and regulatory frameworks that govern these practices (Mota-Rojas et al., 2023; Roldán-Clarà et al., 2014).

Overall patterns in pet keeping demonstrate that Latin American households favor canines and felines. In urban settings such as those studied in Colombia, Costa Rica, and Mexico, pet owners display a proactive attitude toward animal safety, as evidenced by the fact that about 74.5 percent of respondents indicated they would evacuate with their pets during emergencies (Hesterberg, Huertas, and Appleby, 2012). Furthermore, there is a rising interest in nontraditional pets, including reptiles and wild birds. For instance, studies in Brazil have documented the practices of reptile care, showing that reptiles are housed and fed under conditions that may reflect both traditional husbandry and emerging trends in exotic pet keeping (Fuente et al., 2023). Additionally, wild birds continue to be a popular choice in certain regions, particularly in semiarid areas of northeastern Brazil, where local residents engage in keeping wild birds as pets despite potential welfare challenges (Bezerra, Araújo, and Alves, 2017). These findings signal a dual trend: while companionship remains rooted in longstanding traditions, there is also an increased diversification of species kept as pets.

The welfare status of pets in Latin America is a growing concern, particularly given the diversity of practices and the disparate levels of regulatory enforcement. As Mota-Rojas et al. point out, certain legal practices—such as aesthetic surgical procedures and events like bull-, cock-, and dog fighting—are culturally embedded and can substantially impair the quality of life for animals (Mota-Rojas et al., 2023). This regulatory gap is compounded by the fact that many pets, especially wild and exotic species, are kept under conditions that are not optimally designed to meet their physical and psychological needs. Although urban pet owners in Latin America have shown some readiness to incorporate

Migrants in Latin America boarding a train with their pet dog. Many people in Latin American urban areas have indicated they would evacuate with their pets during emergencies. (David Peinado/Dreamstime.com)

their pets into emergency planning (Hesterberg, Huertas, and Appleby, 2012), broader aspects such as routine veterinary care, appropriate enclosure conditions, and behavioral enrichment often lag behind the standards set in more regulated markets. Consequently, the literature calls for improved management policies and more comprehensive research on social organization and owner perceptions to ensure a higher welfare status for companion animals across the region (Roldán-Clarà et al., 2014; Mota-Rojas et al., 2023).

The tradition of keeping pets—ranging from dogs and cats to reptiles and wild birds—is deeply entrenched in Latin America, as illustrated by national surveys in Costa Rica and urban studies in several countries (Drews, 2001; Hesterberg, Huertas, and Appleby, 2012). However, the divergence between popular practices and optimal welfare conditions emphasizes the need for enhanced policy interventions and educational initiatives. Collaborative efforts between researchers, policymakers, and communities remain essential to address these welfare challenges and to align pet keeping practices with the best standards of care and management in the region (Mota-Rojas et al., 2023; Fuente et al., 2023; Bezerra, Araújo, and Alves, 2017).

See also: Community, Pets And

References and Further Reading:

Bezerra, D., H. Araújo, and R. Alves (2017). "Keeping Wild Birds as Pets in a Semiarid Region of Rio Grande Do Norte State, Northeastern Brazil." *El Hornero*, 32 (1): 85–93.

Drews, C. (2001). "Wild Animals and Other Pets Kept in Costa Rican Households: Incidence, Species and Numbers." *Society and Animals*, 9 (2): 107–26.

Fuente, M., B. Araújo, I. Policarpo, H. Pereira, A. Borges, W. Vieira, G. Filho, and Alves, R. (2023). "Keeping Reptiles as Pets in Brazil: Keepers' Motivations and Husbandry Practices." *Journal of Ethnobiology and Ethnomedicine*, 19 (1).

Hesterberg, U., G. Huertas, and M. Appleby. (2012). "Perceptions of Pet Owners in Urban Latin America on Protection of their Animals During Disasters." *Disaster Prevention and Management an International Journal*, 21 (1): 37–50.

Mota-Rojas, D., A. Strappini, A. Whittaker, M. Ghezzi, C. Titto, N. Calderón-Maldonado, A. Orihuela, et al. (2023). "Controversial Topics in Animal Welfare in Latin America: A Focus on the Legislation Surrounding the Human-companion Animal Relationship and Animals Used for Recreational Practices." *Animals*, 13 (9): 1463.

Roldán-Clarà, B., X. López-Medellín, I. Espejel, and E. Arellano (2014). "Literature Review of the Use of Birds as Pets in Latin-America, with a Detailed Perspective on Mexico." *Ethnobiology and Conservation*, 3.

LAWS AND REGULATIONS

Where the care of animals was once dictated by tradition, culture, and economic concerns, today, virtually all nations have at least a basic level of legislation regarding the care and treatment of animals.

In the United States, unlike many other countries with a strong central government, most legislation regarding animals, like most criminal legislation in general, happens at the state level. This means that in the United States, there is very little unity or consensus regarding how much or how little protection animals deserve.

State anti-cruelty laws have traditionally provided the primary framework for protecting companion animals, with each state developing its own set of standards regarding the treatment of these animals. Frasch and Lund (2011) detail that state laws classify animal cruelty in various forms—from infractions to felony offenses—depending on the circumstances and severity. This reflection illustrates the diverse approaches taken across jurisdictions.

In the United States, the largest laws and those that are found at the federal level include the Animal Welfare Act, which provides regulations concerning animals in three areas—entertainment, research, and pets; the Humane Methods of Slaughter Act, which covers the slaughter of livestock; and the Endangered Species Act, which covers wild animals that are threatened or endangered. Most Western nations have similar federal laws, although many go much farther than American laws.

Animal protection organizations have acted as key catalysts in the legislative process by mobilizing public sentiment and lobbying for more robust legal frameworks. These large organizations like the Humane Society of the United States and the American Society for the Protection of Animals, have not only influenced the creation of new statutes that recognize the legal rights of companion animals but also have helped reframe the discourse around animal protection (Favre, 2019). These groups have been pivotal in challenging traditional legal paradigms and driving innovations that better align statutory language with contemporary ethical standards regarding animal welfare.

The growth of animal law as a legal specialty is another significant trend from the last few decades. The increasing recognition of companion animals as entities with legal protection has spurred academic inquiry and practical legal reform. The evolving nature of academic programs and professional discourse surrounding animal law mirrors broader societal shifts, which are increasingly demanding legal accountability for cruelty and neglect and pushing for sophisticated legal frameworks that acknowledge the complex status of companion animals (Favre, 2020; Favre, 2019).

Joyce Tischler's analysis of animal law (2008, 2012) tracks the evolution of this field back to *Jones v. Butz*, a 1973 case that aimed to contest the religious exemptions found in the US Humane Methods of Slaughter Act. Although the case was ultimately dismissed, it signaled the inception of animal law and the use of legal mechanisms to advocate for animal welfare. Notably, this case—and the majority of subsequent cases—originated in the United States. The first animal law courses were also introduced in the United States before extending to other predominantly English-speaking nations.

Other significant developments during the initial phase of animal law, including pioneering journals, conferences, animal law sections within bar associations, and the publication of relevant books, were similarly concentrated in the United States and subsequently in other English-speaking countries. From a pragmatic standpoint, this phenomenon is understandable. The foundation of American law is English common law, which was disseminated to the colonies. Over time, other countries began to establish their own animal law frameworks that align with their unique legal traditions, resulting today in a global mosaic of laws that offer limited protection for specific animals in particular contexts.

Similar to the situation in the United States, most of these regulations safeguard individual categories of animals (typically companion animals) and include extensive exemptions permitting various forms of violence against animals, especially within the agricultural sector. Nonetheless, there has been a comparatively slow evolution of international laws and treaties dedicated to the protection of animals, irrespective of their residing nation or those traveling internationally.

Charlotte Blattner (2019) observes the growing levels of international trade in animals, their derivatives, and related products. While international commerce involving animals is not a new occurrence, it is undoubtedly experiencing rapid expansion, involving billions of animals annually. Although there exist notable international agreements, such as CITES, the Convention on International Trade in Endangered Species of Wild Flora and Fauna, a Universal Declaration for Animal Rights has yet to be established, despite a proposal made to UNESCO in 1978.

In recent years, the movement has achieved significant legislative milestones both within the United States and globally. Despite having a limited number of federal laws addressing animal welfare, the US statutes do not comprehensively cover cruelty, leading individual states to enact their own cruelty laws, many of which, regrettably, do not protect farm animals. Conversely, several other nations have implemented federal laws safeguarding animal rights. For instance, all European Union member states are required to adhere to the European Convention for the Protection of Pet Animals, which offers protection against pain, suffering, and abandonment for companion animals. Notably, all such laws were enacted within the last twenty-five years and provide a level of protection that surpasses any federal legislation in the United States.

Moreover, the twenty-first century has witnessed a surge in new legislation promoting animal welfare. The European Union has either banned or is in the process of phasing out various egregious practices concerning farm animals, such as battery cages for egg-laying hens (2012), gestation crates for pregnant pigs, and veal crates for calves (2006)—all of which remain legal in the United States. The EU has also established EU Directive 2010/63/EU, which enforces the strictest standards for the treatment of animals used in scientific research globally, and by 2013, ceased cosmetic testing on animals within its jurisdiction. In 2001, several nations, including the United Kingdom, Austria, the Netherlands, Switzerland, Croatia, and Bosnia and Herzegovina, enacted bans on fur farming. Additionally, in 2008, Spain introduced legislation bestowing nonhuman primates with rights to life, liberty, and freedom from experimentation.

Beyond Europe, India, where the foundation of animal protection laws dates back to the third century BCE (as detailed in Lahiri, 2015 regarding the Edicts of Ashoka), and which pioneered the cow protection movement in the nineteenth century, became the first nation to prohibit animal testing for cosmetics in 2014. In Latin America, several significant laws have also been enacted; for instance, Bolivia was the first country to outlaw the use of animals in circuses, and Brazil, the only nation to include animal welfare within its constitution, prohibited most cosmetic testing in 2014. Currently, animal rights organizations are pursuing several key legislative goals, all focused on federal legislation.

For example, the Humane Society of the United States (HSUS) has established a political action committee named Humane USA, dedicated to lobbying federal politicians and candidates to pass animal-friendly legislation.

See also: Animal Protection Movement; Animal Welfare Act

References and Further Reading

Blattner, C. E. (2019). "The Recognition of Animal Sentience by the Law." *Journal of animal ethics*, 9 (2): 121–36.
Chiesa, L. E. (2008). "Why Is It a Crime to Stomp on a Goldfish—Harm, Victimhood, and the Structure of Anti-Cruelty Offenses." *Mississippi Law Journal*, 78: 1.
Favre, D. (2019). "Next Steps for Animal Rights." *Derecho Animal Forum of Animal Law Studies*, 10 (1): 21.
Favre, D. (2020). "The Growing Reality of Legal Rights for Companion Animals." Society Register, 3 (3): 143–50.
Frasch, P. and H. Lund (2011). "The Unequal Treatment of Animals By Species and Practice in the United States: A Moral and Legal Dilemma." *Derecho Animal Forum of Animal Law Studies*, 2 3): 1.
Lahiri, N. (2015). *Ashoka in Ancient India*. Cambridge: Harvard University Press.
Park, M. (2024). Animal Cruelty and Punishment in the United States, and Animal Welfare Act in Korea. *Korean Association of Public Safety and Criminal Justice*, 33 (1): 91–130.
Tischler, J. (2008). "The History of Animal Law, Part I (1972–1987)." *Stanford Journal of Animal Law & Policy*, 1: 1.
Tischler, J. (2012). "A Brief History of Animal Law, Part II (1985–2011)." *Stanford Journal of Animal Law & Policy*, 5: 27.

LINK BETWEEN ANIMAL AND HUMAN ABUSE

In recent years, there has been a rising public consciousness regarding the issue of animal cruelty, largely due to the findings shared by psychologists, social workers, and law enforcement professionals who have begun to document a troubling reality. There is an emerging consensus on the notion of "the link," which denotes a significant correlation between acts of violence against animals and violent behaviors directed toward humans.

Historically, early academic inquiry, followed by media attention, primarily highlighted extreme incidents, such as those involving serial killers or school shooters whose violent trajectories often began with the torture of animals. Such cases are notably evident among young perpetrators, including school shooters like Eric Harris and Dylan Klebold from Littleton, Colorado; Nikolas Cruz from Parkland, Florida; as well as teenage offenders such as Andrew Golden, Luke Woodham, and Michael Carneal. Each of these individuals had histories of torturing and mutilating animals—cats, squirrels, cows, and dogs—prior to their violent assaults on classmates, teachers, and family members.

Similarly, adult offenders, including Jeffrey Dahmer, the Boston Strangler, and Ted Bundy, began their violent careers with acts of cruelty toward animals during their formative years. Such extreme case profiles have led the public to draw simplistic conclusions that children who inflict harm upon animals are destined to become murderers in adulthood. However, this perspective fails to encapsulate the multifaceted nature of the issue.

Research indicates that violence toward animals is associated with a broad spectrum of both violent and nonviolent criminal behaviors, and establishing a causal relationship between these behaviors remains a complex challenge. Although the significance of "the link" has garnered attention from scholars, members of law enforcement, and mental health professionals in recent years, the recognition of this issue has been acknowledged by keen observers for centuries.

For instance, in the eighteenth century, English artist Thomas Hogarth illustrated this concept in his series of engravings, "The Four Stages of Cruelty," created in 1751. This art depicted the fictional journey of a character named Tom Nero, whose engagement in increasingly severe acts of cruelty spiraled from childhood into adulthood. While Hogarth was not an animal rights advocate per se, he intended to impart the lesson that cruelty leads to further cruelty, particularly aimed at educating the impoverished populace of England. Current understanding, however, reveals that the graduation hypothesis—the proposition that childhood cruelty toward animals directly translates into adult violence toward people—is not robustly supported by scholarly data. Such research often relies on retrospective testimonials from adult criminal offenders about their childhood experiences.

This reliance on adults' recollections can introduce biases into the findings, alongside the common use of samples populated primarily by violent or criminally involved individuals, rather than representative samples from the general population. Consequently, we lack comprehensive data regarding the number of children who exhibit animal cruelty and later develop into law-abiding, nonviolent citizens. The prevalence of childhood animal abuse indicates that not all children engaged in such behaviors will transition into violent adults. A study conducted by Arluke (2002) found that a significant majority of former childhood animal abusers eventually "grew out" of these behaviors and adopted law-abiding lifestyles, indicating that we cannot reliably predict that a child who mistreats an animal will become a murderer in adulthood. To gain further insights into the phenomenon, it is essential to examine the demographics of those who perpetrate animal violence and the connections between their violent behaviors and their childhood experiences.

Data consistently show that males, particularly boys, exhibit a greater propensity for committing acts of physical cruelty against animals and engaging in animal fighting competitions. Research indicates that approximately 95 percent of individuals charged with animal cruelty cases are male. Conversely, women generally display more favorable attitudes toward animals. Nonetheless, it is noteworthy that a higher proportion of women than men are implicated in hoarding cases, while neglect cases see a more equitable distribution across genders, albeit with men often more frequently identified as culprits.

Regardless of gender, the majority of animal abusers tend to be the guardians or owners of the animals involved. While a definitive profile of a "typical" animal abuser has yet to be established, emerging data indicate that boys constitute the most prevalent demographic among abusers. A study conducted in Italy found that 66 percent of respondents identified as male had committed acts of abuse against animals (Baldry, 2005). It is crucial to recognize that children—irrespective of gender—who engage in animal cruelty tend to experience heightened levels of psychological, emotional, and behavioral issues compared to their peers.

Nonetheless, it is critical to contextualize animal abuse within a broader framework of risk factors, which may include poverty, exposure to trauma, or environments characterized by crime or violence.

Recent investigations into the correlation between childhood animal cruelty and violent later behaviors have yielded noteworthy findings. A 1985 study by Kellert and Felthous identified that 25 percent of violent offenders exhibited previous violence toward animals, compared to merely 6 percent among nonviolent criminals. Additionally, a 2004 study indicated that 36 percent of adult offenders had previously engaged in animal cruelty during childhood (Tallichet and Hensley, 2004). More recently, Levitt, Hoffer, and Loper (2016) revealed that 96 percent of individuals arrested for animal cruelty had histories of either prior or subsequent violent criminal offenses. The American Psychiatric Association currently categorizes animal cruelty as a symptom of conduct disorder, denoting a persistent pattern of antisocial behavior that may continue into adulthood.

Understanding the motivations behind animal abuse among children—primarily boys—reveals that many have often been victims of abuse themselves, including sexual victimization or bullying, or may have been perpetrators of bullying (Baldry, 2005; Henry and Sanders, 2007; Parkes and Signal, 2017). Typically, children who inflict harm on animals come from dysfunctional family environments where other forms of abuse or neglect are prevalent. Research indicates a striking correlation; up to 88 percent of families that experience child abuse simultaneously encounter some level of animal abuse (DeViney, Dickert, and Lockwood, 1983). In a study conducted by Duncan, Thomas, and Miller, (2005), it was found that children who abuse animals are twice as likely to have experienced abuse themselves.

In some cases, children may also witness animal abuse perpetrated by a father or another paternal figure (Browne, Hensley, and McGuffee, 2016). Flynn's (1999) research highlighted that witnesses of animal abuse possess an elevated likelihood of engaging in cruelty toward companion animals, with 8 percent admitting to having killed an animal. Instances of sexual abuse within the home may correlate with sexual abuse of animals as well (Holoyda and Newman, 2016). The impact of witnessing a parent's violent treatment of another parent is a significant factor; children exposed to parental violence often develop a propensity to harm animals.

Studies by Currie (2006) and Baldry (2005) indicated that children who observed domestic violence were three times more prone to inflict harm on animals compared to their peers. Such children may be acting out their own traumas or attempting to regain a sense of power through animal abuse (McDonald et al., 2018). It is plausible that experiences of trauma significantly contribute to childhood cruelty (Bright et al., 2018; Hawkins, Hawkins, and Williams, 2017). The normalization of violence, coupled with disrupted empathy, may play a role in children's behavior toward animals.

Even children untouched by direct abuse or violence may find themselves directing emotional distress stemming from familial conflict toward animals. Conversely, there are instances where children actively intervene to shield their mothers and pets from physical harm, even placing themselves at risk for the sake of animal protection. The interconnectedness is bidirectional: not only do children living in violent households face a heightened risk of committing violence, but they are also are jeopardized if they reside in an environment where either an adult targets another adult or an animal. Regrettably, although research firmly establishes that familial abuse correlates with animal cruelty, legal frameworks often lack mechanisms to ensure that animal abuse is reported to authorities—

except for a few US states where legislation mandates veterinarians to report suspected cruelty. This gap results in the oversight of potential warning signs.

A recent positive development has been the advent of cross-reporting, which entails the collaborative reporting of incidents of animal abuse, child abuse, and, at times, elder abuse among health care providers, child protective services, and animal enforcement agencies. Such collaboration enables professionals who observe abuse across these sectors to share essential information with relevant entities to facilitate appropriate interventions for the victims involved.

Moreover, a well-documented linkage exists between animal cruelty and domestic violence. Numerous studies (Ascione, 1997; Ascione, Weber, and Wood, 1997; Flynn, 2000; Faver and Strand, 2003) have indicated that between 50 percent and 85 percent of women who fled domestic violence situations disclosed that their partners had inflicted abuse on the family pets. In the United States, approximately 1.5 million women experience rape or physical assault by an intimate partner annually, making domestic violence one of the most prevalent forms of abuse faced by women today. The Humane Society of the United States estimates that nearly one million animals are abused or killed in association with domestic violence each year, based on the prevalence of assaults reported by women whose abusers also targeted their pets. This grim reality poses substantial threats to women, children, and their animal companions.

Perpetrators of abuse frequently resort to violence or threats against animal companions as a means of exerting control over their partners. The killing or intimidation of an animal serves to reinforce a woman's sense of entrapment and loss of control over her own existence, often imprisoning her within an abusive relationship. Such acts are executed as punishments, precautions, or deterrents to prevent her from leaving. Though many domestic violence offenders may be women, the vast majority of severe abuses against partners are perpetrated by men, who also constitute the majority of animal abusers. In a pertinent study, Loring and Bolden-Hines (2004) found that a significant proportion of men—41 percent—arrested for domestic violence reported having committed acts of animal cruelty during their adulthood, compared to only 1.5 percent of the general male demographic (Febres et al., 2014).

A significant challenge remains that most domestic violence shelters are unable to accommodate companion animals, even temporarily. This limitation leaves women in a dire situation where they must either abandon their pets to further victimization or remain in environments fraught with danger for both themselves and their animals. As highlighted in a survey of women who left abusive relationships (Faver and Strand, 2003), one in four respondents indicated that concerns for their pets' safety had delayed their entry into a shelter. Fortunately, initiatives such as the SAF-T program (Sheltering Animals and Families Together) are emerging, which provide substantial resources to domestic violence shelters to offer safe housing for pets. Furthermore, a growing number of shelters are collaborating with local animal shelters to create refuge options for animal victims of domestic violence, and many shelters are now increasingly inquiring about pet-related abuse during the intake processes.

Thanks to ongoing efforts from social workers, sociologists, psychologists, and professionals from law enforcement and animal welfare sectors, there is an improved recognition of the substantive connections between violence against animals and humans, albeit with acknowledged complexities. The pressing question evolves into how society can effectively prevent such violence and address it once it arises. Encouragingly, various programs have been developed to counsel and rehabilitate young abusers before their violent impulses escalate further. The National Cruelty Investigation School, a program under the Law Enforcement Training Institute at the University of Missouri, exemplifies a national initiative that educates law enforcement officers on effective cruelty investigations with an emphasis on understanding the relationships between animal cruelty and human violence. Additionally, numerous local jurisdictions have begun establishing specialized units to combat this pressing issue.

As of 2019, a total of sixteen US states require psychological evaluations and, when warranted, treatment for individuals convicted of animal cruelty. Another nineteen states, along with the US Virgin Islands, have established mandates for evaluation and/or treatment for abusers. Nonetheless, it is imperative to note that not all mental health professionals possess training to adequately counsel individuals who have committed such acts of cruelty, presenting another arena where animal protection organizations can step in to contribute toward positive change.

In conclusion, the intricate ties between animal cruelty and broader patterns of interpersonal violence call for comprehensive inquiries and systematic approaches to prevention and intervention. Continued collaboration across sectors is essential to create an environment where both human and animal welfare is prioritized, thereby enhancing safety and health for all.

See also: Cruelty; Laws and Regulations; Neglect; United States, Pets In

References and Further Reading

ArlUKe, A. (2002). "Animal Abuse as Dirty Play." *Symbolic Interaction* 25: 405–30.

ArlUKe, A., and E. Madfis (2014). "Animal Abuse as a Warning Sign of School Massacres: A Critique and Refinement." *Homicide Studies* 18, no. 1: 7–22.

Ascione, F. R., C. Weber, and D. S. Wood (1997). "The Abuse of Animals and Domestic Violence: A National Survey of Shelters for Women Who Are Battered." *Society & Animals* 5 (3): 205–18.

Baldry, A. C. (2005). "Animal Abuse Among Preadolescents Directly and Indirectly Victimized at School and at Home." *Criminal Behaviour and Mental Health* 15, (2): 97–110.

Bright, M. A., M. S. Huq, T. Spencer, J. W. Applebaum, and N. Hardt (2018). "Animal Cruelty as an Indicator of Family Trauma: Using Adverse Childhood Experiences to Look Beyond Child Abuse and Domestic Violence." *Child Abuse & Neglect* 76: 287–96.

Browne, J. A., C. Hensley, and K. M. McGuffee (2016). "Does Witnessing Animal Cruelty and Being Abused During Childhood Predict the Initial Age and Recurrence of Committing childhood Animal Cruelty?" *International Journal of Offender Therapy and Comparative Criminology* 61, (16): 1850–65.

Currie, C. L. (2006). "Animal Cruelty by Children Exposed to Domestic Violence." *Child Abuse & Neglect* 30, (4): 425–35.

DeViney, E., J. Dickert, and R. Lockwood (1983). "The Care of Pets Within Child Abusing Families." *International Journal for the Study of Animal Problems* 4, (4): 321–29.

Duncan, A., J. C. Thomas, and C. Miller (2005). "Significance of Family Risk Factors in Development of Childhood Animal Cruelty in Adolescent Boys with Conduct Problems." *Journal of Family Violence* 20 (4): 235–39.

Faver, C. A., and E. B. Strand. (2003). "Domestic Violence and Animal Cruelty: Untangling the Web of Abuse." *Journal of Social Work Education* 39: 237–53.
Febres, J., H. Brasfield, R. C. Shorey, J. Elmquist, A. Ninnemann, Y. C. Schonbrun, J. R. Temple, P. R. Recupero, and G. L. Stuart. (2014). "Adulthood Animal Abuse Among Men Arrested for Domestic Violence." *Violence Against Women* 20 (9): 1059–77.
Flynn, C. P. (1999). "Exploring the Link Between Corporal Punishment and Children's Cruelty to Animals." *Journal of Marriage and the Family* 61 (4): 971–81.
Flynn, C. (2000). "Woman's Best Friend: Pet Abuse and the Role of Companion Animals in the Lives of Battered Women." *Violence Against Women* 6: 162–77.
Hawkins, R. D., E. L. Hawkins, and J. M. Williams. (2017). "Psychological Risk Factors for Childhood Nonhuman Animal Cruelty: A Systematic Review." *Society & Animals* 25 (3): 280–312.
Henry, B., and C. E. Sanders. (2007). "Bullying and Animal Abuse: Is There a Connection?" *Society & Animals* 15 (2): 107–26.
Holoyda, B. J., and W. J. Newman. (2016). "Childhood Animal Cruelty, Bestiality, and the Link to Adult Interpersonal Violence." *International Journal of Law and Psychiatry* 47: 129–35.
Kellert, S. R., and A. R. Felthous. (1985). "Childhood Cruelty Toward Animals Among Criminals and Noncriminals." *Human Relations* 38 (12): 1113–29.
Levitt, L., T. A. Hoffer, and A. B. Loper. (2016). "Criminal Histories of a Subsample of Animal Cruelty Offenders." *Aggression and Violent Behavior* 30: 48–58.
McDonald, S. E., A. M. Cody, L. J. Booth, J. R. Peers, C. O'Connor Luce, J. H. Williams, and F. R. Ascione (2018). "Animal Cruelty Among Children in Violent Households: Children's Explanations of Their Behavior." *Journal of Family Violence*, 33 (7): 469–80.
Loring, M. T., and T. A. Bolden-Hines (2004). "Pet Abuse by Batterers as a Means of Coercing Battered Women into Committing Illegal Behavior." *Journal of Emotional Abuse*, 4 (1): 27–37.
Parkes, D., and T. Signal. (2017). "Revisiting a Link: Animal Abuse, Bullying and Empathy in Australian Youth." *Human-Animal Interaction Bulletin* 5 (1): 26–40.
Tallichet, S. E., and C. Hensley (2004). "Exploring the Link Between Recurrent Acts of Childhood and Adolescent Animal Cruelty and Subsequent Violent Crime." *Criminal Justice Review*, 29 (2): 304–16.
Verlinden, S., M. Hersen, and J. Thomas. (2000). "Risk Factors in School Shootings." *Clinical Psychology Review* 20 (1): 3–56.

LITERATURE, PETS IN

Animals are one of the most common features in world literature. Often, their presence is only metaphorical in that they are often used to represent humans and human culture. But sometimes, animals are represented as themselves.

Why are animals used so frequently in literature? And why are they almost always used as stand-ins for humans? Are animals ever used in literature as animals? We know that animals have played a major role in the symbolic behaviors of humans for thousands of years, through art, through religion, and through folklore and myth. As new ways of telling stories emerged, such as writing, and, later, film, of course animals continued to play a major role.

Just as animals featured heavily in the myths of the great ancient civilizations, they were prominently featured in the plays and poems of the Greeks and Romans, as well as in the epic writings of India, such as the Vedas and the Puranas. For instance, in the Greek tragedy

The Bacchae, each character is likened to an animal, and the story unfolds as a contest between the hunters and the prey. *Aesop's Fables*, a collection of tales starring animals, each with a moral lesson, is probably the most famous of Greek writings involving animals. In the Indian epic poem the *Ramayana*, a number of animals play prominent roles, such as Hanuman, the monkey god, and Jatayuvu, a mythical bird, and a key battle is won for Rama with the help of an army of monkeys, bears, and other animals. Homer's *Odyssey* features a dog named Argos, who patiently awaits his master's return from battle; after a decade of waiting, Ulysses returns and Argos dies, having fulfilled his mission. The loyalty of Argos has been a common theme in stories in Western literature involving dogs for thousands of years.

With the rise of Christianity in Medieval Europe, much of the literature that was available was religious; it would not be until after the invention of the printing press in the fifteenth century that secular writing was more widely available. One of the earliest types of literature involving animals was the Medieval bestiary, in which used animals as allegorical devices to convey moral lessons to readers of the time. Each animal included in the bestiary was described in terms of its personality and moral traits; the partridge, for example, was said to steal the eggs of others, and its "immorality" was held up as a characteristic to avoid. Most Europeans at the time could not read, so bestiaries were primarily used by clergy to prepare their sermons, while illiterate church members could follow along by looking at the pictures.

Animals were used in the Age of Enlightenment as well, but at this time, with the waning influence of religion, animals were often used to parodied or satirized humans, rather than to impart a moral lesson. For example, the eighteenth-century English writer Jonathan Swift's *A Modest Proposal* (2000) is a satirical essay arguing for the breeding and sale of poor children as food in order to reduce the number of starving Irish. Swift employed language generally used to refer to animals to make his point, discussing their feeding habits, their breeding, and how much they could be sold for.

During the Romantic Age of the nineteenth century, as the concept of conservation of nature was emerging, animals began to be treated differently in literature. Writers and poets like William Wordsworth, Lord Byron, and John Keats wrote about wild animals and their beauty and how humans could be inspired by that wildness. With the rise of the Victorian era came the rise of natural history writing on animals, beginning with Charles Darwin, and at the same time, domesticated animals became a much more prominent feature of English and American households. With the rise of pet keeping came a new form of literature, primarily aimed at children, which once again used animal characters as a way to teach children morality and other positive traits, such as kindness to children.

In the twentieth and twenty-first centuries, animals continue to be major characters in literature in the West. In this firmly post-Darwinian era, where in which the human/animal border continues to crumble, writers have been grappling with a number of different issues through which animals can play a role. For instance, twentieth-century writers like Franz Kafka use animality as a way to understand what it is that makes us human. In his 1915 short story "The Metamorphosis" (2013), Gregor, the protagonist of the story, wakes up one morning to find that he is now an insect. As his appearance and behavior change, his family begins to shun him, and eventually he dies of starvation and an injury caused when his father threw apples at him.

Another Kafka story, "A Report to the Academy" (1979), has a chimpanzee narrating a story about how he became human. The chimpanzee, Red Peter, had found himself in a cage in the cargo hold of a ship and realized that his only way out of the cage was to be human: "There was nothing else for me to do, provided always that freedom was not to be my choice" (258). In both texts, Kafka's use of animals serves to destabilize the concept of humanity.

Animals also feature heavily in graphic novels (Chaney 2011; Brown 2012; Herman 2018). While the best known of such novels is Art Spiegelman's *Maus* series, but there are numerous other graphic novels that have something interesting to say about animals or their relationships with us. Such works include *Animosity Vol. 1: The Wake* by Marguerite Bennett and *Duncan the Wonder Dog: Show One*, both of which are about what happens when animals wake up one day able to speak. This is a theme covered in other novels as well, such as *The Rabbi's Cat*. As Lisa Brown notes:

> Comics can address animals in a way that is unique by providing an alternate perspective on how we humans believe animals think and behave, and also how we treat them as a result. By providing other animals an outlet for their voices, artists simultaneously allow them a forum to air their grievances and provide humans an arena to hear what they might say. (2012: 87)

The twentieth century has seen a great deal of writing by women involving animals. African American poet Maya Angelou's autobiography, *I Know Why the Caged Bird Sings* (1997), which takes its title and theme from poet Paul Laurence Dunbar's poem "Sympathy," evokes a caged bird, beating its wings fruitlessly against the bars of its cage, crying to be freed. Another African American writer, Alice Walker, wrote the short story "Am I Blue" (1988) about a horse named Blue, and her feelings of loss when her companion horse, Brown (who was brought to stay with her specifically to make her pregnant), was taken away. "Am I Blue" can be read, on the one hand, as an allegory about slavery, because of the way that the humans use Blue and other animals without regard for their own needs, and their lack of recognition that horses create meaningful bonds with other animals and mourn when those bonds are broken. But "Am I Blue" does more than make animals into rhetorical devices: Walker empathizes heavily with Blue and wants the reader to recognize that animals have feelings and that those feelings need to be acknowledged and met. Walker's narrator, for example, at the end of the story, Walker's narrator spits out a mouthful of steak that she was eating because she recognized that in eating meat, she was "eating misery." But she also recognizes that when humans can treat animals in this way, they can treat humans in this way as well.

See also: Film, Animals In; Language, Pets And

References and Further Reading

Adams, R. (1972). *Watership Down*. London: Rex Collings.
Angelou, M. (1997). *I Know Why the Caged Bird Sings*. New York: Bantam.
Armstrong, W. H. 1969. *Sounder*. New York: Harper & Row.

Auster, P. (1999). *Timbuktu*. New York: Henry Holt.
Beierl, B. H. (2008). "The Sympathetic Imagination and the Human—Animal Bond: Fostering Empathy Through Reading Imaginative Literature." *Anthrozoös*, 21 (3): 213–20.
Berger, J. (2000). *King: A Street Story*. New York: Vintage.
Bettelheim, B. (2010). *The Uses of Enchantment: The Meaning and Importance of Fairy Tales*. New York: Vintage.
Brown, L. (2012). "The Speaking Animal: Non-Human Voices in Comics." In *Speaking for Animals: Animal Autobiographical Writing*. Routledge Advances in Sociology, ed. M. DeMello, 82–7. New York: Routledge.
Burnford, S. E. 1960. *The Incredible Journey*. London: Hodder & Stoughton.
Chaney, M. A. (2011). "Animal Subjects of the Graphic Novel." *College Literature*, 38 (3): 129–49.
Copeland, M. (2012). "'Straight from the Horse's Mouth': Equine Memoirs and Autobiographies." In *Speaking for Animals: Animal Autobiographical Writing*. Routledge Advances in Sociology 80, 186–99. New York: Routledge.
Cosslett, T. (2017). *Talking Animals in British Children's Fiction*, 1786–1914. London: Routledge.
DeMello, M., ed. (2013). *Speaking for Animals: Animal Autobiographical Writing*. Routledge Advances in Sociology 80. New York: Routledge.
Gipson, F., and J. Weiss (1956). *Old Yeller*. New York: Harper & Row.
Herman, D. (2018). "Introduction: More-Than-Human Worlds in Graphic Storytelling." In D. Herman (ed.), *Animal Comics: Multispecies Storyworlds in Graphic Narratives*, 1–25. London: Bloomsbury.
Jacobs, J. (1966). *The Fables of Aesop*. 1894. New York: Schocken.
Jaques, Z. (2015). *Children's Literature and the Posthuman: Animal, Environment, Cyborg*. London: Routledge.
Kafka, F. (1979). *A Report forto an Academy*. Pretoria, SA: Pretoria M and M Productions.
Kafka, F. (2013). *The Metamorphosis*. New York: Modern Library Classics.
Kendall-Morwick, K. (2014). "Mongrel Fiction: Canine Bildung and the Feminist Critique of Anthropocentrism in Virginia Woolf's *Flush*." *Modern Fiction Studies*, 60 (3): 506–26.
Kipling, R. (1987). *The Jungle Book*. 1894. New York: Viking.
Lockley, R. M. (1976). *Private Life of the Rabbit: An Account of the Life History and Social Behaviour of the Wild Rabbit*. London: Corgi.
Malay, M. (2018). *The Figure of the Animal in Modern and Contemporary Poetry*. Palgrave Studies in Animals and Literature. Cham, Switzerland: Palgrave Macmillan.
Martel, Y. (2010). *Beatrice and Virgil*. Edinburg: Canongate Books.
McHugh, Susan. (2011). *Animal Stories: Narrating across Species Lines*. Minneapolis: University of Minnesota Press. *Animal Stories: Narrating Across Species Lines*. Minneapolis: University of Minnesota Press.
Melson, G. F. (2001). *Why the Wild Things Are: Animals in the Lives of Children*. Cambridge: Harvard University Press.
Norris, M. (1985). *Beasts of the Modern Imagination: Darwin Nietzsche, Kafka, Ernst and Lawrence*. Baltimore, MD: Johns Hopkins University Press.
Orwell, G. (1946). *Animal Farm*. New York: Harcourt, Brace and Company.
Parry, C. (2017). *Other Animals in Twenty-First Century Fiction*. Basingstoke, UK: Palgrave Macmillan.
Rohman, C. (2009). *Stalking the Subject: Modernism and the Animal*. New York: Columbia University Press.
Sewell, A. (1877). *Black Beauty*. New York: Scholastic Paperbacks.
Sfar, J. (2005). *The Rabbi's Cat*. New York: Pantheon Books.
Smith, C. (2002). "Across the Widest Gulf: Nonhuman Subjectivity in Virginia Woolf's Flush". *Twentieth Century Literature*, 48 (3): 348–61.
Spiegelman, A. (1991). *The Complete Maus* 1. New York: Pantheon.

Swift, J., and P. Magee (2000). *A Modest Proposal*. North Carolina: Project Gutenberg.
Walker, A. (1988). "Am I Blue? Ain't These Tears in These Eyes Tellin' You?'" In I. Zahava (ed.), *Through Other Eyes: Animal Stories by Women*, 1–6. Freedom, CA: Crossing Press.
White, E. B. (1980). *Charlotte's Web*. 1952. New York: Scholastic.
Wolfe, C. (2014). "Commentary: Apes Like Us." In U. Chaudhuri and H. Hughes (Eds), *Animal Acts: Performing Species Today*, 156–62. Ann Arbor: University of Michigan Press.
Woolf, V. (1998). *Flush*. New York: Oxford University Press.

M

MENTAL HEALTH, PETS AND

The relationship between pet ownership and mental health has been widely studied, yielding both positive and negative implications for psychological well-being. Numerous studies suggest that the companionship provided by pets significantly enhances psychological health, particularly among individuals experiencing social isolation or heightened stress. For instance, it has been reported that pets, especially dogs, can serve as facilitators for social connections and reduce feelings of loneliness, making challenging life situations seem more manageable for their owners (Needell and Mehta-Naik, 2016; Fayyaz, 2023). The companionship and unconditional affection offered by pets can bolster emotional health by increasing feelings of happiness and reducing symptoms of depression and anxiety (Wells, 2009; Manero et al., 2023).

The benefits of pet ownership extend beyond mere companionship; they also include physiological health advantages, such as improved cardiovascular health and better dietary habits, which contribute to overall mental well-being (Zhou, Yin, and Gao, 2020). Indeed, research suggests that interactions with pets promote regular exercise and a structured routine that can enhance the psychological state of their owners, particularly during stressful times like the Covid-19 pandemic (Amiot, Gagné, and Bastian, 2022; Bussolari et al., 2021).

However, it is essential to note that the impact of pet ownership on mental health is not universally positive. Some studies indicate that pet ownership can lead to increased distress, particularly among individuals who feel overwhelmed by the responsibilities associated with caring for a pet (Hawkins, Hawkins, and Tip, 2021). Additionally, attachment to pets, while often beneficial, may complicate emotional health by potentially acting as a substitute for human relationships, particularly for those who are depressed or socially isolated (Purewal et al., 2017). Conflicting results in the literature highlight that the psychological benefits of pet ownership can vary significantly among individuals, influenced by factors such as pre-existing mental health conditions and the nature of the individual's relationship with their pet (Herzog, 2011; Parslow et al., 2004).

Moreover, the literature indicates that outcomes can be context-dependent. For example, during periods of enforced isolation like lockdowns, pet owners often reported a greater sense of purpose and reduced feelings of loneliness (Chopik et al., 2023), Antonacopoulos and Pychyl, 2010). Yet, some researchers emphasize that these benefits could be short-lived, with the possibility of diminishing returns as individuals acclimatize to their pets post-adoption (Chopik et al., 2023).

In summary, while the presence of pets generally correlates with improved mental health outcomes such as enhanced mood and reduced anxiety, there are nuanced dynamics at play. The intricacies of pet attachment, the responsibilities of pet ownership, and individual differences significantly influence these outcomes, highlighting the need for a more individualized understanding of the human–animal bond in the context of mental health.

See also: Bond, Human-Animal; Disability, Pets And; Grief and Pet Loss; Human Health, Pets And

References and Further Reading

Amiot, C., C. Gagné, and B. Bastian (2022). "Pet Ownership and Psychological Well-being During the COVID-19 Pandemic." *Scientific Reports*, 12 (1).

Antonacopoulos, N. and T. Pychyl (2010). "An Examination of the Potential Role of Pet Ownership, Human Social Support and Pet Attachment in the Psychological Health of Individuals Living Alone." *Anthrozoös*, 23 (1): 37–54.

Bussolari, C., J. Currin-McCulloch, W. Packman, L. Kogan, and P. Erdman (2021). "'I Couldn't Have Asked for a Better Quarantine Partner!': Experiences with Companion Dogs During COVID-19." *Animals*, 11 (2): 330.

Chopik, W., J. Oh, R. Weidmann, J. Weaver, R. Balzarini, G. Zoppolat, and Slatcher, R. (2023). "The Perks of Pet Ownership? The Effects of Pet Ownership on Well-Being During the Covid-19 Pandemic." *Personality and Social Psychology Bulletin*, 51 (6): 928–48.

Fayyaz, F. (2023). Pet-Keeping and its Role on the Psychological Well-being of Adolescents in Khyber Pakhtoon Khwah (KP). *Journal of Development and Social Sciences*, 4 (III).

Hawkins, R., E. Hawkins, and L. Tip (2021). "I Can't Give Up When I Have Them to Care for": People's Experiences of Pets and Their Mental Health. *Anthrozoös*, 34 (4): 543–62.

Herzog, H. (2011). "The Impact of Pets on Human Health and Psychological Well-being." *Current Directions in Psychological Science*, 20 (4): 236–9.

Manero, R., J. Dechusa, R. Guanzon, and E. Ceballo (2023). "Pawsitive Connection: Widowers' Life Experiences on Therapeutic Value of Owning Domestic Pets." *Journal of Education Society and Behavioural Science*, 36 (8): 97–110.

Needell, N. and N. Mehta-Naik (2016). "Is Pet Ownership Helpful in Reducing the Risk and Severity of Geriatric Depression?" *Geriatrics*, 1 (4): 24.

Parslow, R., A. Jorm, H. Christensen, B. Rodgers, and P. Jacomb (2004). "Pet Ownership and Health in Older Adults: Findings from a Survey of 2,551 Community-based Australians aged 60–64." *Gerontology*, 51 (1): 40–7.

Purewal, R., R. Christley, K. Kordas, C. Joinson, K. Meints, N. Gee, and C. Westgarth (2017). "Companion Animals and Child/adolescent Development: A Systematic Review of the Evidence." *International Journal of Environmental Research and Public Health*, 14 (3): 234.

Wells, D. (2009). "Associations Between Pet Ownership and Self-reported Health Status in People Suffering from Chronic Fatigue Syndrome." *The Journal of Alternative and Complementary Medicine*, 15 (4): 407–13.

Zhou, Z., D. Yin, and Q. Gao (2020). "Sense of Presence and Subjective Well-being in Online Pet Watching: The Moderation Role of Loneliness and Perceived Stress." *International Journal of Environmental Research and Public Health*, 17 (23): 9093.

MIDDLE EAST, PETS IN

The literature on pet keeping in the Middle East, particularly concerning dogs in Muslim countries and cats in Turkey, reveals a complex interplay between cultural attitudes, health implications, and individual psychological benefits associated with pet ownership.

In many Muslim countries, which make up the bulk of countries in the region, societal attitudes toward dogs can be historically ambivalent due to Islamic teachings regarding purity and contamination. However, recent trends indicate an increased acceptance of dog ownership, particularly in urban areas. For instance, the rising number of individuals keeping dogs stems from shifts toward modern lifestyles where pets contribute to social status and companionship. Research indicates that pet keeping among children, including dogs, correlates positively with psychological well-being. Specifically, studies demonstrate that pet keepers tend to have higher levels of mental health benefits compared to non-pet owners (Fayyaz, 2023). Notably, pet ownership, including dogs, has shown protective effects against asthma and allergies in children, suggesting a "healthy pet-keeping effect" that contradicts traditional avoidance behaviors associated with allergenic responses (Medjo et al., 2013).

In Turkey, cats hold a particularly esteemed cultural status, being viewed not only as companions but also as important parts of urban life. Turkish cities like Istanbul showcase a unique coexistence between humans and street cats, leading to community-based cat care emphasizing caretaking and responsibility toward these animals. Cats in Turkey are often regarded as spiritual symbols, aligning with the country's broader cultural heritage. Pet keeping behaviors in Turkey highlight a blend of affection toward animals and social responsibilities, which differ significantly from traditional views of dog ownership in many Muslim contexts.

Moreover, the literature points out that the psychological impacts of owning pets—whether dogs in predominantly urban Islamic regions or cats in Turkey—are substantial. Participants in studies expressed that companionship from pets fosters a sense of belonging and emotional stability, significantly impacting their mental health (Fayyaz, 2023).

Furthermore, the increase in exotic pet keeping, including domestic cats, is becoming more prevalent across the Middle East. This trend, fueled by social media and global travel, emphasizes the region's demand for companionship alongside traditional pets. However, it raises various ethical and health concerns regarding zoonotic diseases and the implications of wildlife trade, documented as a rising concern (Bush, Baker, and Macdonald, 2014; Hooper, 2025).

The underlying dynamics of pet keeping, including cultural perceptions, health implications, and psychological effects, are significantly influenced by regional attitudes toward animals. In conclusion, while Muslim countries exhibit a growing acceptance of dogs, particularly in urban domains, Turkey's cultural embrace of cats reflects a deep-rooted relationship with these animals, demonstrating diverse influences that shape pet keeping in the Middle East.

See also: Cats; Dogs; Exotic Pets; Free-Roaming Cats; Religious Institutions, Pets And

References and Further Reading

Bush, E., S. Baker, and D. Macdonald (2014). "Global Trade in Exotic Pets 2006–2012." *Conservation Biology*, 28 (3): 663–76.

Fayyaz, F. (2023). "Pet-Keeping and its Role on the Psychological Well-being of Adolescents in Khyber Pakhtoon Khwah (KP)." *Journal of Development and Social Sciences*, 4 (III).

Hooper, J. (2025). "The Conservation Costs of Commercial Breeding of Civet Pets in Indonesia." *Discover Animals*, 2 (1): 6.

Medjo, B., M. Atanaskovic-Markovic, D. Nikolic, B. Spasojevic-Dimitrijeva, P.Ivanovski, and S. Djukic (2013). "Association Between Pet-Keeping and Asthma in School Children." *Pediatrics International*, 55 (2): 133–7.

N

NATURAL DISASTERS AND PETS

Since 2005, when the United States experienced Hurricane Katrina, a hurricane that caused an incredible amount of devastation throughout the Gulf of Mexico region, we have become much more attuned to the importance of including pets in emergency planning and policy. This disaster has catalyzed significant changes in emergency management strategies concerning pets, revealing the profound emotional bonds between humans and their companion animals, and leading to a new focus on pet welfare throughout the stages of an emergency.

Research conducted since Katrina indicates that many pet owners—both in the United States and elsewhere—are hesitant to evacuate in a disaster if their pets cannot accompany them, as we saw during Katrina, where many individuals prioritized their pets' safety over their own evacuation orders (Adams, 2015; Farmer, DeYoung, and Wachtendorf, 2016; Babcock and Smith, 2020). This sentiment is supported by findings showing that nearly all surveyed pet owners believed animals should be included in evacuation efforts, underscoring the critical role pets play in disaster planning (Tanaka et al., 2019).

Animal protection organizations have emerged as key advocates for the integration of pets into emergency management frameworks. Following Hurricane Katrina, the passage of the Pets Emergency and Transportation Standards (PETS) Act in the United States highlighted the necessity of including animals in disaster preparedness strategies (Glassey, 2018). Efforts from animal welfare organizations have led to the establishment of pet-friendly shelters and trained volunteer networks to ensure animals are cared for post-disaster, which helps minimize panic and confusion among pet owners during crises (Gunter et al., 2022). Moreover, organizations have continued to promote public awareness regarding disaster preparedness for pets, recognizing that educational outreach can significantly influence pet owner behaviors in crisis situations (Chadwin, 2017).

A member of a North Dakota humane society takes care of a dog whose owner's home was flooded. The owner could not take the dog with him to the shelter where he was staying. (Photo by Patsy Lynch/FEMA, National Archives)

The aftermath of Hurricane Katrina also showed that the psychological toll on individuals who lost animals during the disaster was profound, leading to increased depression and post-traumatic stress (Wijnker, Leinenga, and Lipman, 2018). This emphasizes the need for understanding human–animal relationships, particularly during emergencies where those relationships can influence decision-making and overall community resilience (Pimentel et al., 2021). Furthermore, the emotional significance of pets is reflected in public health discussions, advocating for a comprehensive strategy to include pets in emergency planning, not only for animal welfare but also for the mental well-being of pet owners (Babcock and Smith, 2020; Heath and Linnabary, 2015; DeYoung et al., 2020).

Recent studies, including those conducted in response to Hurricane Irma in 2017, further illustrate the continuous evolution of emergency response protocols that prioritize animal welfare. Recognizing the impact of pets on evacuation behavior has led to improvements in local and county-level emergency management to create inclusive plans that accommodate both human and animal needs (Spain et al., 2017; Iacobucci, 2019). This understanding is critical in fostering an environment where pet owners feel more secure and willing to evacuate when faced with natural disasters, thereby enhancing the safety and effectiveness of disaster response initiatives.

In summary, the literature indicates that the challenges posed by pet ownership during natural disasters necessitate a multifaceted approach involving improved emergency planning, community education, and enhanced support systems. The lessons learned from Hurricane Katrina, alongside subsequent disasters, have guided significant reforms in animal emergency management, emphasizing the integral role of pets in modern disaster preparedness strategies (Berger, 2021; Kajiwara, 2020).

See also: Animal Protection Movement; Bond, Human-Animal; Laws and Regulations; Rescue Groups; War, Pets And

References and Further Reading

Adams, H. (2015). "Why Populations Persist: Mobility, Place Attachment and Climate Change." *Population and Environment,* 37 (4): 429–48.

Babcock, S. and D. Smith (2020). "Pets in Comprehensive Disaster Planning: the Post–Hurricane Katrina Experience." *American Journal of Public Health,* 110 (10): 1500–1.

Berger, M. (2021). "All Creatures Safe and Sound: The Social Landscape of Pets in Disasters." *World Medical & Health Policy*, 15 (2).

Chadwin, R. (2017). "Evacuation of Pets During Disasters: A Public Health Intervention to Increase Resilience." *American Journal of Public Health*, 107 (9): 1413–7.

DeYoung, S., A. Farmer, Z. Callaro, and S. Naar (2020). "Disaster Preparedness Among Service Dog Puppy- Raisers (Human Subject Sample)." *Animals*, 10 (2): 246.

Farmer, A., S. DeYoung and T. Wachtendorf (2016). "Pets and Evacuation: An Ongoing Challenge in Disasters." *Journal of Homeland Security and Emergency Management*, 13 (4).

Glassey, S. (2018). "Did Harvey Learn From Katrina? Initial Observations of the Response to Companion Animals During Hurricane Harvey." *Animals*, 8 (4): 47.

Gunter, L. M., R. J. Gilchrist, E. M. Blade, J. L. Reed, L. T. Isernia, R. T. Barber,... and C. D. Wynne (2022). "Emergency Fostering of Dogs from Animal Shelters During the COVID-19 Pandemic: Shelter Practices, Foster Caregiver Engagement,and Dog Outcomes." *Frontiers in Veterinary Science*, 9: 862590.

Heath, S. and R. Linnabary (2015). "Challenges of Managing Animals in Disasters in the U.S." *Animals*, 5 (2): 173–92.

Iacobucci, D. (2019). "A Challenge Within Macromarketing: Global Disasters." *Journal of Macromarketing*, 39 (3): 334–8.

Irvine, L. (2009). *Filling the Ark: Animal Welfare in Disasters*. Philadelphia: Temple University Press.

Kajiwara, H. (2020). *Surviving with Companion Animals in Japan: Life After a Tsunami and Nuclear Disaster*. Cham: Palgrave Macmillan.

Pimentel, L., A. May, J. Iskander, R. Banks, and J. Gibbins (2021). "Assessment of One Health Knowledge, Animal Welfare Implications, and Emergency Preparedness Considerations for Effective Public Health Response." Public Health Reports, 137 (5): 964–71.

Spain, C., R. Green, L. Davis, G. Miller, and S. Britt (2017). "The National Capabilities for Animal Response in Emergencies (Ncare) Study: An Assessment of US States and Counties." *Journal of Homeland Security and Emergency Management*, 14 (3).

Tanaka, A., J. Saeki, S. Hayama, and P. Kass (2019). "Effect of Pets On Human Behavior and Stress in Disaster." *Frontiers in Veterinary Science*, 6: 113.

Wijnker, J., S. Leinenga, and L. Lipman (2018). "Farm Evacuation Coefficient: A Novel Indicator in Veterinary Crisis Management." Veterinary Record, 182 (16): 462–2.

NEGLECT

Animal neglect, particularly prevalent during economic hardship, is influenced by various socioeconomic dynamics and legal frameworks. In the United States, animal welfare laws aim to prevent such neglect, but their efficacy varies significantly depending on the social and economic circumstances faced by animal owners. Animal neglect is typically viewed within the context of animal cruelty and is often included in the same legislation, but the difference is that cruelty is active, involving the application of harm to an animal, while neglect is passive and involves the lack of care rather than intentional harm.

Research indicates a connection between socioeconomic status and animal welfare practices. Individuals from lower socioeconomic backgrounds are more likely to engage in neglectful practices, potentially due to financial constraints that hinder proper pet care (Shih, Paterson, and Phillips, 2019). Studies show that economic hardships often lead to the relinquishment of pets, with financial challenges cited as a primary reason for this trend (Rah and Choi, 2023; Coe et al., 2014). This relationship underscores how socioeconomic factors diminish the ability of owners to provide adequate care and may contribute to a decline in general attitudes toward animal welfare (Ly et al., 2021).

Laws designed to protect animals, such as the Animal Welfare Act in the United States, are foundational but often fail to address the underlying causes of neglect, particularly in economically disadvantaged communities. Where enforcement occurs, it has been observed that socioeconomic vulnerabilities, such as financial instability and lack of access to veterinary care, increase the risk of neglect (Ly et al., 2021; McDowall et al., 2023). Investigations into companion animal abuse suggest that while socioeconomic challenges do not exempt owners from responsibility, they complicate their circumstances, necessitating a nuanced understanding of animal welfare interventions (Monsalve et al., 2023).

Furthermore, specific types of animals, such as horses, are particularly vulnerable during economic downturns; owners may struggle to afford their care, resulting in neglect or abandonment (Pritchard, Upjohn, and Hirson, 2018; Seck et al., 2024). The compromises made in animal welfare issues reflect not only legal shortcomings but also broader societal attitudes influenced by economic realities (Pritchard, Upjohn, and Hirson, 2018). Effective interventions, therefore, require not just legal compliance but also community-based strategies that tackle structural barriers to accessing veterinary services (Hawes et al., 2021; Dyer and Milot, 2019).

Addressing community welfare is crucial for improving animal care standards. Programs that successfully mitigate structural barriers hindering access to pet care services can enhance community perceptions of animal welfare and reduce rates of neglect (Hawes et al., 2021). Education also plays a vital role; initiatives aimed at raising awareness of responsible pet ownership can foster a culture of care that transcends economic limitations (McDowall et al., 2023; Dyer and Milot, 2019).

While legal frameworks exist to combat animal neglect, including for equines, their effectiveness in the United States, as in other countries, is significantly undermined by socioeconomic conditions. Economic hardship serves as a catalyst for neglect, highlighting the need for a multifaceted approach that incorporates both legal and educational strategies tailored to the socioeconomic context of communities.

See also: Cruelty; Socioeconomic Class, Pets And

References and Further Reading

Coe, J., I. Young, K. Lambert, L. Dysart, L. Borden, and A. Rajić (2014). "A Scoping Review of Published Research on the Relinquishment of Companion Animals." *Journal of Applied Animal Welfare Science*, 17 (3): 253–73.

Dyer, J. and L. Milot (2019). "Social Vulnerability Assessment of Dog Intake Location Data as a Planning Tool for Community Health Program Development: A Case Study in Athens-Clarke County, GA, 2014–2016." *Plos One*, 14 (12): e0225282.

Hawes, S., T. Hupe, J. Winczewski, K. Elting, A. Arrington, S. Newbury, and K. Morris (2021). "Measuring Changes in Perceptions of Access to Pet Support Care in Underserved Communities." *Frontiers in Veterinary Science*, 8.

Ly, L., E. Gordon, and A. Protopopova (2021). "Exploring the Relationship Between Human Social Deprivation and Animal Surrender to Shelters in British Columbia, Canada." *Frontiers in Veterinary Science*, 8.

McDowall, S., S. Hazel, C. Chittleborough, A. Hamilton-Bruce, R. Stuckey, and T. Howell (2023). "The Impact of the Social Determinants of Human Health on Companion Animal Welfare." *Animals*, 13 (6): 1113.

Monsalve, S., J. Hammerschmidt, M. Ribeiro, M. Caleme, M. Marconcin, S., G. Filius, and R. Garcia (2023). "A One Welfare Approach to Identify Socioeconomic Vulnerability in Families During Investigations into Companion Animal Abuse in Pinhais, Brazil." *Animal Welfare*, 32.

Pritchard, J., M. Upjohn, and T. Hirson (2018). "Improving Working Equine Welfare in 'Hard-Win' Situations, Where Gains are Difficult, Expensive or Marginal." *Plos One*, 13 (2): e0191950.

Rah, H. and S. Choi (2023). "Are Veterinary Costs and Socioeconomic Status Risk Factors for Companion Animal Relinquishment in the Republic of Korea?" *Animals*, 13 (21): 3406.

Seck, M., G. Carder, J. Wathan, M. Randau, K. Fletcher, and L. Proops (2024). "Working Horse Welfare in Senegal is Linked to Owner's Socioeconomic Status, their Attitudes and Belief in Horse Sentience." *Plos One*, 19 (10): e0309149.

Shih, H., M. Paterson, and C. Phillips (2019). "Socioeconomic Influences on Reports of Canine Welfare Concerns to the Royal Society for the Prevention of Cruelty to Animals (RSPCA) in Queensland, Australia." *Animals*, 9 (10): 711.

P

PET DEATH INDUSTRY

The pet death industry has evolved significantly in recent years, reflecting changes in societal attitudes toward pet ownership and the acknowledgment of the emotional bond between pets and their human companions. This transformation is demonstrated through the emergence of new practices and methods for dealing with the aftermath of a pet's death, particularly concerning the grief experienced by pet owners.

Research indicates that the grief experienced after the loss of a pet parallels that of losing a human loved one, highlighting the profound emotional attachment that many owners develop toward their animals (Morris, 2012; Packman, Carmack, and Ronen, 2012). This emotional response has led to a demand for more comprehensive end-of-life care options, such as veterinary hospice services and pet cemeteries, which provide families with a range of respectful and dignified ways to memorialize their pets (Bishop et al., 2016). Online platforms and support groups have also arisen, allowing grieving pet owners to share their experiences and coping mechanisms, which are crucial for managing disenfranchised grief—a term used to describe grief that is not recognized or validated by societal norms (Packman et al., 2014).

In addition to the emotional aspects, the logistics of pet death care are being redefined. The availability of cremation services has risen, with many providers emphasizing respect and compassion during the process, similar to human funeral services (Chur-Hansen, 2011). Moreover, guidelines from veterinary associations aim to educate pet owners about their options and promote discussions regarding the ethical considerations around euthanasia and end-of-life care (Bishop et al., 2016). As awareness of human–animal bonds grows, the need to integrate such services into general healthcare discussions becomes evident, advocating for acknowledgment of pet loss in societal bereavement frameworks (Wong et al., 2015).

Furthermore, the industry is increasingly recognizing the importance of addressing the needs of marginalized groups, including the homeless population who often find companionship in pets (Rhoades, Winetrobe, and Rice, 2014; Labrecque and Walsh, 2011). Studies show that pets offer significant psychological benefits, serving as a source of comfort for individuals facing homelessness or other challenging life circumstances (Schmitz, Carlisle, and Tabler, 2021). However, these individuals also face additional vulnerabilities related to pet loss, as accessing services that accommodate pets in shelters can significantly impact their willingness to seek help (Garland-Lewis et al., 2024).

Today, the pet death industry is undergoing a transformative phase influenced by a greater understanding of the emotional and psychological significance of pets in human lives. This evolution is mirrored by the expansion of services that cater not only to the logistical needs associated with a pet's death but also address the grief and emotional stress that often accompany this loss. As the societal context continues to change, the industry must adapt to better support grieving pet owners, particularly among underserved populations, emphasizing the importance of inclusivity and acknowledgment of the profound bonds shared between humans and their pets.

See also: Attachment; Bond, Human-Animal; Cemeteries, Pet; Grief and Pet Loss; Healthcare Spending on Pets; Veterinary Care

References and Further Reading:

Bishop, G., K. Cooney, S. Cox, R. Downing, K. Mitchener, A. Shanan, N. Soares, B. Stevens, and T. Wynn (2016). "2016 AAHA/IAAHPC End-of-Life Care Guidelines." *Journal of the American Animal Hospital Association*, 52 (6): 341–56.

Chur-Hansen, A. (2011). "Cremation Services Upon the Death of A Companion Animal: Views of Service Providers and Service Users." *Society and Animals*, 19 (3): 248–60.

Garland-Lewis, G., E. Scott, V. Ramirez, and P. Rabinowitz (2024). "Shifting the Focus: A Photovoice Exploration of the Benefits and Barriers of Having a Pet While Experiencing Homelessness." *Plos One*, 19 (3): e0295588.

Labrecque, J. and C. Walsh (2011). "Homeless Women's Voices On Incorporating Companion Animals into Shelter Services." *Anthrozoös*, 24 (1): 79–95.

Morris, P. (2012). "Managing Pet Owners' Guilt and Grief in Veterinary Euthanasia Encounters." *Journal of Contemporary Ethnography*, 41 (3): 337–65.

Packman, W., B. Carmack, and R. Ronen (2012). "Therapeutic Implications of Continuing Bonds Expressions Following the Death of a Pet." *Omega - Journal of Death and Dying*, 64 (4): 335–56.

Packman, W., B. Carmack, R. Katz, F. Carlos, N. Field, and C. Landers (2014). "Online Survey as Empathic Bridging for the Disenfranchised Grief of Pet Loss." *Omega - Journal of Death and Dying*, 69 (4): 333–56.

Rhoades, H., H. Winetrobe, and E. Rice (2014). "Pet Ownership Among Homeless Youth: Associations with Mental Health, Service Utilization and Housing Status." *Child Psychiatry & Human Development*, 46 (2): 237–44.

Schmitz, R., Z. Carlisle, and J. Tabler (2021). "'Companion, Friend, Four-Legged Fluff Ball': The Power of Pets in the Lives of LGBTQ+ Young People Experiencing Homelessness." *Sexualities*, 25 (5–6): 694–716.

Wong, P., K. Lau, L. Liu, G. Yuen, and P. Wing-Lok (2015). "Beyond Recovery." *Omega - Journal of Death and Dying*, 75 (2): 103–23.

PET FOOD

The evolution of pet food as a commercial product has undergone significant transformations since its inception. Originally, pets consumed the same food as their owners, primarily scraps and leftovers. With the rise of the pet industry in the late nineteenth century, particularly in Western countries, specialized pet foods began to emerge. This development was driven by the growing perception of pets as family members and companions, rather than mere animals.

Research indicates that modern pet owners increasingly view their pets as integral to their family structure, leading to a demand for quality pet foods that align with this newfound familial status (Vinassa et al., 2020; Kumcu and Woolverton, 2014; Buff et al., 2014). Pet food companies have responded by marketing products that mimic human food trends, highlighting aspects like "natural" and "premium" ingredients, which resonate strongly with pet owners looking to provide optimal nutrition (Vinassa et al., 2020; Prata, 2022; Samant et al., 2021; Meeker and Meisinger, 2015).

This shift toward specialized pet foods has played a critical role in expanding pet ownership. The burgeoning pet food industry, now valued at billions globally, not only reflects but also reinforces the attachment people have to pets (Okin, 2017; Kumar et al., 2024). In fact, pet ownership has been associated with enhanced social well-being, especially among children, indicating a reciprocal relationship where pets contribute positively to human health and happiness (Purewal et al., 2017). The expansion of pet ownership has also driven advancements in pet food innovation, including the introduction of raw food diets and alternative products aimed at catering to specific dietary preferences and health needs of pets (Swanson et al., 2013; (Dhakal et al., 2024).

In recent years, there has been a notable trend toward the "humanization" of pet food, where ingredients and formulations align more closely with human dietary preferences (Kumcu and Woolverton, 2014; Dhakal et al., 2024). This trend has included a rise in vegetarian and vegan diets, which are the subject of a great deal of controversy, especially for cats, and raw food diets, which proponents argue provide more natural nutrition for pets, although concerns regarding food safety and pathogen risks, such as Salmonella, remain crucial (Dhakal et al., 2024). Additionally, the pet food industry has seen the emergence of pet food delivery services, reflecting the convenience-driven consumer culture and the demand for personalized nutrition plans for pets (Kumar and Sharma, 2024). These services often offer tailored meal plans based on a pet's specific needs, which further illustrates the shift toward customized pet nutrition.

Overall, the landscape of pet food continues to evolve, with increasing attention paid to sustainability, health implications, and the role of consumer preferences in shaping product offerings. The pet food industry is expected to continue leveraging human nutritional trends, to meet the evolving needs and interests of pet owners (Kumar and Sharma, 2024; Buchanan et al., 2011).

In summary, pet food has developed from simple, repetitive feeding practices into a complex commercial industry that caters to consumers' desires for quality, safety, and the nutritional needs of their pets. As the industry advances, it is essential to balance product quality with environmental considerations and public health implications.

See also: Family, Pets As; Pet Industry; Socioeconomic Class, Pets And; Veterinary Care

References and Further Reading

Buchanan, R., R. Baker, A. Charlton, J. Riviere, and R. Standaert (2011). "Pet Food Safety: A Shared Concern." *British Journal of Nutrition*, 106 (S1): S78–S84.

Buff, P., R. Carter, J. Bauer, and J. Kersey (2014). "Natural Pet Food: A Review of Natural Diets and their Impact on Canine and Feline Physiology." *Journal of Animal Science*, 92 (9): 3781–91.

Dhakal, J., L. Cancio, A. Deliephan, B. Chaves, and S. Tubene (2024). "Salmonella Presence and Risk Mitigation in Pet Foods: A Growing Challenge with Implications for Human Health." *Comprehensive Reviews in Food Science and Food Safety*, 23 (6).

Kumar, R., and A. Sharma (2024). "Review of Pet Food Packaging in the US Market: Future Direction Towards Innovation and Sustainability." *Annual Research & Review in Biology*, 39 (6): 16–30.

Kumcu, A., and A. Woolverton (2014). "Feeding Fido: Changing Consumer Food Preferences Bring Pets to the Table." *Journal of Food Products Marketing*, 21 (2): 213–30.

Meeker, D., and J. Meisinger (2015). "Companion Animals Symposium: Rendered Ingredients Significantly Influence Sustainability, Quality, and Safety of Pet Food." *Journal of Animal Science*, 93 (3): 835.

Okin, G. (2017). "Environmental Impacts of Food Consumption by Dogs and Cats." *Plos One*, 12 (8): e0181301.

Prata, J. (2022). "Survey of Pet Owner Attitudes on Diet Choices and Feeding Practices for their Pets in Portugal." *Animals*, 12 (20): 2775.

Purewal, R., R. Christley, K. Kordas, C. Joinson, K. Meints, N. Gee, and C. Westgarth (2017). "Companion Animals and Child/Adolescent Development: A Systematic Review of the Evidence." *International Journal of Environmental Research and Public Health*, 14 (3): 234.

Samant, S., P. Crandall, S. Arroyo, and H. Seo (2021). "Dry Pet Food Flavor Enhancers and their Impact on Palatability: A Review." *Foods*, 10 (11): 2599.

Swanson, K., R. Carter, T. Yount, J. Aretz, and P. Buff (2013). "Nutritional Sustainability of Pet Foods." *Advances in Nutrition*, 4 (2): 141–50.

Vinassa, M., D. Vergnano, E. Valle, M. Giribaldi, J. Nery, L. Prola, D. Bergero and A. Schiavone (2020). "Profiling Italian Cat and Dog Owners' Perceptions of Pet Food Quality Traits." *BMC Veterinary Research*, 16 (1): 131.

PET INDUSTRY

The pet industry has experienced significant growth over the last decade, largely driven by an increase in pet ownership and an evolving relationship between humans and their pets. Each component of this industry, including veterinary care, food, travel, boarding, daycare, clothing, and luxury products, collectively contributes to the economic dynamics that underscore this burgeoning market.

The pet food segment, as one of the largest components, is experiencing rapid expansion, valued at approximately $42 billion within the United States alone. This growth is, in part, driven by an increasing pet population accelerated by the pandemic, which has resulted in a larger pet ownership base in the United States than any other country globally (White, 2023; Moses et al., 2024). This increased ownership corresponds with a rising consumer tendency toward humanized pet food, as owners seek products that not only cater to their pet's health but also reflect their values concerning sustainability and transparency (Kumar and Sharma, 2024). Furthermore, there exists a growing awareness among consumers regarding the environmental impact of pet food production, prompting the industry to innovate and address concerns through sustainable practices Okin (2017) and alternative protein sources (Kumar and Goswami, 2024).

In addition to food, pet travel has emerged as a notable component of the industry, with a growing segment of tourists opting to travel with their pets. This niche market is recognized for its economic potential, as accommodations increasingly adapt to meet the needs of pet owners (Buhalis and Chan, 2023). Empirical research indicates that pets are perceived as family members, leading to heightened expectations for pet-friendly services (Bernardi, Edelheim, and Crossley, 2022). As tourist accommodations evolve, their willingness to welcome pets can significantly enhance their marketability, as a considerable portion of pet owners are prepared to spend generously on comfortable arrangements for their pets (Larrea and Smith, 2024).

The services sector within the pet industry, including boarding and daycare facilities, has similarly expanded to cater to pet owners' needs during travel. The incorporation of such services plays a crucial role in supporting mobility for tourists while ensuring their pets are cared for. The rationale for utilizing these facilities stems from pet owners' desire for peace of mind during travel, which is reflected in their spending behaviors on premium pet amenities (Bernardi, Edelheim, and Crossley, 2022).

Furthermore, the rise of luxury pet products signifies a deeper societal trend where animals are increasingly viewed as companions but also as status symbols. This has led to the proliferation of high-end services and products, including pet attire and accessory markets, where personalized and luxury options are becoming mainstream. Consequently, the pet industry is no longer just a niche market; it encompasses a broad array of sectors catering to ever-evolving consumer demands.

In summary, the pet industry exhibits robust growth due to various interrelated components, from food and travel to luxury products and services. This growth trajectory reflects a comprehensive trend toward greater integration of pets into family dynamics, prompting both ethical considerations and economic opportunities across multiple market segments.

See also: Family, Pets As; Pet Food; Veterinary Care

References and Further Reading

Bernardi, C., J. Edelheim, and É. Crossley (2022). "The Role of Catteries and Boarding Kennels in Enabling Tourist Mobility." *Tourism Culture & Communication*, 22 (4): 387–92.

Buhalis, D., and J. Chan (2023). "Traveling with Pets: Designing Hospitality Services for Pet Owners/parents and Hotel Guests." *International Journal of Contemporary Hospitality Management*, 35 (12): 4217–37.

Kumar, R., and M. Goswami (2024). "Harnessing Poultry Slaughter Waste for Sustainable Pet Nutrition: A Catalyst for Growth in the Pet Food Industry." *Journal of Dairy Veterinary & Animal Research*, 13 (1), 31–3.

Kumar, R. and A. Sharma (2024). "Review of Pet Food Packaging in the US Market: Future Direction Towards Innovation and Sustainability." *Annual Research & Review in Biology*, 39 (6): 16–30.

Larrea, G. and S. Smith (2024). "Bear E. Drooley: Achieving Business Growth with the Help of Pets-in-Residence and Resort Mascots." *Journal of Hospitality & Tourism Cases*, 13 (1): 3–11.

Moses, I., Á. Ribeiro, T. Valiatti, F. Santos, R. Cayô, and A. Gales (2024). "First Detection of International High-risk Blakpc-2-Harbouring Escherichia Coli Pandemic Lineage ST648 in Pet Food Packages." *Transboundary and Emerging Diseases*, 2024 (1).

Okin, G. (2017). "Environmental Impacts Of Food Consumption by Dogs and Cats." *Plos One*, 12 (8): e0181301.

White, B. (2023). "Insights-driven Development of Humanized Foods for Pets." *Meat and Muscle Biology*, 6 (3): 14397.

PIGS

Pigs have increasingly become popular as pets, particularly in the form of miniature pigs like pot-bellied pigs. The rise of pot-bellied pigs in the 1980s served as a catalyst for the popularity of these smaller breeds to emerge as companions in households across the Western world. Initially prized for their compact size and perceived intelligence and friendliness, miniature pigs quickly gained traction, leading to an increase in breeding operations catering to this demand. In contrast to traditional livestock, which face tight regulations and practical farming constraints, pot-bellied pigs presented an appealing alternative that fit the trend toward unconventional pets. Over the past few decades, this niche has grown, leading to a lack of standardization in breeding practices that has inevitably raised welfare concerns (Godyń, Nowicki, and Herbut, 2019).

While many owners are drawn to the idea of having a pet pig due to their intelligence and social nature, significant welfare issues have arisen from how these animals are bred and marketed. Many breeders do not adhere to ethical breeding practices, with a focus on profit over the health and well-being of the animal (Grant, Montrose, and Wills, 2017). For example, pigs are often sold under the label "miniature," which can lead to misunderstandings of their size at full maturity. Many of these pigs may grow larger than expected, resulting in abandonment and surrender to shelters when owners can no longer manage them (Godyń, Nowicki, and Herbut, 2019). Moreover, the marketing of these animals frequently highlights their cute aesthetic rather than adequately addressing their comprehensive animal care needs, which include proper housing, socialization, and veterinary care (Grant, Montrose, and Wills, 2017).

Welfare issues also arise from the living conditions that many pet pigs endure. Pigs kept in inadequate spaces, often alone or with little environmental enrichment, may experience significant behavioral problems and health-related issues (Godyń, Nowicki, and Herbut, 2019)—especially once the pig grows beyond babyhood. Furthermore, many owners are ill-prepared for the responsibilities that come with pig ownership, including their dietary and space requirements, leading to neglect or inadequate care (Godyń ,Nowicki, and

Pot-bellied pigs have become increasingly popular as pets due to their small size, social nature, and intelligence. (Irina Kozorog/Dreamstime.com)

Herbut, 2019). The need for sufficient space and social engagement is crucial for pigs' mental health; without it, they can develop stress-related behaviors that further complicate their welfare (Godyń ,Nowicki, and Herbut, 2019).

The trend of keeping pigs as pets has raised numerous welfare concerns due to irresponsible breeding and marketing practices, alongside inadequate care from owners. The popularity of miniature pigs, specifically pot-bellied pigs, emerged from their perceived cuteness and companionship qualities, leading to a rise in demand that has not been matched by adherence to ethical breeding standards. One result of this has been the rise, in the past twenty years, of animal sanctuaries that focus entirely on former pet pigs.

See also: Exotic Pets; Rescue Groups

References and Further Reading

Godyń, D., J. Nowicki, and P. Herbut (2019). "Effects of Environmental Enrichment on Pig Welfare—A Review." *Animals*, 9 (6): 383.

Grant, R., V. Montrose, and A. Wills (2017). "Exnotic: Should we be Keeping Exotic Pets?" *Animals*, 7 (6): 47.

POCKET PETS

The concept of "pocket pets" or "children's pets" encompasses a variety of small animals typically kept as pets, often for children. This includes species such as hamsters, guinea pigs, rabbits, and small reptiles, which are often perceived as simpler to care for compared to dogs or cats. However, there are significant welfare concerns associated with these pets due to misconceptions about their care requirements and the responsibilities involved in pet ownership.

A critical area of concern is the attitude of pet owners toward small animals. For instance, studies indicate that pet rabbits may be regarded as inexpensive and replaceable, which can lead to inadequate veterinary care (Rioja-Lang et al., 2019). This lack of care may result in relinquishment or in reduced life expectancy for these animals. Furthermore, children often bear the brunt of care responsibilities, which may not be appropriate for their maturity level, leading to compromised welfare outcomes for the pets involved (Muldoon, Williams, Lawrence, 2014). This dynamic raises alarm bells regarding the appropriate education of both parents and children about animal welfare and the true nature of the responsibilities entailed in pet ownership.

Moreover, the welfare of exotic pets, including reptiles, has been brought into focus given the unique challenges presented by their specialized care needs, which are frequently misunderstood by owners (Goins and Hanlon, 2021). Many believe these pets require less interaction or care, potentially resulting in distress and poor welfare outcomes. This misconception also extends to the broader category of small mammals, where inappropriate housing and care standards prevail. Research emphasizes that inadequate living environments, such as undersized cages, significantly jeopardize animal welfare by limiting their natural behaviors and compromising their health (Bläske et al., 2022).

Socioeconomic factors also play a notable role in pet welfare. Communities characterized as animal welfare deserts—regions lacking access to veterinary care and animal welfare resources—exhibit stark disparities in animal welfare outcomes (Reese and Li, 2023). Residents in such areas may face additional challenges, contributing to higher rates of pet abandonment (Clark, Gruffydd-Jones, Murray, 2012). Thus, educational initiatives that promote awareness of proper care and available resources are vital for improving outcomes for pocket pets in under-resourced areas.

The attachment of children to pets may foster empathy and compassionate behavior but also requires guidance to ensure that such relationships are healthy and beneficial for both the pets and the children involved (Hawkins and Williams, 2017). Understanding these influences on child development highlights the necessity for structured educational programs aimed at teaching children responsible pet ownership in conjunction with their emotional development.

See also: Children and Pets; Classroom Pets; Rabbits; Rodents; Socioeconomic Class, Pets And

References and Further Reading

Bläske, A., A. Schwarzer, M. Ebner, H. Gerbig, S. Reese, M. Erhard, and A. Wöhr (2022). "Evaluation of Small Mammal Pet Supplies Offered in German Retail under Animal Welfare Aspects." *Plos One*, 17 (2): e0262658.

Clark, C., T. Gruffydd-Jones, and J. Murray (2012). "Number of Cats and Dogs in UK Welfare Organisations." *Veterinary Record*, 170 (19): 493–3.

Goins, M. and A. Hanlon (2021). "Exotic Pets in Ireland: 1. Prevalence of Ownership and Access to Veterinary Services." *Irish Veterinary Journal*, 74 (1).

Hawkins, R. and J. Williams (2017). "Childhood Attachment to Pets: Associations Between Pet Attachment, Attitudes to Animals, Compassion, and Humane Behaviour." *International Journal of Environmental Research and Public Health*, 14 (5): 490.

Muldoon, J., J. Williams, and A. Lawrence (2014). "'Mum Cleaned It and I Just Played With It': Children's Perceptions of Their Roles and Responsibilities in the Care of Family Pets." *Childhood*, 22 (2): 201–16.

Reese, L. and X. Li (2023). "Animal Welfare Deserts: Human and Nonhuman Animal Inequities." *Frontiers in Veterinary Science*, 10.

Rioja-Lang, F., H. Bacon, M. Connor, and C. Dwyer (2019). "Rabbit Welfare: Determining Priority Welfare Issues for Pet Rabbits Using a Modified Delphi Method." *Veterinary Record Open*, 6 (1).

PRESIDENTIAL PETS

The presence of pets in the lives of American presidents has been a notable aspect of their public personas and reflects broader cultural attitudes toward animals throughout the history of the United States.

Approximately thirty US presidents have owned pets during their time in office, with dogs being the most common companion animals. Noteworthy examples include President Franklin D. Roosevelt's Scottish Terrier, Fala, and President Richard Nixon's Cocker Spaniel, Checkers. Both pets played significant roles in shaping the public's perception of their owners; for instance, Checkers was the most crucial element in Nixon's famous 1952 "Checkers" speech, which humanized him during a time of political contention, thereby bolstering his public favor and saving his political career. Similarly, Fala's presence underscored Roosevelt's approachable personality, contributing to his affectionate image among the electorate, especially during the Second World War.

Richard Nixon and family with their dog Checkers at a New Jersey beach, 1953. Nixon's 1952 "Checkers" speech helped humanize him in the eyes of the public, saving his political career. (National Archives)

The types of pets owned by presidents have varied, comprising predominantly dogs but also including cats, birds, and even more exotic pets such as reptiles. Notably, some presidents, including William Henry Harrison, James K. Polk, and Calvin Coolidge, either had no pets or did not maintain the tradition of pet ownership during their presidencies. This absence can sometimes lead to a perception of aloofness or detachment from the public, contrasting with those presidents who actively displayed their affection for their pets.

Richard Nixon and family with their dog Checkers at a New Jersey beach, 1953. Nixon's 1952 "Checkers" speech helped humanize him in the eyes of the public, saving his political career. (National Archives)

In modern political discourse, the relationship between presidents and their pets continues to be relevant; for instance, Donald Trump's lack of any pets during his presidency raised eyebrows and sparked conversations about the implications of not owning a pet. Observers have speculated that his lack of a pet might hinder his ability to connect with the public in a way that feels personable and relatable. Moreover, the presence of pets in political imagery and speeches often serves to soften a politician's image and enhance their relatability, a tactic that Trump opted out of, potentially impacting his public reception.

In regard to public sentiment, pets symbolize warmth and family values, which are critical components in shaping a political figure's image in the eyes of the electorate. This phenomenon can be seen in the successful campaigns of various presidents, where their partnerships with pets played a key role in endearing them to voters. Pets not only provide companionship but can also serve as strategic assets in political communication, establishing trust and familiarity that might otherwise be absent in the political arena.

In summary, presidential pets not only reflect personal choices but also play pivotal roles in shaping public perception, establishing emotional connections, and aiding presidential image management.

See also: Diplomacy and Pets; Family, Pets As

References and Further Reading

Blankfield, B. (2017). "Political Animals: Prosopopoeia in the 1944 Presidential Election." *Rhetoric Society Quarterly*, 47 (4): 335–58.

Hoewe, J., D. Zulli, P. Arceneaux, and A. Lusvardi (2025). "Political Pets: The Pawsitive Effect of Dogs in Politicians' Social Media Posts." *Journal of Information Technology & Politics*, 1–13.

Maltzman, F., J. H. Lebovic, E. N. Saunders, and E. Furth (2012). "Unleashing Presidential Power: The Politics of Pets in the White House." *PS: Political Science & Politics*, 45 (3): 395–400.

Mutz, D. C. (2010). "The Dog That Didn't Bark: The Role of Canines in the 2008 Campaign." *PS: Political Science & Politics*, 43 (4): 707–12.

PUPPY MILLS

The concept of "puppy mill" has gained significant recognition among the American public, largely due to the advocacy efforts of numerous national and local animal welfare organizations, as well as investigations by major media outlets, including national newspapers, magazines, and television programs such as "Dateline," "Hard Copy," and Atlantic Monthly.

Puppy mills are defined as extensive breeding enterprises, often lacking proper licensing, and generally located in rural regions. They are characterized by unsanitary conditions in which a large number of puppies are bred, predominantly with the aim of selling them to pet retailers through intermediaries. Notably, those facilities that sell directly to consumers are not mandated to obtain licenses. In these establishments, dogs typically endure their entire lives confined to wire cages or limited dog runs. They are subjected to repeated breeding cycles, producing successive litters until their health is depleted. Consequently, the puppies originating from these mills frequently suffer from health issues stemming from inadequate medical care and substandard living conditions.

An inherent challenge with the term "puppy mill" lies in the fact that breeders seldom classify their operations using this label, and pet stores typically deny sourcing their puppies from such environments. According to the United States Department of Agriculture (USDA) regulations, any commercial breeding site generating over $500 annually, possessing more than three female breeding dogs, and engaging in wholesale transactions must hold a USDA license and is subject to inspections by the Animal and Plant Health Inspection Service (APHIS). Nevertheless, there is a significant shortfall of USDA inspectors, with fewer than one hundred available to oversee compliance across all licensed breeders (approximately 3,500), dealers (approximately 1,100), and transporters, not to mention the thousands of research facilities and animal exhibitions in operation, accounting for more than 9,000 total establishments.

While the animals in licensed facilities must adhere to standards set forth in the Animal Welfare Act (AWA), the AWA does not delineate what constitutes a puppy mill, complicating the ability of the public to ascertain the existence of such facilities and the origins of puppies sold in local pet stores. Commercial breeders—designated as Class "A" dealers by the USDA—breed dogs, cats, rabbits, and other animals for wholesale purposes. These facilities may vary in scale, with some housing over a thousand animals and utilizing single or multiple breeds, and even various species. Commercial breeding establishments generating over $500 annually are required to obtain USDA licensing and to adhere to the AWA, meeting specific criteria related to space, shelter, nutrition, water, sanitation, and veterinary care.

Wholesale breeders typically distribute their animals to brokers or dealers, who then facilitate the distribution to pet stores (or laboratories) that ultimately retail the animals. The cost of puppies at the wholesale level can range from $50 to $200, while those same puppies may be sold in pet stores for prices between $500 and $2,000. For rabbits, wholesale prices may start as low as $3, with retail prices ranging from $25 to $30.

The conditions prevalent in large commercial breeding facilities can vary considerably from acceptable to appalling, especially if one considers the confinement of adult animals to cages throughout their lives as "acceptable." At the lower end of the spectrum, there exist countless documented cases of dogs or rabbits living in appalling conditions, characterized by filthy living environments, matted fur, sores on limbs, and various health disorders (e.g., parvovirus, kennel cough, hip dysplasia, deafness, cataracts, epilepsy, glaucoma, and oral diseases). These animals can endure extreme temperature fluctuations, from oppressive summer heat to frigid winter climates. Instances have been reported where dogs are subjected to debarking procedures, and routine breeding occurs with little regard for the well-being of the animals involved.

Female dogs are generally bred twice yearly, producing litters until approximately six or seven years of age; however, in poor conditions, they may cease producing viable litters at an earlier age. Conversely, some breeders continue breeding their dogs until the animals ultimately succumb to health declines. Cat breeders, on average, produce one litter per year, but due to their reproductive cycle, cats may be bred more frequently, especially in profit-driven operations, leading to increased litter frequencies. Once the reproductive viability diminishes, these animals are often discarded.

Puppy mills have become a critical concern within the realm of animal welfare, particularly with respect to the regulatory oversight of dog breeding practices and associated ethical ramifications. The AWA serves as the primary federal legislation intended to safeguard the humane treatment of animals through regulatory frameworks governing breeding and care conditions. The USDA is tasked with the administration and inspection of licensed breeding facilities under the AWA; however, criticisms have emerged regarding the enforcement of these regulations, as many violations go unresolved or receive inadequate penalties (Novotný, 2020). The enforcement of existing animal welfare laws remains inconsistent.

Research corroborates the existence of numerous puppy mills operating with minimal oversight, particularly those that circumvent regulations or exploit legislative loopholes (Novotný, 2020). Reports of violations highlight the urgent necessity for more stringent penalties and enhanced enforcement mechanisms under the AWA to deter unethical breeding practices. The USDA faces numerous challenges in effectively monitoring these facilities, which permits many puppy mills to continue operations with inadequate adherence to humane standards. Statistical evidence demonstrates an upward trend in the acquisition of puppies from unethical sources, a situation aggravated by spikes in demand, particularly during the Covid-19 pandemic (Brand et al., 2022; Packer et al., 2023).

Animal advocates and researchers have revealed that many puppies sold in pet stores receive prophylactic antibiotic treatments—often a consequence of inadequate breeding practices. The pandemic-induced shifts in puppy purchasing behaviors led many prospective pet owners to acquire animals without thoroughly vetting breeders, further compounding the demand cycle for puppies from mills (Merritt et al., 2025; Packer et al., 2021). An observable increase in impulsive pet purchases has underscored the public's prioritization of immediate acquisition over responsible breeding practices (Packer et al., 2021).

In conclusion, while the AWA constitutes a critical regulatory framework for safeguarding animal welfare in breeding operations, the efficacy of its enforcement by the USDA remains in question. The emergence of the "Pandemic Puppies" phenomenon exemplifies how sociocultural factors can drive puppy demand, resulting in detrimental implications for animal welfare. Addressing these challenges necessitates not only improved regulatory oversight but also public education focusing on ethical pet ownership and the ramifications of supporting puppy mills (Packer et al., 2023; Brand et al., 2022).

See also: Covid-19; Dogs; Laws and Regulations; Pet Industry; Selective Breeding; United States, Pets In

References and Further Reading

Brand, C., D. O'Neill, Z. Belshaw, C. Pegram, K. Stevens, and R. Packer (2022). "Pandemic Puppies: Demographic Characteristics, Health and Early Life Experiences of Puppies Acquired During the 2020 Phase of the COVID-19 Pandemic in the UK. *Animals*, 12 (5): 629.

Costa, A., T. Nielsen, E. Grande, J. TUKe, and S. Hazel (2020). "Regulatory Compliance in Online Dog Advertisements in Australia." *Animals*, 10 (3): 425.

Merritt, B., D. O'Neill, C. Brand, Z. Belshaw, F. Dale, C. Pegram, and Packer, R. (2025). "Lasting Lockdown Love? Problem Behaviour and Pandemic and Non-pandemic Related Risk Factors Influencing the Owner-Dog Relationship in a UK Cohort of Dogs Reaching Early Adulthood." *Plos One*, 20 (2): e0316166.

Novotný, L. (2020). "How to Fight Puppy Mills: Toughening the Sentences for Animal Abuse in the Post-Communist Region." *Animals*, 10 (6): 1020.

Packer, R., C. Brand, Z. Belshaw, C. Pegram, F. Dale, K. Stevens, and O'Neill, D. (2023). "Is UK Puppy Purchasing Suffering a Long Covid Effect? Ongoing Negative Impacts of the COVID-19 Pandemic upon Puppy Purchase Motivations and Behaviours in 2021." *Animals*, 13 (13): 2186.

Packer, R., C. Brand, Z. Belshaw, C. Pegram, K. Stevens, and D. O'Neill (2021). "Pandemic Puppies: Characterising Motivations and Behaviours of UK Owners who Purchased Puppies During the 2020 COVID-19 Pandemic." *Animals*, 11 (9): 2500.

R

RABBITS

The Romans first controlled wild rabbits in "leporaria" (or rabbit gardens) and used them for meat and fur. True domestication, in which rabbits were bred for size and form, began somewhere between 500 and 1000 CE in French monasteries. The monks initially became interested in keeping wild rabbits because they thought that rabbit fetuses and newborn babies, called laurices, were aquatic and so could be eaten during Lent and other fasts when meat was forbidden (Rogers 1994).

While domesticated pet rabbits share the same genus and species *(Oryctolagus cuniculus)* as their wild counterparts, during domestication some of the rabbits' genes were subtly altered, making domestic rabbits, for example, less timid and more friendly than wild rabbits (Carneiro et al. 2014). Early domestication attempts probably focused on breeding for tameness and size, since meat was the primary concern, but not for color. Breeding for color probably began around the middle of the sixteenth century; records from that time specifically mention piebald, white, and black rabbits (Fox 1994). One hundred years later, monks and others began breeding for fur, and deliberately creating white, black, "blue," and the distinctly collared Dutch rabbit for pelts.

By 1700, seven different colors had been deliberately created (Ibid.). The rediscovery of Mendel's work on dominance and selection and gene combinations fueled rabbit breeding in the late nineteenth and early twentieth centuries. Since then, dozens of rabbit breeds have been created for looks and for fur, meat, or wool. In the last hundred years, rabbits have also been bred to be pets. The exact history of this practice is hard to find. This is partly because the history of rabbit breeding is so heavily dominated by commercial interests and partly because women, who cared for rabbits historically and were probably most likely to develop bonds with them, had fewer opportunities to write than did men. But it seems clear that some rabbits were cherished as personal companions—at least by women—even before the nineteenth century.

Roman rules forbade residents of the British Isles to eat rabbits (as well as hares, chickens, and geese) but Julius Caesar noted that the Britons could "keep these animals for pleasure" (Ezpeleta 1996: 22). Many centuries later, in 1387, William of Wykeham exhorted his nuns to "presume henceforward to bring to church no birds, hounds, rabbits or other frivolous things that promote indiscipline" (Flux 1990a: 9), which implies that the nuns had a personal bond with their small charges.

By the eighteenth century, when Europeans had introduced the first fancy rabbit breeds, people were beginning to see rabbits as more than just food. The English lop, with its extraordinarily long ears, for example, is among the oldest breeds of domestic rabbit and has been found in images and accounts dating back to at least the early eighteenth century (Crook 1985). Pet rabbits—or at least meat rabbits with whom the family felt a bond—also appear in artwork throughout the nineteenth century, illustrating the growth of the human-rabbit bond. Rabbits also became (thanks in part to their long symbolic association with women and children) the ultimate "children's pet" (Davis and DeMello 2003).

During this same period, people were developing and buying more and more fancy breeds of rabbits. The nineteenth century also saw the development of organized rabbit shows and specialty clubs, giving "fancy" rabbit breeders an opportunity to show off their rabbits and further spreading the breeds throughout the world. That at least some of these animals were pets seems clear. Today, more than one hundred breeds of rabbits exist, and thousands of rabbit breeders worldwide cater to the pet market, the show community, and the commercial rabbit industries. But a pet rabbit is not necessarily a family member. For at least two hundred years, pet rabbits have been most commonly kept outdoors, living alone in wood and wire hutches, with little protection from the elements, and little to no companionship or comfort. (These hutches are remnants of the practice of families breeding their own rabbits for meat; rabbits were bred and kept in backyard hutches until the time that they were slaughtered. The fact that pet rabbit keeping emerged directly out of the raising of meat rabbits is one of the most unique facets of the pet rabbit; pet dogs and cats, for example, did not start out as meat animals.) For the most part, rabbits were fed and watered once a day, and occasionally taken out by the children to play, with little other exercise or interaction with humans or other rabbits. In England, where the keeping of pet rabbits first emerged, we have a handful of early and notable pet rabbit lovers. William Cowper, the eighteenth-century English poet, wrote movingly about living with three hares he named Puss, Tiney, and Bess (Milford 1975). Author and illustrator Beatrix Potter was another famous rabbit lover. In 1889, she purchased her first rabbit, a Belgian hare she named Benjamin Bouncer; Peter Piper followed soon afterward. She described Benjamin as "an impudent, cheeky little thing," and both lived with her at least part-time in the house and went with her on walks on a leash. When Peter died at nine, Potter wrote, "Whatever the limitations of his intellect or outward shortcomings of his fur, and his ears and toes, his disposition was uniformly amiable and his temper unfailingly sweet. An affectionate companion and a quiet friend" (Taylor 1986, p. 61).

Ironically, while Potter truly loved Benjamin and Peter, her books, which were based on her rabbits' antics, did not lead to a change in the prevailing attitudes toward rabbits. In fact, rabbits were seen, perhaps more than ever before, as a children's pet, and as such,

Historically, pet rabbits were kept outdoors in wood and wire hutches, but house rabbits have become more common in recent decades. (Allenfive5/Dreamstime.com)

continued to be trivialized. It took another century for rabbits to make the leap from children's pet to true family member.

The publication of Marinell Harriman's *House Rabbit Handbook* in 1985, the first book on living with a "house rabbit," changed the way in which rabbits began to be kept as pets. The book, featuring a dozen stories of people who lived with their rabbits indoors, often without cages, inspired tens of thousands of people to adopt rabbits as house pets and inspired Harriman to found House Rabbit Society (HRS), the first and only international rabbit advocacy group of its kind. In the thirty years since the first edition of the Handbook was released, the idea that rabbits could be treated on the same level as dogs or cats—that they could live indoors as part of a human family, that they could receive veterinary care, that they could be spayed or neutered and live with companions of their own species—has gone from being laughable to being almost commonplace. Thanks to the *Handbook* and House Rabbit Society, tens of thousands of people around the world who now live with house rabbits have taken the rabbit–human relationship to an entirely new level. Rabbits don't just share the homes of house rabbit lovers—they occupy an enormous part of people's lives and have influenced the creation of an entirely new cottage industry that is dedicated not to the selling of house rabbits, but to providing for their needs, from rabbit toy makers to specialized food and hay producers to rabbit condominium manufacturers to healthy snack companies.

According to the American Veterinary Medical Association (AVMA, 2022, U.S. Pet Ownership Statistics), as of 2022, there are 2.2 million pet rabbits in the United States. American Pet Products Association (APPA) also discovered in their 2013–14 survey that 25 percent of all American pet rabbits were adopted from animal shelters or rescue groups, and 14 percent were purchased from pet stores, with the rest coming from other sources. Over the past few years, the APPA has found that the number of rabbits purchased in pet stores as opposed to being adopted decreases, indicating that more people are thinking of rabbits as beings, as opposed to commodities.

In a 2015 study (Welch 2015) of both Canadian and American rabbit owners, the author found that rabbits were acquired because the owners wanted to have the companionship of a rabbit, and/or wanted a companion for their current rabbit, and most rabbits came through adoption rather than purchase. The author also found that the likelihood of getting one's rabbit neutered went down with the number of rabbits owned. For adopters, the study found, saving an animal was the most important factor in adoption, rather than breed, size, or other considerations. The results of this study mark a major change in the ways in which rabbits have long been thought of and treated.

Perhaps because rabbits have been domesticated for a much shorter period of time than dogs or even cats (and did not co-evolve alongside humans, as did dogs), rabbits are not the easiest animals to integrate into a human household. They chew, sometimes voraciously, leading house rabbit guardians to devise creative means of keeping phone and electrical cords, baseboards, walls, and furniture safe from sharp teeth. They are curious and spend more time than one might imagine climbing, digging, and jumping into and onto furniture, crevices, and counters, often causing much trouble in the process. Rabbits do not always come when called, obey their caretakers, or behave in an appropriate "pet-like" fashion. They are intelligent, using tools like toys, bowls, furniture, or cage bars to communicate their desires to humans and other animals. And they are messy, not so much with their waste (they are relatively easy to litterbox train) but with their toys, as well as with whatever they choose to claim as their own. All of these assessments must be understood not as judgments but rather as measures of how distinct rabbit psychology and behavior are from more traditional domesticated pets—and how much house rabbit guardians must compromise in order to provide a safe and happy home for them. Thanks to the work of rabbit advocates and the ingenuity of thousands of rabbit lovers, issues like these are now managed through "bunnyproofing" the house, through purchasing and creating toys out of cardboard, wood, wicker, and wire, and through creatively reimagining housing for rabbits inside of human homes. In addition, rabbit veterinarians have learned how to safely treat rabbit illnesses and to spay and neuter them, allowing for the possibility of rabbit–rabbit relationships and making living with rabbits in the home a cleaner proposition.

The result is that not only have tens of thousands of domesticated rabbits enjoyed the benefits of their new status as house rabbits, and their guardians likewise reaped the benefits of developing intimate relationships with these intelligent, playful, curious, and willful animals, but we have begun to gain some entrée into the psyche of these once-inscrutable animals. Adopting a house rabbit not only changes one's living conditions, it alters the way in which rabbits are perceived and thus treated.

The house rabbit movement has, in creating the concept of the "house rabbit," given rabbits some measure of personhood and has changed them from objects used as food or fur to subjects of a life. On the other hand, rabbits, both in the United States and elsewhere, are now surrendered to animal shelters or abandoned in greater numbers than ever before. The increase in popularity of the rabbit as a house pet, combined with the challenges of keeping a rabbit as a pet (along with the lack of education for many owners), has led to rabbits becoming one of the most surrendered animals in the United States. For this reason, the United States, where the majority of house rabbits live, is home to the largest number of rabbit rescue organizations (over ten dozen) in the world.

Cook and McCobb (2012) completed a six-year study of rabbit intakes and adoptions at four American shelters on the east coast and found that anywhere from 100 to just over 4,000 rabbits came into each of the shelters each year, with the vast majority of these rabbits (77 percent) being surrendered by their owners, and the second most being strays. (A Canadian study [Ledger 2010] found that 48 percent of all incoming rabbits were owner surrenders, with 38 percent coming in as strays.) Unlike the majority of House Rabbit Society members' rabbits, most (81 percent) of the American rabbits were unaltered (compared to just under 50 percent of all dogs surrendered to American animal shelters [Salman et al. 1998]), and most were surrendered because the owners stated that they could not care for them. The Canadian study also found that most (95 percent) of the rabbits were surrendered because their owners felt they could no longer keep them; of those who surrendered their rabbits for behavioral reasons, education about rabbit care and behavior would have mitigated many of the problems (such as, e.g., destructive chewing or house soiling). In fact, Ledger (2010) suggests that more education on the specific needs of rabbits is necessary in order to reduce the number of rabbits entering Canadian shelters. These results indicate that rabbits still have a way to go before they are thought of on the same level as dogs and cats.

Most countries do not offer up easy statistics on rabbit ownership. The House Rabbit Society, the world's largest rabbit advocacy organization, counts members all over the world, with 98.7 percent located in the United States and Canada. Only 0.63 percent of the organization's members are based in Europe. The rest of the members are from Asia (0.15 percent apiece in China and Japan) and 0.34 percent in Australia and New Zealand, showing that the English-speaking countries have the greatest representation in terms of HRS membership.

There is no way to know how many pet rabbits live in Europe, although there are house rabbit rescue groups in the Netherlands, France, Switzerland, Finland, Germany, Denmark, Norway, Belgium, Spain, Italy, Russia, Poland, Croatia, Serbia, Turkey, and Greece. These groups both rescue homeless rabbits, but, like HRS, also educate the public on house rabbit care. In the UK alone, there are five dedicated rabbit rescue groups and sanctuaries. The presence of these groups indicates at least an interest in keeping of rabbits as house pets. Rabbits have been kept as pets in the UK for longer than in most countries, having been brought to the islands by the Romans as a source of meat. The Pet Food Manufacturers' Association in the UK conducts an annual survey and found that 1 million rabbits lived with 3 percent of all UK households in 2013, making rabbits the third most popular companion mammal, just as in the United States. Other estimates put the numbers higher, at 2 million. While UK rabbit owners are similar to American owners in many ways, two main differences have to do with where pet rabbits are procured and how they are housed. While the adoption of house rabbits from rabbit rescue groups is certainly becoming more popular, it is still extremely common for British citizens to purchase their rabbits at pet stores or to get them from friends and acquaintances. Once purchased, many of those rabbits live outdoors (at least part-time), either in a hutch or loose in the garden. Mullan and Main (2007) found that only 4 percent of the rabbits surveyed were house rabbits; the rest lived outdoors most or all of the time.

The UK's largest rabbit welfare association, the Rabbit Welfare Association & Fund, works to educate British citizens on the importance of giving rabbits a better life than can be found in a hutch in their "A Hutch Is Not Enough" campaign (Rabbit Welfare Association, n.d.). According to a 2011 study (Edgar and Mullan 2011), 81 percent of rabbit owners in the UK conduct research prior to getting a pet rabbit, but even with that research, most still had limited knowledge of rabbits' dietary and social needs; over a third of all British rabbits, for example, received no hay, even though hay should be the most important part of a rabbit's diet, and almost half of all guardians fed their pets a diet made up in part of grains, which can cause serious digestive problems. In addition, those who did no research at all (19 percent) were less likely to get their rabbit altered, while those who did get their rabbit altered planned to find a companion for their rabbit after the surgery was complete.

According to an earlier study (Mullan and Main 2006), well over half (57.84 percent) of pet rabbits in the UK are not neutered. These unneutered rabbits can never have a companion rabbit, unless they are to be bred, which means that most British rabbits live alone. (Rabbits are social animals and thrive in the companionship of their own kind.) In fact, another study, this one commissioned by the Royal Society for the Prevention of Cruelty to Animals (RSPCA) and carried out by the University of Bristol (Rooney et al. 2014) found that 60 percent of all pet rabbits in the UK lived alone, without a companion.

The UK, then, with the longest history of pet rabbits, still has work to do in terms of seeing rabbits as deserving the same treatment as other companion animals. The Netherlands has about 1 million pet rabbits, according to a 2009 study (Schepers, Koene, and Beerda, 2009). Thanks to this research, we know that 58 percent of Dutch rabbits live outdoors, with only 35 percent living as indoor-only pets, and just less than half lived alone. As with the RSPCA survey, most of the rabbits surveyed showed signs of poor welfare—small, inappropriate cages, poor diets, among other issues. The result was not only that these rabbits lived less long than they should, with a life expectancy of only 4.2 years, 8 but that they did not appear happy, healthy, or well socialized. In a similar study conducted in Switzerland in 2009, 45 percent of the rabbits surveyed were house rabbits, with 55 percent living outdoors (Leyvraz, McBride, and Bishop, 2009). Both the Swiss and Dutch numbers indicate that continental pet rabbits are far more likely to live indoors than in the UK. This may speak to the long history of raising rabbits for meat in outdoor hutches in the UK, or the longstanding notion that rabbits are properly children's pets. On the other hand, the Swedish University of Agricultural Sciences conducted a research study (Edbom 2011) which showed that most Swedish rabbits (the country's third most popular pet) do live outdoors throughout the year. Pet rabbits in Sweden are also very likely to live alone, and while no numbers are available on how many Swedish rabbits are altered, we can assume, given the fact that most are both living outdoors and alone, that most are not. Another study (Ulfsdotter 2013) looked at how people both procure and find homes for rabbits. Rabbits are extremely easy to purchase in pet stores and are not surrendered to, or adopted from, animal shelters in nearly the quantity that we see in the United States; in fact, many shelters in Sweden do not accept rabbits. Instead, they are likely to be rehomed by the owners themselves.

In Sweden, people who no longer want their rabbits—like in the United States and Canada, Swedish owners tend to give up their rabbits because they cannot keep them—tend to put them up for sale online. One thing that may help bring down the number of rabbits who are purchased on a whim and later rehomed in Sweden is the new practice of "renting" summer rabbits. With this program, Swedish families can borrow a rabbit from a rabbit breeder and return it at the end of the summer if they no longer want it. While these programs are extremely popular (there is even a waiting list for individual rabbits), there is no data on the care that these "rented" rabbits receive or on what happens to them after they are no longer rented. Another fact that makes Sweden somewhat unusual is that many Swedish people buy rabbits for competition (rabbit jumping is a popular sport in northern Europe, which began in Sweden in 1987), rather than for companionship.

Rabbits are becoming extremely popular pets in Asia, a region of the world that has a very different relationship with these animals than in Europe or North America. Unlike Europeans and Euro-Americans, who first began keeping rabbits for food and only later as pets, rabbit meat is not commonly eaten in Asia. For this reason, keeping rabbits as pets is a relatively new phenomenon in Asian countries. In Japan, where rabbits, cats, and small dogs are extremely popular pets, modern pet keeping did not emerge until the twentieth century (Veldkamp 2009), and it wasn't until after the Second World War and the rapid urbanization that accompanied its end that small pets became so prevalent in Japan. Today, with the birth rate dropping, more and more Japanese people have chosen to have pets over children—there are over 22 million pets in Japan and only 16.6 million children. While dogs and cats are still the most common pets, ten rabbits are becoming increasingly popular as well, as can be seen by the rising number of rabbit-only pet stores, rabbit-only pet magazines, and the rise in Japanese members of HRS. No other small animal has garnered this kind of attention in Japan.

In addition, Japan has a number of "rabbit cafes" in which patrons pay an hourly fee so that they can drink tea and play with rabbits. Rabbit cafés are a subcategory of Japanese animal cafés, but rabbit cafés have experienced a boom in recent years. Finally, Okunoshima (known as Rabbit Island), a small island in the south of Japan, hosts a population of approximately one thousand feral rabbits, who attract huge numbers of tourists each day. Another reason for the popularity of rabbits in Japan comes from the term kawaii: they are cute. In fact, dwarf rabbits are among the most popular breeds in Japan, both because of their small size (Japan is very crowded and living space is tight) and because of their cuteness. Rabbits are much more likely to live indoors, as house rabbits, in Japan than in Europe.

Most Japanese people buy their pets, whether dogs, cats, or rabbits; adopting from animal shelters is rare in Japan. Instead, rabbits, like other animals, are often purchased for hundreds of yen at pet stores throughout the island. Like dogs and cats, these rabbits are often lavished with attention, clothing, and sometimes taken outdoors for walks in baby carriages. Also like dogs and cats, they are memorialized with cremation or burial after death. Rabbits are popular house pets in other Asian countries as well. Here, as in Japan, they are purchased for hundreds of dollars apiece at pet stores, but in Singapore, Taiwan, and Hong Kong, they can be adopted from animal shelters or rabbit rescue groups as well.

The most popular rabbits are small fancy breeds, probably because, as in Japan, most people live in relatively small apartments. Unfortunately, while rabbits benefit from living indoors in, for example, Singapore, they are also are subject to the same issues that they face in Europe and the United States. Here, they are given away when their owners cannot or no longer want to keep them anymore.

House Rabbit Society Singapore (HRSS), for example, reports that they are inundated with emails from people requesting rehoming assistance, and take in ten rabbits per month from the public. These requests mirror those found among rabbit rescuers in other countries: the children lost interest, someone got pregnant, the family is moving, and so on. However, this problem may be on the decline, thanks to a 2013 law that prevents children under sixteen from buying animals and mandates that pet stores provide care checklists and minimal adoption screening with each animal purchase. Rabbit rescue and animal welfare groups in both Singapore and Taiwan launched educational campaigns in 2011, the last Year of the Rabbit (which occurs every twelve years in the Chinese calendar), to discourage people from buying rabbits to celebrate the holiday.

Rabbits are thought of as lucky animals in many Asian countries and thus are purchased and later abandoned during Year of the Rabbit celebrations. HRSS reported that in the 1999 Year of the Rabbit, rabbits in Singapore were abandoned at that nation's shelters at twice the rate they were in 1998. That same year, CNN reported that dead rabbits were appearing in mailboxes throughout China, as the Chinese bought many rabbits for the celebration from online vendors, and the rabbits took up to five days to reach the purchasers, often dying on the journey. Australia Rabbits are also popular pets in Australia, although they face serious obstacles there as well, and there is no way to know how many rabbits live as pets in the country.

Two-thirds of all Australians live with a companion animal, but there have been no surveys to determine how many rabbits live in Australian homes. However, a recent article published in the Journal of Applied Animal Welfare Science (Howell, Mornement, and Bennett 2015) looked at the conditions under which rabbits in Victoria live and found that 35 percent of the rabbits were altered, and approximately two-thirds of all respondents had a single rabbit, with one-third (presumably the same people with altered rabbits) having a pair. Most rabbits were acquired as babies, and a third of all the rabbits were acquired for free. Most (76 percent) of the respondents considered their rabbits to be members of their family, and yet most (68 percent) of all rabbits spend most of their time outdoors. Of those outdoor rabbits, 13 percent of guardians allow their rabbits to run freely, which correlates with another finding: 12 percent of the respondents said that they had lost a rabbit in the last five years. And finally, nearly half of all respondents keep their rabbits caged for almost their entire lives, letting them out to exercise less than one hour per week.

It is clear that Australian rabbits, even with the benefit of four rabbit rescue and education organizations in the country, suffer from a lack of good quality care, just like those in the other countries surveyed. In addition, rabbits in Australia face another obstacle—they are also seen as pests throughout the country.

The British introduced domestic rabbits to Australia (and New Zealand) throughout the 1770s and 1780s to be used for hunting. For about seventy years, those rabbits pretty much stayed within the colony settlements, probably because the land hadn't been cultivated and natural predators could easily wipe out domestic rabbits that wandered into the bush (Flux 1990b). But in the mid-nineteenth century, colonists developed "acclimatization societies" aimed at importing and then releasing game, fish, and songbirds in Australia and New Zealand, including wild rabbits. The goal, according to an 1882 report by the society in Victoria, was "to enrich (the country) by stocking its broad territory with the choicest products of the animal kingdom borrowed from every temperate region on the face of the globe" (ibid., p. 147).

The country did indeed become enriched with wild rabbits. By 1879, there were so many rabbits in western Victoria that farmers had to abandon their land. By 1887, farmers killed 10 million rabbits in New South Wales in just eight months, and by 1888, an Intercolonial Commission on Rabbits had declared that the rabbit was a pest and the country should be rid of the species (Hinds et al. 1996). The rapid increase in Australia's rabbit population led to an increase in the European red fox and feral cats (also introduced species), which in turn contributed to the extinction or a reduction in the number of several indigenous birds and mammals.

Rabbits in Australia also cost the agricultural industry some $600 million a year, as they eat cereals and other crops, and reduce the forage available for sheep and cows (it is estimated that ten rabbits eat as much herbage as one sheep; ibid.). Since the turn of the twentieth century, scientists and government officials have tried a variety of methods to control the rabbit population, including building a transcontinental rabbit-proof fence, encouraging citizens to kill rabbits on their own, trapping, poisoning, bringing in predator species, and developing biological agents, such as the Myxoma virus, to kill off the rabbits. Unfortunately for Australian farmers, none of these methods has been effective; unfortunately for the rabbits, myxomatosis, after it jumped from the target population, is now the most dangerous disease affecting wildlife in Europe (and kills pet rabbits as well).

Another result of the "plague" of wild rabbits in Australia is that they are extremely vilified there. It is illegal to keep rabbits as pets in Queensland, although rabbit lovers do advocate a lifting of the ban. In 2014, for example, the Queensland government once again rejected calls for legalizing pet rabbits. Those "outlaws" who keep rabbits in Queensland risk penalties of $44,000 and six months in prison—far more than some forms of violent crime. Queensland is now the only location on earth where it is illegal to live with a pet rabbit. Throughout the rest of the country, however, rabbits are available both through pet shops and from animal shelters and the four rabbit rescue organizations.

It is clear that rabbits are, while not yet threatening dogs or cats, becoming more popular pets around the world. It is also clear that they are suffering from many of the same problems that plague cats and dogs—abandonment, impulse purchases, and neglect. In fact, because fewer people are aware of the rabbit's particular needs, they arguably suffer more—they are more likely to live outdoors, they are less likely to be spayed or neutered (and thus have a companion), and they are far more likely to spend their lives in cages.

In addition, rabbit meat is experiencing a Renaissance in the United States, United Kingdom, and other places today, and with the rise in popularity of rabbit meat comes new challenges for pet rabbits and rabbit lovers. Because rabbits are used for a variety of purposes, they often lack the legal protections that other animals enjoy; for instance, in the United States, rabbits are excluded from the protections afforded by the Humane Methods of Slaughter Act, and, when raised for meat, are not protected by state or local anti-cruelty laws. Ultimately, we will see rabbits gain ever more popularity as pets, thanks to their intelligence, curiosity, mild temperament, and, of course, their cuteness. What is yet to be determined is if rabbits will, at the same time, gain more respect and better treatment as well.

See also: Animal Cafes; Japan, Pets In; Rescue Groups; United States, Pets In

References and Further Reading

American Veterinary Medical Association. (2007). *U.S. Pet Ownership and Demographics Sourcebook*. Schaumburg, IL: Center for Information Management of the American Veterinary Medical Association.

Carneiro, M., C. J. Rubin, F. Di Palma, F. W. Albert, J. Alföldi, A. M. Barrio, et al. (2014). "Rabbit Genome Analysis Reveals a Polygenic Basis for Phenotypic Change During Domestication." *Science*, 345 (6200): 1074–9.

Clutton-Brock, J. (1987). *A Natural History of Domesticated Mammals*. Cambridge: Cambridge University Press.

Cook, A. J., and E. McCobb (2012). "Quantifying the Shelter Rabbit Population: An Analysis of Massachusetts and Rhode Island Animal Shelters." *Journal of Applied Animal Welfare Science*, 15 (4): 297–312.

Crook, S. (1985). *Lop Rabbits as Pets*. Neptune City, NJ: TFH Publications.

Davis, S., and M. DeMello (2003). *Stories Rabbits Tell: A Natural and Cultural History of A Misunderstood Creature*. New York: Lantern Books.

Edbom, M. (2011). *Kaninhållning för SällskapsbrUK (Rabbit group behavior)*. First cycle, G2E. Skara: SLU, Department of Animal Environment and Health.

Edgar, J. L., and S. M. Mullan (2011). "Knowledge and Attitudes of 52 UK Pet Rabbit Owners at the Point of Sale." *Veterinary Record-English Edition*, 168 (13): 353.

Ezpeleta, A. (1996). Rabbits Everywhere. New York: Harry N. Abrams Publishers.

Flux, J. E. C. (1990a). "The Hares and Jackrabbits." In J. A. Chapman and J. E. C. Flux (Eds), *Rabbits, Hares, and Pikas*. Gland, Switzerland: International Union for the Conservation of Nature and Natural Resources.

Flux, J. E. C. (1990b). "Exotic populations." In J. A. Chapman and J. E. C. Flux (Eds), *Rabbits, Hares, and Pikas*. Gland: International Union for the Conservation of Nature and Natural Resources.

Fox, R. (1994). "Taxonomy and Genetics." In P. J. Manning, D. H. Ringler, and C. E. Newcomer (Eds), *The Biology of the Laboratory Rabbit*. New York: Academic Press.

Grier, K. (1999). "Childhood Socialization and Companion Animals: United States 1820–1870." *Society and Animals*: *Journal of Human-Animal Studies*, 7 (2): 95–120.

Harriman, M. (1985 [1995]). *House Rabbit Handbook: How to Live With an Urban Rabbit*. Alameda, CA: Drollery Press.

Hinds, L. A., B. D. Cooke, T. J. Robinson, and C. K. Williams (1996). "Rabbits Prospects for Long Term Control: Mortality and Fertility Control." *Prime Minister's Science and Engineering Council*, 13.

Howell, T. J., K. Mornement, and P. C. Bennett (2015). "Companion Rabbit and Companion Bird Management Practices Among a Representative Sample of Guardians in Victoria, Australia." *Journal of Applied Animal Welfare Science* , 18 (3): 287–302.

Ledger, R. A. (2010). "The Relinquishment of Rabbits to Rescue Shelters in Canada." *Journal of Veterinary Behavior: Clinical Applications and Research*, 5 (1): 36–7.

Leyvraz, A. M., A. McBride, and F. L. Bishop (2009). "Management and behaviour of pet rabbits in Switzerland." In *Proceedings of the 14th Annual Meeting of the European Society of Veterinary Clinical Ethology & 5th Annual Meeting of the European College of Veterinary Behavioural Medicine-Companion Animals*. European Society of Veterinary Clinical Ethology, 42.

Milford, H. S. (Ed.) (1975). *The Poetical Works of William Cowper*. Oxford: Oxford University Press.

Mullan, S. M., and D. C. Main (2006). "Survey of the Husbandry, Health and Welfare of 102 Pet Rabbits." *The Veterinary Record*, 159 (4): 103–9.

Mullan, S. M., and D. C. Main (2007). "Behaviour and Personality of Pet Rabbits and their Interactions with their Owners." *The Veterinary Record*, 160 (15): 516–20.

Podberscek, A. L., E. S. Paul, and J. A. Serpell (2005). "*Companion Animals and Us: Exploring the Relationships Between People and Pets*. Cambridge: Cambridge University Press.

Rabbit Welfare Association. (n.d.). *A Hutch Is Not Enough*. http://www.rabbitwelfare.co.UK/ahutchisnotenough.htm

Rogers, P. M. (1994). "The Rabbit in Continental Europe." In H. Thompson (ed.), *The European Rabbit: The History and Biology of a Successful Colonizer*. New York: Oxford University Press.

Rooney, N. J., E. J. Blackwell, S. M. Mullan, R. Saunders, P. E. Baker, J. M. Hill, et al. (2014). "The Current State of Welfare, Housing and Husbandry of the English Pet Rabbit Population." *BMC Research Notes*, 7: 942.

RSPCA. (2015). "Is your rabbit a happy bunny?" *RSPCA News*. February 3.

Salman, M. D., J. G. New Jr., J. M. Scarlett, P. H. Kris, R. Ruch-Gaille and S. Hetts (1998). "Human and Animal Factors Related to the Relinquishment of Dogs and Cats in 12 Selected Animal Shelters in the United States." *Journal of Applied Animal Welfare Science*, 1 (3): 207–26.

Schepers, F., P. Koene, and B. Beerda (2009). "Welfare Assessment in Pet Rabbits." *Animal Welfare*, 18 (4): 477–85.

Taylor, J. (1986). *Beatrix Potter: Artist, Storyteller, and Countrywoman*. London: Frederick Warne.

Ulfsdotter, L. (2013). *Rehoming of Pet Rabbits in Sweden*. Master thesis, Presented to the Swedish University of Agricultural Sciences, Department of Animal Environment and Health.

Veldkamp, E. (2009). "The Emergence of 'Pets as Family' and the Socio-Historical Development of Pet Funerals in Japan." *Anthrozoös: A Multidisciplinary Journal of The Interactions of People & Animals*, 22 (4): 333–46.

Welch, T. (2015). *Motivations for and Thoughts Toward Rabbit Ownership and Factors Contributing to Companion-Rabbit Owners' Knowledge*. Master thesis, The University of Guelph, Guelph, ON, Canada.

RELIGIOUS INSTITUTIONS, PETS AND

Contemporary religions have a somewhat uneasy relationship with other animals. As symbols, animals help us to understand important religious concepts such as purity, sacrifice, morality, and creation. As such, they play important roles in the myths of cultures around the world. Some religions hold certain animals to be sacred, and some worship certain animals or make certain animals taboo for religious purposes. Other cultures hold that some animals are the mythic founders of a group or clan and have set those animals aside as totems. Finally, religious rituals often make use of animals directly, usually as a sacrifice to the gods or ancestors. But even though animals figure into every world religion, that does not mean that those religions necessarily offer anything to the animals themselves.

Religions, through their myths and practices, encode a worldview that is specific to that religion. In this sense, a worldview is an orientation to the world that includes assumptions about the natural world and our place in it, our relationship to the sacred realm, and notions about the construction of reality. Within a worldview is often, either explicitly or implicitly, a set of assumptions about humanity's relationship with nature and with other animals.

For instance, the monotheistic religions—Judaism, Christianity, and Islam—are often called the "religions of the book" because they all share the Old Testament as a source of God's wisdom, share some similarities in their views toward animals as well as some differences. In the Old Testament, God created man in his own image and the animals and plants of the world, over which man was given dominion (Gen. 1:26). Although this can be interpreted to mean that humans were given rights of dominance over animals, some interpret it to mean a stewardship in which humans have fundamental obligations toward animals. In any case, it is clear that in the Old Testament, animals and humans are separate creations, with the latter being given authority over the former.

Christianity, because it shares with Judaism the Old Testament, holds the idea that humans were created separately from animals with the power and responsibility of dominion over them. In the New Testament's book of the Christians, very little is written about animals. Most of what we know about Christian views toward animals comes later, starting in the fourth century with the writings of St. Augustine (Augustinus, Rotelle, and Hill 2000), who stated that only man has a rational soul, created by the breath of God, as opposed to animals, who have only sensation and cannot attain eternal life. It is Augustine's view (which itself borrows from Greek thought about the inferiority of animals)—later expanded by the thirteenth-century theologian Thomas Aquinas (2007), who held that man had no moral obligations toward animals—that sets out what would become the dominant Christian view of animals. In this view, because only humans were created in God's image, only humans have an immortal soul, and because God became a human being (in the body of Jesus), animals and humans belong on different ontological levels. In other words, they are radically different beings who don't deserve the same consideration as humans. For example, the Papal Encyclical "Evangelium Vitae," though opposing needless suffering for animals, does so because it is contrary to human dignity, not because it is harmful to animals (John Paul II 1995). The encyclical also says that humans should not spend money on animals that should go toward relieving human suffering. And the 1994 Catechism of the Catholic Church tells us that animals are "destined for the common good of past, present and future humanity" (1994, 516, 2,415).

In Islam, the third religion of the book, all creatures were made by Allah, who loves all creatures. The Prophet Muhammad, too, was said to love animals, and he features in a number of stories that relate his caring. The Quran explicitly states, however, that animals can be used for human benefit. "It is God who provided for you all manner of livestock, that you may ride on some of them and from some you may derive your food. And other uses in them for you to satisfy your heart's desires. It is on them, as on ships, that you make your journeys" (Quran 40:79–80). Like the previous traditions, however, in Islam, although animals should be treated with kindness and compassion, they exist for the benefit of human beings.

Buddhism and Hinduism, the primary religions of India, differ substantially from Judaism, Christianity, and Islam in their worldview regarding animals. Hinduism is a polytheistic religion with a great many traditions and thus has no single view regarding animals. However, many Hindu gods take on animal forms, such as Ganesh, who has the head of an elephant, or Hanuman, who takes the shape of a monkey; in addition, cows are sacred to Hindus. The god Vishnu was incarnated as a fish, a tortoise, a boar, and as half lion/half man; he is worshipped today in all of those forms. Because of these beliefs, Hindus do not eat cows, and villagers will not harm monkeys; both animals are protected by Indian law (Krishna 2010). Many Hindus are vegetarian because of the concept of *ahimsa*, which means avoidance of harm. Finally, Hindus believe in reincarnation, which means that one is reborn into another life after death. Because both animals and humans have souls and exist together in the cycle of rebirth, humans can be reborn as animals and vice versa. But though humans and animals exist in the same continuum of life, humans are considered to be the apex of what life should be; they are thus superior to animals, and accruing negative karma means that one will be reborn as an animal, which is considered to be an unhappy fate. Although many Hindu texts and beliefs promote the humane treatment of animals and the coexistence of humans and animals, many Hindu practices are, nevertheless, harmful toward animals.

Buddhists also share with Hindus the desire to do no harm, along with the concept of reincarnation (or *samsara*, through which both animal and human lives are recycled) and karma; Buddhists should avoid cruelty of all kinds. Like Hindus, many Buddhists are vegetarians. Finally, animals and humans both have "Buddha-nature," which is the potential for reaching enlightenment. However, in Buddhism, as in Hinduism, being reborn as an animal is seen as negative because animals aren't seen as having the ability to improve themselves, making it harder to work off bad karma and ultimately reach enlightenment. Clearly, both Hinduism and Buddhism see animals in a complex way.

Where do pets fit into these religious traditions? Each of the world's major religions—excepting Mormonism, which was founded in 1830—is thousands of years old, and thus they predate the contemporary preoccupation with pets. So while all of them have something to say regarding the moral status of animals, they are not terribly useful when it comes to pets.

However, some religions have, in recent years, begun to speak more directly about pets and our relationship with them. Within Christianity, St. Francis of Assisi, theologian Albert Schweitzer, and writer Isaac Bashevis Singer are examples of deeply religious men who read scripture as encouraging compassion toward animals. The Quakers (known as the Religious Society of Friends), a Christian movement that arose in the seventeenth century, were unusual for their time in that they felt strongly that God's creatures deserved consideration and reverence. George Fox, the founder of the Quakers, abhorred hunting and other practices, and in 1891, the Friends' Anti-Vivisection Society was founded in England in order to oppose the use of animals in science (Glaholt 2012).

St. Francis, known as the patron saint of animals, was said to preach to the birds and believed that it was the duty of man to protect nature and animals. He is the creator of the Christmas Nativity scene, or crèche, and because of his belief that animals are God's creations just as man is, he included animals in the scene of Jesus' birth. Today, St. Francis remains an inspiration to Christians who work on behalf of animals (Hughes 1996; Barbiero 2016; Sorrell 1988). Many churches now hold blessings of the animals rituals on October 3, which is St. Francis' feast day. At these events, members of the church bring in their pets for the priest or monk to bless them. These blessings might have originated in pre-Christian times but evolved with Christianity. The earliest records of such events go back not to St. Francis' feast day but to St. Anthony Abbot, whose feast day is on January 17 and who is also a patron saint of animals. Like St. Francis, St. Anthony was known during his lifetime to heal, bless, and care for animals. As early as the fifteenth century, people might have brought their farm animals to church for blessings. Today, however, companion animals make up the majority of animals brought to blessings like this, as their importance has increased in modern society.

The deep attachment that humans have to companion animals can be seen in the various ways in which people imagine that they will share an afterlife with the animals. We can interpret 14,000-year-old burials of dogs as evidence that not only did humans care deeply about their dogs but that they might have buried them with religious rituals generally used for people. For example, as noted earlier, Egyptians included their mummified pet cats in their family tombs, where the cats were expected to be reunited with the family in the afterlife. And Romans not only buried dogs under marble headstones but also had their companion animals killed after their own death so that their fates might be shared (Collier 2016). With the rise of Christianity, however, animals and humans were separated after death; Christians who wanted to be buried with their companions were not allowed to do so by the church. Pet cemeteries, where pets could be buried, often in ritual fashion, emerged in the late nineteenth century as an alternative to other ways of disposing of beloved pets (Howell 2002).

Roman Catholic dogma says that animals do not possess an immortal soul and as such will not go to Heaven after death. For practicing Catholics who are animal lovers, this raises a serious question: what happens to animals after they die? Many animal lovers whose faiths do not permit animals to join humans in an afterlife must either reconcile themselves to the fact that they will never see their beloved companion animals again or must come up with a new concept of an afterlife that includes animals. For many people, that afterlife is called the Rainbow Bridge. This refers to a meadow where pets (but not other animals) go after death, where they play together until the death of their owners, who then walk over the bridge to be met by their pets, and they all progress to Heaven together. (Other versions of the Rainbow Bridge describe it as a permanent place for deceased animals to play, but they do not rejoin their owners, nor do they go to Heaven.) Some Christians interpret a handful of biblical verses in such a way that suggests that animals will be found in Heaven. For example, Isa. 65:25 says, "the wolf and the lamb will feed together, and the lion will eat straw like the ox," which some interpret to mean that animals will go to Heaven. (Others, however, say that this passage refers to animals metaphorically as representing the end of conflict.) Still others have interpreted Heaven to be a place in which both humans and animals (or at least some animals) are welcome. In most interpretations of this belief system, Heaven is considered to be a place of "perfect happiness," so if one person's idea of perfect happiness includes their companion animals, the pets will be there as a sort of heavenly prop. Finally, in 2009, a new business model emerged for evangelical Christians who believe in the Rapture—an event in which faithful Christians will be "caught up" to Heaven to join Jesus, while the sinners will remain on earth to await Armageddon. Those Christians who have pets can now elect to pay companies to find "sinners" to care for their pets after they have been taken up in the Rapture.

See also: India, Pets In

References and Further Reading

Aftandilian, D., ed. (2007). *What Are the Animals to Us? Approaches from Science, Religion, Folklore, Literature, and Art*. Knoxville: University of Tennessee Press.
Aquinas, T. (2007). *Compendium of the Summa Theologica of St. Aquinas*. Eastbourne: Gardners Books .
Augustinus, A., J. Rotelle, and E. Hill. (2000). *The Works of St. Augustine. A Translation for the 21st Century, Part I: Books*. Volume 13. New York: New City Press.
Barbiero, G. (2016). "Francis of Assisi and the Wolf: Nonviolence as a Moral Value of Biophilia." *Visions for Sustainability* 5: 28–31.
Church, C. (1994). *Catechism of the Catholic Church*. Vatikan: Libreria Editrice Vaticana.
Collier, I. (2016). "More than Just a Bag of Bones: A History of Animal Burials." In M. DeMello (ed.), *Mourning Animals: Rituals and Practices Surrounding Animal Death*. East Lansing: Michigan State University Press.
Glaholt, H. R. (2012). "Vivisection as War: The 'Moral Diseases' of Animal Experimentation and Slavery in British Victorian Quaker Pacifist Ethics." *Society & Animals* 20 (2): 154–72.
Hobgood-Oster, L. (2008). *Holy Dogs and Asses: Animals in the Christian Tradition*. Urbana: University of Illinois Press.
Howell, P. (2002). "A Place for the Animal Dead: Pets, Pet Cemeteries and Animal Ethics in Late Victorian Britain." *Ethics, Place & Environment* 5 (1): 5–22.
Hughes, J. D. (1996). "Francis of Assisi and the Diversity of Creation." *Environmental Ethics* 18 (3): 311–20.
John Paul II, P. (1995). "Evangelium vitae." *Origins* 24 (42): 689–91.
Krishna, N. (2010). *Sacred Animals of India*. New Delhi: Penguin.
Masri, B. A. (1989). *Animals in Islam*. Petersfield: Athene Trust.
Massaro, A. (2016). "The Issue of Animals' Souls Within the Anglican Debate in the Eighteenth and Nineteenth Centuries." In M. DeMello (ed), *Mourning Animals: Rituals and Practices Surrounding Animal Death*, 31–7. East Lansing: Michigan State University Press.
Perlo, K. W. (2009). *Kinship and Killing: The Animal in World Religions*. New York: Columbia University Press.
Sorrell, R. D. (1988). *St. Francis of Assisi and Nature: Tradition and Innovation in Western Christian Attitudes Toward the Environment*. New York: Oxford University Press.
Toperoff, S. P. (1995). *Animal Kingdom in Jewish Thought*. Northvale, NJ: Jason Aronson

REPTILES

Reptiles have gained significant popularity as pets in recent years, with approximately 4 percent of households owning reptiles in the United States, and comparable trends observed in other parts of the world, particularly in Europe and Australia (Sollom and Baron, 2023; Valdez, 2021). The motivations for keeping reptiles as pets are diverse, including emotional connections, convenience, and a desire for unique companions that differ from traditional pets such as dogs and cats (Fuente et al., 2023; Lambert, Carder, and D'Cruze, 2019). This popularity is also reflected in trade data, which indicate a steady rise in interest toward exotic pet species, including reptiles (Valdez, 2021; Azevedo et al., 2021).

The acquisition of pet reptiles primarily stems from two main sources: wild-caught specimens and captive-bred individuals. Captive breeding has increasingly become the preferred method, as it alleviates some of the ecological pressures associated with the extraction of reptiles from their natural habitats (Valdez, 2021; Meletiadis et al., 2022). Captive-bred reptiles tend to be more manageable and better suited to captivity, which is crucial for ensuring animal welfare. However, the trade of wild-caught reptiles persists, often driven by high demand for certain species that are not easily bred in captivity (Lambert et al., 2019; Azevedo et al., 2021).

In the context of regulations, various laws protect reptile species across the United States. The Convention on International Trade in Endangered Species of Wild Fauna and Flora (CITES) governs the international trade of endangered species; however, many reptiles are not listed under this convention, leading to a largely unregulated domestic market (Dipineto et al., 2012). This lack of regulation has prompted concerns about both the legality and ethics of wild-caught reptiles—a factor exacerbated by zoonotic diseases, such as Salmonella, which pose a risk to both reptile owners and public health (Morrison and Rubin, 2020; Chen et al., 2010).

Welfare issues for reptiles kept as pets are substantial and varied. Captive reptiles often face challenges due to inadequate husbandry practices that fail to meet their species-specific needs. Many reptiles require particular environmental conditions, including specific temperature ranges, UVB lighting, and proper nutrition that, if unmet, can lead to stress and health problems (Azevedo et al., 2021; Chan et al., 2024). Furthermore, the commercial trade often prioritizes profit over animal welfare, subjecting these creatures to poor handling and transportation conditions that can exacerbate their physiological and psychological distress (D'Cruze et al., 2020). Despite increasing awareness and campaigns aimed at promoting better husbandry practices, significant gaps in knowledge among pet owners persist, further complicating welfare issues in the reptile pet trade (Chan et al., 2024).

Addressing the welfare challenges inherent in keeping reptiles as pets will require a multifaceted approach involving regulation, education, and the promotion of responsible ownership practices.

See also: Exotic Pets; Health Risks of Pet Ownership; Veterinary Care; Zoonoses

References and Further Reading

Azevedo, A., L. Guimarães, J. Ferraz, M. Whiting, and M. Magalhães-Sant'Ana (2021). "Pet Reptiles—are we Meeting their Needs?" *Animals*, 11 (10): 2964.

Chan, S., W. Fang, B. LePage, and S. Wang (2024). "Assessing the Current Landscape of Reptile Pet Ownership in Hong Kong: A Foundation for Improved Animal Welfare and Future Research Directions." *Animals*, 14 (12): 1767.

Chen, C., W. Chen, S. Chin, Y. Lai, K. Tung, C. Chiou, Y. M. Hsu, and C. Chang (2010). "Prevalence and Antimicrobial Susceptibility of Salmonellae Isolates from Reptiles in Taiwan." *Journal of Veterinary Diagnostic Investigation*, 22 (1): 44–50.

Dipineto, L., M. Capasso, M. Maurelli, T. Russo, P. Pepe, G. Capone, A. Fioretti, G. Cringoli, and L. Rinaldi (2012). "Survey of Co-infection by Salmonella and Oxyurids in Tortoises." *BMC Veterinary Research*, 8 (1).

D'Cruze, N., S. Paterson, J. Green, D. Megson, C. Warwick, E. Coulthard, J. Norrey, M. Auliya, and Carder, G. (2020). "Dropping the Ball? the Welfare of Ball Pythons Traded in the EU and North America." *Animals*, 10 (3): 413.

Fuente, M., B. Araújo, I. Policarpo, H. Pereira, A. Borges, W. Vieira, G. Filho, and R. Alves (2023). "Keeping Reptiles as Pets in Brazil: Keepers' Motivations and Husbandry Practices." *Journal of Ethnobiology and Ethnomedicine*, 19 (1).

Lambert, H., G. Carder, and N. D'Cruze (2019). "Given the Cold Shoulder: A Review of the Scientific Literature for Evidence of Reptile Sentience." *Animals*, 9 (10): 821.

Meletiadis, A., C. Biolatti, D. Mugetti T. Zaccaria, R. Cipriani, M. Pitti, P. Acutis, et al. (2022). "Surveys on Exposure to Reptile-associated Salmonellosis (Ras) in the Piedmont Region—Italy." *Animals*, 12 (7): 906.

Morrison, B. and J. Rubin (2020). "Detection of Multidrug-resistant Gram-negative Bacteria from Imported Reptile and Amphibian Meats." *Journal of Applied Microbiology*, 129 (4): 1053–61.

Sollom, H. and H. Baron (2023). "Clinical Presentation and Disease Prevalence of Captive Central Bearded Dragons (Pogona Vitticeps) at Veterinary Clinics in Australia." *Australian Veterinary Journal*, 101 (5): 200–7.

Valdez, J. (2021). "Using Google Trends to Determine Current, Past, and Future Trends in the Reptile Pet Trade." *Animals*, 11 (3): 676.

RESCUE GROUPS

Animal rescue groups are typically privately funded groups, often made up of volunteers, that rescue domesticated animals and place them up for adoption. The animals can either be surplus animals from public or private animal shelters, unwanted pets from the general public, or stray (former) pets.

While individuals have been finding and keeping stray animals for probably hundreds of years, organized animal rescue groups are a relatively recent invention, dating to the early nineteenth century in England. Here, in 1824, a handful of animal lovers formed the Society for the Prevention of Cruelty to Animals, the first SPCA around the world, later renamed the Royal Society for the Prevention of Cruelty to Animals (RSPCA).

In the United States, the first formal animal welfare group was formed by Henry Bergh, who founded the American Society for the Prevention of Cruelty to Animals (ASPCA) in 1866. The group was formed not to rescue individual animals, but initially to protect animals like carriage horses in New York City and to fight other forms of cruelty. While

Bergh's efforts were primarily focused on anti-cruelty campaigns, resulting in the nation's first anti-cruelty law in New York in 1866, the ASPCA (which took over New York City's animal control contract in 1894, becoming one of the first non-municipal animal shelters in the country) became the model for the country's first animal rescue groups, many of whom still use SPCA in their names today.

The nineteenth century saw the formation of other animal rescue groups in the United States and England, some of which still exist today. The methods of these groups varied, but all were founded in order to help alleviate the suffering of companion animals, usually cats and dogs. Often focused on taking in stray cats and dogs, providing medical care, and attempting to find them homes, many animal rescue groups operated (and in many areas of the world today, still operate) in the absence of any formalized, municipally run animal control agency. These small, often volunteer-run organizations, then, were and often are the only source of help for stray, sick, abused, or starving domestic animals.

Today, in most locations in the developed world, animal rescue groups operate alongside city- and county-run shelters. Many groups are species or breed specific, rescuing only rabbits, Great Danes, or Chihuahuas. As nontraditional pets become mainstream, and as the public purchases and then discards those animals, rescue groups are popping up to handle every species from turtle to parrot to rat.

Animal rescuers with a relationship with their local shelter are generally contacted by staff at the shelter when an animal meeting the breed or species requirement is brought in, and the rescuer will then pick the animal up. Rescuers both aid their local shelters by cutting down on the volume of animals the shelter must deal with and are often better able to find a suitable adoptive home for animals who, because of their breed, species, or temperament, may be difficult for the shelter to place (Huss 2006). Many rescuers operate foster homes that provide permanent sanctuary care to animals who, by virtue of their age, health, or temperament, are deemed unadoptable. Others specialize in certain kinds of animals, such as disabled animals, or seniors, often keeping them as sanctuary animals, but often offering them for adoption.

Rescue groups that focus on rescuing and placing nontraditional pets are faced with some unique challenges. Many so-called exotic pets, for example, are in fact wild animals who are not at all domesticated and should not even be pets to start with because of the damage that the exotic pet industry does to wild habitats and species, and because of these animals' unique behavioral and physical needs. These groups must attempt to get birds, reptiles, and other non-domesticated animals adopted into new homes while at the same time must also discourage people who can't adequately provide for their needs from attaining these animals in the first place. In addition, animal shelters often have to rely on these rescue groups to take their nontraditional animals because they are often ill-equipped to deal with the specific needs of exotic animals, and their shelter adoption rates are consequently much lower than cats and dogs.

Some animal rescue groups specialize in rescuing animals, both domestic and wild, from disasters. In the United States, Hurricane Katrina in 2005 demonstrated the need for comprehensive disaster plans that included provisions for animals to be rescued alongside people. By 2007, and the devastating Southern California wildfires, local and national officials recognized the necessity of providing for animals and joined with numerous animal rescue groups to provide for hundreds, if not thousands of companion animals during that disaster.

Animal rescue groups are funded primarily by private donations from the public. Those which have charitable status can offer tax benefits to their donors, but many groups are operated by well-meaning individuals who have not taken the steps to incorporate or get tax-deductible status. Many groups offer newsletters to their supporters, and most engage in fundraising efforts such as walkathons, merchandise sales, or services such as boarding, veterinary care, or grooming. Groups that are staffed entirely by volunteers and that operate out of a network of foster homes will have lower operating costs than those groups with paid staff and/or a permanent facility such as an animal shelter. Those groups must do additional fundraising in order to meet their expenses. Some groups, like the ASPCA in New York, operate their city's animal control contracts and thus are paid in part by the city. Other groups may run their own private shelters, taking in animals from the public (often for a fee) but not collecting stray animals or responding to cruelty calls. While most animal rescue groups do not euthanize animals (except for health reasons), some, particularly those that operate shelters, do.

While municipal animal shelters are funded through a combination of city or county taxes and private donations, all other rescue groups are responsible for raising their own money. Because municipal shelters contract with the local government to provide animal control services, that is picking up stray animals, taking in unwanted animals, and dealing with cruelty investigations and nuisance animals, at least part of their annual budget is guaranteed. Animal rescuers, whether they are independent or work within an organization, do not have this type of safety net and thus are often forced to pay for rescue activities with their own money.

This illustrates one of the major conflicts within the pet world: while the pet industry makes money by producing pets and then selling all of the supplies necessary to keep a pet healthy and happy, the industry takes no responsibility for the problems produced by it—either the health problems created through intensive breeding or the "surplus" pets produced by it. That means that the rescue world—which is made up overwhelmingly of women—must undertake the enormous job of rescue, rehabilitation, and adoption on their own.

Animal rescue volunteers who work on the front lines rescuing abandoned and unwanted animals collectively spend millions of dollars per year on everything from spaying/neutering and other medical costs, food, and other animal maintenance costs, and the other expenses involved in rescuing animals. Not surprisingly, rescuers see themselves fighting to save animals against a never-ending tide of breeders who breed too many animals, pet stores that sell animals to the public with no screening or education, and the general public that abandons animals. No matter how hard rescuers and animal advocates work, the animals just seem to keep coming, with no end in sight (Neumann 2010; Samdahl 2015; Guenther 2017).

For these volunteers, sometimes the fatigue of knowing how many animals continue to be abandoned and euthanized feels overwhelming. Known as compassion fatigue, animal rescuers are at risk of being overwhelmed and traumatized by the constant animal suffering, and the knowledge that what they do is never enough. Many rescuers are depressed and deal with that depression in unhealthy ways. Many, for example, use food or alcohol or drugs to self-medicate (Figley and Roop 2006; Rank, Zaparanick, and Gentry, 2009; Fournier and Mustful 2019).

Another difficulty faced by animal rescue volunteers is that they often have an antagonistic relationship with the workers at the animal shelters that they assist. Rescue volunteers often feel that they care more about animals than shelter workers, that they are more knowledgeable about the particular breed or species of animal that they rescue, and they sometimes look critically at shelter workers whose jobs require that they must euthanize animals. In return, shelter workers often see rescue volunteers as untrained, unprofessional, and unaware of the realities of working in a shelter and the hard realities of, and need for, euthanasia (Guenther 2017).

Many animal rescue groups have foster homes that provide permanent sanctuary care to animals who, by virtue of their age, health, or temperament, are deemed unadoptable. Other groups specialize in certain kinds of animals, such as disabled animals, or seniors, often keeping them as sanctuary animals, but often offering them for adoption.

Prior to the age of the Internet, animal rescue groups were primarily local operations—with a network of local volunteers, a relationship with their local shelters, and a list of local supporters to provide funding. In the 1980s, in the United States, Project BREED (Breed Rescue Efforts and Education) was founded to provide a resource for animal shelters, the public, and rescue groups. The Project BREED directory, still published today, lists thousands of breed-specific (for dogs) and species-specific rescue groups, as well as specific information on the breed or species, to aid people who are interested in adopting a particular kind of animal.

Today, with the Internet, not only are there numerous websites that provide such information, but rescue groups are able to operate nationwide and even worldwide. House Rabbit Society (HRS), a rabbit rescue organization, is one such group. Originally based in the San Francisco Bay area, the group developed a nationwide mailing list, attracting members and new volunteers from around the country, leading to chapters or volunteers in almost every state. With the Internet, HRS has gone international, with representatives in Europe, Asia, Canada, and Australia.

Most rescue groups today, in fact, use the Internet for public outreach, to advertise their adoptable animals, to fundraise, and to attract supporters. One important website for many groups is Petfinder, which allows those groups without their own websites (or without the technology to easily update a website) to quickly and easily list their adoptable animals for the public to see, along with photos and now videos.

Animal rescue groups have, especially compared to breeders and pet stores, strict adoption procedures that typically include an adoption application, an interview, and often a home visit. These requirements are often stricter than those found at animal shelters and are put in place to ensure that the animals who are adopted end up in a safe, loving, permanent home.

Animal rescue groups can be found around the world today, although developing nations, which often have greater problems with stray animals, tend to have fewer groups with fewer resources.

See also: Animal Protection Movement; Humane Societies; Sanctuaries; Shelters

References and Further Reading

Figley, C. R., and R. G. Roop (2006). *Compassion Fatigue in the Animal-Care Community*. Humane Society Press.

Fournier, A. K., and B. Mustful (2019). Loss and grief in animal-care agents. In *Pet Loss, Grief, and Therapeutic Interventions* 99–123. Routledge.

Guenther, K. M. (2017). "Volunteers' Power and Resistance in the Struggle for Shelter Animal Survival." In *Sociological Forum* 32, (4): 770–92.

Huss, R. J. (2006). "Rescue me: Legislating Cooperation Between Animal Control Authorities and Rescue Organizations." *Conn. L. Rev*, 39, 2059.

Neumann, S. L. (2010). "Animal Welfare Volunteers: Who Are They and Why Do They Do What They Do?" *Anthrozoös*, 23 (4), 351–64.

Rank, M. G., T. L. Zaparanick, and J. E. Gentry (2009). "Non Human-Animal Care Compassion Fatigue." *Best Practices in Mental Health*, 5 (2): 40–61.

Samdahl, D. M. (2015). "'It's [Not] All About the Dogs': Volunteers and Pet Rescue." In *Domestic Animals and Leisure* 89–108. Palgrave Macmillan, London.

Winograd, N. J. (2007). *Redemption: The Myth of Pet Overpopulation and the No Kill Revolution in America.* Cambridge, MA: Almaden Books.

ROBOTIC PETS

Robotic pets are robotic toys that are created to simulate the movements and behaviors of companion animals—primarily dogs. The evolution of robotic pets has gained increasing prominence within the context of elder care, particularly as demographic shifts in many countries result in fewer children available to provide familial support to aging populations. Robotic pets, often designed to mimic emotional support animals, emerged distinctly in the late twentieth and early twenty-first centuries, with several models developed specifically for therapeutic applications in elder care, such as PARO and AIBO (Moshayedi et al., 2023; Robinson, MacDonald, and Broadbent, 2013). These developments reflect the recognition of loneliness and social isolation as significant concerns among older adults, exacerbated by diminishing family structures (Zhang, Luo, and Wang, 2024; Kanamori et al., n.d.).

Target demographics for robotic pets largely encompass elderly populations facing challenges such as loneliness and mental health issues. Research indicates that older adults exhibit a preference for less anthropomorphic designs in robotic pets, finding comfort in their pet-like appearance rather than human-like characteristics (Chiu, Hsieh, and Li, 2021; Lazar et al., 2016). The effectiveness of these robotic companions in fostering emotional benefits is well-documented; they decrease anxiety and enhance overall well-being, particularly in individuals with dementia (Petersen et al., 2016; Abbott et al., 2019). For instance, PARO, a robotic seal, has been shown to profoundly impact the psychological state of users, indicating that robotic pets serve as viable alternatives to traditional live pets within care settings (Jung, Leij, and Kelders, 2017).

In Japan, where the shrinking birthrate has led to a significant demographic shift, the deployment of robotic dogs has become particularly relevant. With fewer children and a rapidly aging population, the integration of robotic pets addresses not only the companionship needs of the elderly but also alleviates the societal pressure on family caregiving (Ihamäki and Heljakka, 2021; Marchetti et al., 2019). Studies conducted in

A girl plays with an AIBO robotic dog designed and manufactured by Sony. Robotic pets have been used to help improve mental health, particularly in elder care. (Maria Argutinskaya/Dreamstime.com)

Japan confirm the high acceptance rates of robotic pets among the elderly, signifying their effectiveness in promoting social interaction and emotional engagement (Zhang, Luo, and Wang, 2024; Kim et al., 2021). Furthermore, interventions involving robotic pets have demonstrated potential in improving social skills and empathy among seniors, further validating their role in communal living settings and senior care facilities (Ihamäki and Heljakka, 2021; Huang et al., 2020).

The complexity of interactions facilitated by these robotic companions has been a focal point of research. Beyond merely providing emotional support, studies are exploring how features like touch sensitivity and speech recognition could enhance engagement and cater to the cognitive needs of users (Jung, Leij, and Kelders, 2017). Ethically, while there are reservations regarding possible deception or diminished human contact due to reliance on robots, the clinical benefits observed in therapy and companionship contexts have largely outweighed these concerns (Sharkey and Sharkey, 2010).

See also: Family, Pets As; Japan, Pets In; Virtual Pets

References and Further Reading:

Abbott, R., N. Orr, P. McGill, R. Whear, A. Bethel, R. Garside, K. Stein, and J. Coon (2019). "How Do "Robopets" Impact the Health and Well-Being of Residents in Care Homes? A

Systematic Review of Qualitative and Quantitative Evidence." *International Journal of Older People Nursing*, 14(3).
Chiu, C., S. Hsieh, and C. Li (2021). "Needs and Preferences of Middle-Aged and Older Adults in Taiwan for Companion Robots and Pets: Survey Study." *Journal of Medical Internet Research*, 23 (6): e23471.
Ihamäki, P., and K. Heljakka (2021). "Robot Pets as "Serious Toys"- Activating Social and Emotional Experiences of Elderly People." *Information Systems Frontiers*, 26 (1): 25–39.
Jung, M., L. Leij, and S. Kelders (2017). "An Exploration of the Benefits of an Animallike Robot Companion with More Advanced Touch Interaction Capabilities for Dementia Care." *Frontiers in Ict*, 4.
Kim, J., S. Kim, S. Kim, E. Lee, Y. Heo, C. Hwang, Y. Choi, et al. (2021). "Companion Robots for Older Adults: Rodgers' Evolutionary Concept Analysis Approach." *Intelligent Service Robotics*, 14 (5): 729–39.
Lazar, A., H. Thompson, A. Piper, and G. Demiris (2016). "Rethinking the Design of Robotic Pets for Older Adults." In *Proceedings of the 2016 ACM Conference on Designing Interactive Systems*, 1034–46. New York: Association for Computing Machinery.
Marchetti, E., W. Juel, R. Langedijk, L. Bodenhagen, and N. Krüger (2019). "The Penguin – On the Boundary between Pet and Machine. An Ecological Perspective on the Design of Assistive Robots for Elderly Care." In: Zhou, J., Salvendy, G. (ed.), *Human Aspects of IT for the Aged Population. Social Media, Games and Assistive Environments. HCII 2019. Lecture Notes in Computer Science()*, vol 11593, 425–43. Springer: Cham.
Moshayedi, A., N. Uddin, X. Zhang, and M. Andani (2023). "Exploring the Role of Robotics in Alzheimer's Disease Care: Innovative Methods and Applications." *Robotic Intelligence and Automation*, 43 (6): 669–90.
Petersen, S., S. Houston, H. Qin, C. Tague, and J. Studley (2016). "The Utilization of Robotic Pets in Dementia Care." *Journal of Alzheimer's Disease*, 55 (2): 569–74.
Robinson, H., B. MacDonald, and E. Broadbent (2013). "Physiological Effects of a Companion Robot on Blood Pressure of Older People in Residential Care Facility: A Pilot Study." *Australasian Journal on Ageing*, 34 (1): 27–32.
Sharkey, A., and N. Sharkey (2010). "Granny and the Robots: Ethical Issues in Robot Care for the Elderly." *Ethics and Information Technology*, 14 (1): 27–40.
Zhang, Y., L. Luo, and X. Wang (2024). "Aging with Robots: A Brief Review on Eldercare Automation." *Interdisciplinary Nursing Research*.

RODENTS

Rodents have gained considerable popularity as pets in recent years, evidenced by a significant rise in households owning these animals. Reports indicate an 11 percent increase in rodent ownership in the United States between 2007 and 2012, reflecting a broader trend observed globally (Lankau et al., 2015). Among various rodent species, certain types stand out as particularly favored among pet owners. For instance, species such as the guinea pig, Syrian hamster, and common rat have become staples in the pet trade (Pereira et al., 2024). The appeal of these rodents often lies in their manageable size, alleged ease of care, and sociable nature, making them especially popular among families with children (James and Wills, 2025). Furthermore, guinea pigs are notably recognized for their friendly interactions and docility, which contribute to their categorization among the top choices (Júnior et al., 2024).

Despite their popularity, the welfare considerations for pet rodents are a matter of significant concern. Rodents can harbor various zoonotic diseases that pose risks to humans, such as leptospirosis and infections caused by parasites and bacterial pathogens (Mori et al., 2017; Matthews and Ausman, 2021). These health risks necessitate robust biosecurity measures in pet shops and a strong commitment from owners toward responsible pet care practices to mitigate the transmission of zoonotic pathogens (Panti–May et al., 2017; Halsby et al., 2014). Additionally, veterinary care for pet rodents can be less systematic than for cats or dogs, often resulting in underdiagnosis of health issues due to their innate ability to conceal illness (Wentz et al., 2020). Owners must be educated about the specific welfare needs of their rodents, which include proper housing, diet, social and mental enrichment, and health monitoring, to enhance the overall well-being of these pets (Chen, Wu, and Lin, 2024).

Moreover, underlying health issues prevalent in pet rodents, including the frequent development of tumors, highlight the importance of regular veterinary check-ups (Pereira et al., 2024). It has been observed that conditions such as urinary tract infections can be problematic, particularly in guinea pigs, and veterinary intervention is crucial for managing such issues (Roberts-Steel et al., 2019). The increasing trend of rodent ownership also raises questions about the environmental impacts of the pet trade, specifically regarding invasive species and their management in different ecosystems (Mori, Zozzoli, and Menchetti, 2018). The merging of public health with pet ownership underscores the necessity for awareness and preventive measures surrounding zoonotic threats inherent to keeping rodents as pets (Chen et al., 2021; Gwenzi et al., 2021).

In sum, while the popularity of rodents as pets continues to grow, ensuring their welfare involves addressing significant health risks and educating owners on appropriate care practices. This multifaceted approach is essential for fostering a safe environment for both pet rodents and their human companions.

See also: Children and Pets; Classroom Pets; Pocket Pets; Zoonoses

References and Further Reading

Chen, J., W. Wang, Y. Lin, L. Sun, N. Li, Y. Guo, M. Kvac, et al. (2021). "Genetic Characterizations of Cryptosporidium Spp. from Pet Rodents Indicate High Zoonotic Potential of Pathogens from Chinchillas." *One Health*, 13: 100269.

Chen, Y., J. Wu, and W. Lin (2024). "Common Spontaneous Tumors and Tumor-Like Lesions in 70 Pet Rodents and Negative Mmtv Detection in Mammary Tumors." *Animals*, 14 (10): 1469.

Gwenzi, W., N. ChaUKura, N. Muisa-Zikali, C. Teta, T. Musvuugwa, P. Rzymski, and A. Abia (2021). "Insects, Rodents, and Pets as Reservoirs, Vectors, and Sentinels of Antimicrobial Resistance." *Antibiotics*, 10 (1): 68.

Halsby, K., A. Walsh, C. Campbell, K. Hewitt, and D. Morgan (2014). "Healthy Animals, Healthy People: Zoonosis Risk from Animal Contact in Pet Shops, a Systematic Review of the Literature." *Plos One*, 9(2), e89309.

James, L., and A. Wills (2025). "Perception and Utilisation of Veterinary Services by Rodent Owners in the United Kingdom." *Veterinary Record*, 196 (8).

Júnior, F., A. Moreira, N. Medeiros, M. Carmo, and M. Lima (2024). "Sporotrichosis in Guinea Pig (cavia porcellus) - Case Report]." *Arquivo Brasileiro De Medicina Veterinária E Zootecnia*, 76 (3).

Lankau, E., J. Sinclair, B. Schroeder, G. Galland, and N. Marano (2015). "Public Health Implications of Changing Rodent Importation Patterns - United States, 1999–2013." *Transboundary and Emerging Diseases*, 64 (2): 528–37.

Matthews, C., and S. Ausman (2021). "Rat-Bite Fever: Taking the Bite out of a Textbook Case of Streptobacillus Moniliformis Blood Stream Infection." *Idcases*, 25: e01199.

Mori, E., R. Zozzoli, and M. Menchetti (2018). "Global Distribution and Status of Introduced Siberian Chipmunks Eutamias Sibiricus." *Mammal Review*, 48 (2): 139–52.

Mori, M., P. Bourhy, M. Guyader, M. Esbroeck, Z. Djelouadji, A. Septfons, A. Kodjo, and M. Picardeau (2017). "Pet Rodents as Possible Risk for Leptospirosis, Belgium and France, 2009 to 2016." *Eurosurveillance*, 22 (43).

Panti–May, J., L. Caraveo-Centeno, S. Hernández-Betancourt, M. Robles, and C. Machaín-Williams (2017). "Survey of Intestinal Helminths Collected from Pet Rodents in México." *Parasitology Research*, 116 (11): 3239–42.

Pereira, R., T. Lima, R. Oliveira, S. Fonseca, N. Wicpolt, R. Farias, R. Lucena, et al. (2024). "Clinical, Pathological and Immunohistochemical Characterization of Spontaneous Neoplasms in Pet Rodents in Northeastern Brazil." *Pesquisa Veterinária Brasileira*, 44.

Roberts-Steel, S., J. Oxley, . Carroll, A. Wills (2019). "Frequency of Owner-Reported Bacterial Infections in Pet Guinea Pigs." *Animals*, 9 (9): 649.

Wentz, M., M. Bianchi, L. Mello, C. Pietzsch, M. Alievi, D. Driemeier, L. Sonne, and S. Pavarini (2020). "Neoplasms in Domestic Hamsters in Southern Brazil: Epidemiological and Pathological Aspects of 40 cases." *Pesquisa Veterinária Brasileira*, 40 (12): 1029–38.

RUSSIA, PETS IN

Pet keeping in Russia has undergone significant transformations since the dissolution of the Soviet Union in 1991, transitioning from a state-affording privilege to a widespread personal choice that reflects socioeconomic changes, urban dynamics, and cultural values.

Barker (1999) provides valuable insights into the pet keeping landscape during the post-Soviet era, describing the revival of pet ownership as a means to embrace personal freedoms previously constrained under the Soviet regime. Pet keeping emerged as a form of self-expression, contributing to a burgeoning consumer culture that aligns with Western ideals, and reflecting a shift toward individualistic values.

In the Soviet era, Nelson (2006) delineates the problems associated with pets in the Soviet Union, highlighting how state policies oscillated between repression and allowance of companion animals. Pets were often viewed as both a means of emotional support and a potential disruptive element to collective ideology.

In the modern era, Nelson (2016) observes a rehabilitation of pet keeping practices within urban settings, indicating a new understanding of animals from mere property to essential members of modern family units. In a contemporary context, pets are increasingly perceived as companions rather than commodities, reshaping social interactions and household dynamics. This evolution is particularly pronounced in urban areas, as discussed by Bekova and Makenov (2018), who investigate dog-keeping practices in cities, illustrating how urban environments influence pet ownership trends and responsibilities. With growing urban populations, a variety of dog ownership styles have emerged, necessitating an understanding of pet welfare, training, and public policies regulating animal care.

Beyond traditional companion animals, Shukhova and MacMillan (2020) delve into the fascination with exotic pets in Russia, illustrating the motivations behind keeping such animals. This trend is linked to unique identity formations, social status, and a reflection of personal interests. The authors report that the choice to keep exotic species provides a sense of novelty and distinction, yet also raises ethical and ecological concerns, prompting discussions on responsible pet ownership and biodiversity conservation.

Current empirical studies confirm that pets in Russia are not only companions but also integral to understanding social dynamics and individual identities within Russian society. Moving forward, ongoing scholarly attention to the implications of pet ownership practices will be critical in shaping policy frameworks and fostering ethical attitudes toward animals in an ever-evolving urban context.

See also: Communism, Pets And; Europe, Pets In; Exotic Pets; Socioeconomic Class, Pets And Status Pets

References and Further Reading

Barker, A. M. (1999). "Going to the Dogs: Pet Life in the New Russia." In *Consuming Russia: Popular Culture, Sex and Society since Gorbachev*, 266–80. Durham: Duke University Press.

Bekova, S., and M. Makenov (2018). "Two Sides of the City: Dog-Keeping Practices in Russian Urban Areas." *Anthrozoös*, 31 (4): 423–32.

Nelson, A. (2006). "A Hearth for a Dog: The Paradoxes of Soviet Pet Keeping." In Lewis H. Siegelbaum (ed.), *Borders of Socialism: Private Spheres of Soviet Russia*, 123–44. New York: Palgrave Macmillan US.

Nelson, A. (2016). "Bringing the Beast Back in: The Rehabilitation of Pet Keeping in Soviet Russia." *Companion Animals in Everyday Life: Situating Human-Animal Engagement within Cultures*, 43–58.

Shukhova, S., and D. C. MacMillan (2020). "From Tigers to Axolotls: Why People Keep Exotic Pets in Russia." *People and Nature*, 2 (4): 940–9.

S

SAFE HAVENS

Safe havens are practices that allow for victims of domestic or interpersonal violence to bring their pets to safety after fleeing a violent home. Traditionally, domestic violence shelters in the United States do not accept pets; this means that for many victims, they must choose between getting themselves (and their children) to safety and ensuring that their pets are not harmed.

Pets often serve as critical emotional support for individuals experiencing intimate partner violence (IPV), creating a strong bond that is often difficult to sever. The role of pets in such contexts cannot be underestimated, as research indicates that many victims delay seeking help out of concern for their animals' safety (Kogan et al., 2004; Stevenson, Fitzgerald, and Barrett, 2017; Ascione et al., 2007). A significant body of research shows that victims frequently report that their abuser has threatened or harmed their pets, which complicates the decision to leave abusive situations (Simmons and Lehmann, 2007; Krienert et al., 2012; Stevenson, Fitzgerald, and Barrett, 2017). Moreover, studies reveal that pets are often perceived as family members, and their welfare is closely tied to the mental health and emotional well-being of their owners (Ascione et al., 2007; Hardesty et al., 2013).

Historically, there has been a lack of accommodation in domestic violence shelters for pets. Most shelters have treated pets as property or possessions rather than integral family members, which has led to barriers that prevent victims from taking refuge along with their animals (Fitzgerald et al., 2020; Ascione et al., 2007). The Sheltering Animals and Families Together (SAF-T) initiative illustrates a shift toward understanding the necessity of pet-inclusive shelter solutions, promoting the establishment of models where victims and their pets can coexist safely (Giesbrecht, 2021). Despite this progress, there remains a pressing need for comprehensive legislation to ensure that federal funding is adequately available for these initiatives.

Legislative efforts to support the creation of pet-friendly domestic violence shelters are crucial. Many advocates emphasize the need for increased funding and more flexible regulations to accommodate the needs of pet owners fleeing situations of IPV. Research suggests that providing safe spaces for both victims and their pets can enhance recovery efforts, as it alleviates the stress and anxiety associated with leaving pets behind (Giesbrecht, 2021; Stevenson, Fitzgerald, and Barrett, 2017). This trend is gaining traction, with several domestic violence programs actively integrating pet sheltering capabilities into their services. Studies have shown that shelters with such offerings report higher rates of victims successfully leaving abusive situations, underscoring the importance of these programs in facilitating safe transitions (Simmons and Lehmann, 2007; Stevenson , Fitzgerald, and Barrett, 2017).

Furthermore, understanding the connections between animal abuse and human violence can inform the interventions and support systems available to domestic violence victims. Violent behavior toward pets by abusers often serves as a method of control, reinforcing the need for comprehensive approaches that address both human and animal victims within a single framework of support (Simmons and Lehmann, 2007; Krienert et al., 2012). The recognition that animal welfare is a critical element in addressing the full scope of domestic violence reflects a growing understanding of the complexities involved in these situations (Stevenson, Fitzgerald, and Barrett, 2017; November, 2023). Future legislative measures and support programs will be vital in ensuring the safety and emotional well-being of both victims and their pets, thus paving the way for a more holistic approach to addressing domestic violence.

See also: Cruelty; Laws and Regulations; Link between Animal and Human Abuse

References and Further Reading

Ascione, F., C. Weber, T. Thompson, J. Heath, M. Maruyama, and K. Hayashi (2007). "Battered Pets and Domestic Violence." *Violence Against Women*, 13 (4): 354–73.

Fitzgerald, A., B. Barrett, A. Gray, and C. Cheung (2020). "The Connection between Animal Abuse, Emotional Abuse, and Financial Abuse in Intimate Relationships: Evidence from a Nationally Representative Sample of the General Public." *Journal of Interpersonal Violence*, 37 (5–6): 2331–53.

Giesbrecht, C. (2021). "Animal Safekeeping in Situations of Intimate Partner Violence: Experiences of Human Service and Animal Welfare Professionals." *Journal of Interpersonal Violence*, 37 (17–18): NP16931-NP16960.

Hardesty, J., L. Khaw, M. Ridgway, C. Weber, and T. Miles (2013). "Coercive Control and Abused Women's Decisions about Their Pets When Seeking Shelter." *Journal of Interpersonal Violence*, 28 (13): 2617–39.

Kogan, L., S. McConnell, R. Schoenfeld-Tacher, and P. Jansen-Lock (2004). "Crosstrails." *Violence Against Women*, 10 (4): 418–34.

Krienert, J., J. Walsh, K. Matthews, and K. McConkey (2012). "Examining the Nexus between Domestic Violence and Animal Abuse in a National Sample of Service Providers." *Violence and Victims*, 27 (2): 280–96.

November, R. (2023). "Protection for Pets and People: The Critical Need for Increased Federal Funding for Pet-Friendly Domestic Violence Services." *Family Court Review*, 62 (1): 243–56.

Simmons, C., and P. Lehmann (2007). "Exploring the Link between Pet Abuse and Controlling Behaviors in Violent Relationships." *Journal of Interpersonal Violence*, 22 (9): 1211–22.
Stevenson, R., A. Fitzgerald, and B. Barrett (2017). "Keeping Pets Safe in the Context of Intimate Partner Violence." *Affilia*, 33 (2): 236–52.

SANCTUARIES

Animal sanctuaries are places which take in abandoned animals, often those who have been subjected to exploitation or hardship, and keep them for the duration of their lives. The development of these sanctuaries can be traced back to the late twentieth century, as activists and organizations sought to create safe havens for animals originally rescued from abuse or neglect, particularly in the context of rising societal awareness regarding animal rights and welfare issues. The modern sanctuary movement gained momentum around the 1980s, paralleling a broader shift in public consciousness toward humane treatment and rehabilitation of both farmed and wild animals (Paik, 2017; Jackson, Lewis, and Fultz, 2024; Rosenfeld, 2023), and for the need to provide permanent refuge for animals who cannot or should not be adopted out or otherwise released.

These sanctuaries typically serve several purposes: they offer lifelong care to animals that cannot be reintroduced to the wild, facilitate rehabilitation for those that may be eligible for release, and often engage in educational activities aimed at raising public awareness concerning animal welfare issues (Trayford and Farmer, 2012; Kuhar et al., 2011; Abrell, 2019). Most animal sanctuaries operate under a model that emphasizes a non-exploitative relationship with their residents. This contrasts with traditional zoos or commercial facilities, which may prioritize profit and entertainment over animal welfare (Rosenfeld, 2023; Clegg, 2021). Importantly, animal sanctuaries frequently collaborate with animal shelters and rescue groups to address challenges posed by overpopulation, particularly by providing a permanent residence for animals that would otherwise face euthanasia (Paik, 2017; Speiran, 2025).

The impact of animal sanctuaries on society extends beyond the immediate benefits to rescued animals. Sanctuaries play a crucial role in shaping public perceptions of animal rights, engaging communities through outreach programs, and often fostering a greater appreciation for biodiversity (Trayford and Farmer, 2012; Islam et al., 2018; Abrell, 2019). The educational components of sanctuaries serve to inform the public about the realities of animal exploitation and the importance of conservation efforts. By bridging the gap between animal welfare and environmental conservation, these institutions promote social activism and drive forward the discourse on ethical considerations in human–animal relations (Moorhouse et al., 2015; Abrell, 2019). Moreover, sanctuaries also impact local economies through tourism and volunteer opportunities, contributing to both community development and conservation (Trayford and Farmer, 2012; Rosenfeld, 2023).

In summary, animal sanctuaries in the United States emerged from a context of growing advocacy for animal welfare, collaborating closely with shelters and rescue organizations to provide essential care for vulnerable animals. Their multifaceted societal impacts illustrate

A wolf at the Wild Animal Sanctuary, a 720-acre animal refuge housing more than 350 large animals near Keenesburg, Colorado. The sanctuary specializes in rescuing and caring for large predators that have being ill-treated. (Library of Congress)

their significance not only in fostering a compassionate ethos toward animals but also in engaging communities to consider the societal implications of animal welfare (Paik, 2017; Trayford and Farmer, 2012; Abrell, 2019).

See also: Animal Protection Movement; Humane Societies; Rescue Groups; Shelters

References and Further Reading

Abrell, E. (2019). "Sanctuary-Making as Rural Political Action." *Journal for the Anthropology of North America*, 22 (2): 109–11.

Clegg, I. (2021). "What Does the Future Hold for the Public Display of Cetaceans?." *Journal of Applied Animal Ethics Research*, 3 (2): 240–78.

Islam, M., R. Aktar, M. Nahiduzzaman, B. Barman, and M. Wahab (2018). "Social Considerations of Large River Sanctuaries: A Case Study from the Hilsa Shad Fishery in Bangladesh." *Sustainability*, 10 (4): 1254.

Jackson, R., R. Lewis, and A. Fultz (2024). "The Interface between Veterinary and Behavioral Management of Chimpanzees (Pan Troglodytes) in a United States Sanctuary Demonstrated by Two Clinical Cases." *Veterinary Sciences*, 11 (11): 523.

Kuhar, C., T. Bettinger, K. Lehnhardt, B. Cartwright, and D. Cress (2011). "Education Program Evaluation at Multiple Primate Sanctuaries in Equatorial Africa." *International Journal of Primatology*, 33 (1): 208–17.

Moorhouse, T., C. Dahlsjö, S. Baker, N. D'Cruze, and D. Macdonald (2015). "The Customer Isn't Always Right—Conservation and Animal Welfare Implications of the Increasing Demand for Wildlife Tourism." *Plos One*, 10 (10): e0138939.
Paik, A. (2017). "Abolitionist Futures and the US Sanctuary Movement." *Race & Class*, 59 (2): 3–25.
Rosenfeld, H. (2023). "Deviant Accumulation at Farmed Animal Sanctuaries." *Environment and Planning E Nature and Space*, 7 (1): 271–87.
Speiran, S. (2025). "The 'Sanctuary Gap': Reviewing the Research on Captive Wildlife Sanctuary Tourism." *Animals*, 15 (4): 496.
Trayford, H., and K. Farmer (2012). "Putting the Spotlight on Internally Displaced Animals (IDAS): A Survey of Primate Sanctuaries in Africa, Asia, and the Americas." *American Journal of Primatology*, 75 (2): 116–34.

SELECTIVE BREEDING

The literature on selective breeding in pets, particularly dogs, reflects a significant evolution over time in both purpose and consequences. Initially, breeding practices were driven by functional needs, focusing on traits that enhanced a dog's work performance, such as hunting or guarding abilities (Menor-Campos, 2024; Smith and Laitman, 2020). However, there has been a notable shift toward breeding for aesthetic traits that adhere to breed standards, often to the detriment of health and functionality (Collins et al., 2010; Mäki et al., 2005).

This change has sparked considerable debate within the scientific and veterinary communities regarding its implications for animal welfare. Selective breeding for specific physical attributes—such as brachycephalic features—illustrates this concern vividly, as these traits can lead to a range of health issues, including respiratory problems and decreased longevity (Collins et al., 2010; Selba et al., 2018; Åsbjer, Hedhammar, and Engdahl, 2024). The popularity of breeds like Bulldogs and Pugs, which are often chosen for their physical appearance, has highlighted the serious health conditions associated with such traits, including Brachycephalic Obstructive Airway Syndrome (Selba et al., 2018; Åsbjer, Hedhammar, and Engdahl, 2024). Research indicates that breeding practices prioritizing aesthetics often overlook the potential for inherited defects, exacerbating health issues in contemporary dog breeds (Menor-Campos, 2024; Subramanian and Kumar, 2024; Packer, Murphy, and Farnworth, 2017).

Moreover, inbreeding remains a significant factor linked to various health problems in purebred dogs, as it tends to increase the prevalence of deleterious genetic variants (Subramanian & Kumar, 2024). The relationship between high inbreeding coefficients and increased morbidity has been well-documented. Breeding for desirable traits often results in a limited gene pool, which exacerbates health problems and reduces genetic diversity—a crucial aspect for the long-term viability of breed populations (Felska-Błaszczyk and ŁASKA, 2024). This has led to urgent calls from experts for new breeding approaches that prioritize health and robustness over appearance (Mäki et al., 2005; Leighton et al., 2018).

Ongoing discussions focus on reshaping selective breeding strategies, including the implementation of more comprehensive health screening processes and consideration of mixed-breed dogs, which may offer healthier genetic profiles. The enthusiasm for "designer breeds" and combining breeds in a more health-oriented manner holds promise for improving welfare outcomes for dogs (Proschowsky et al., 2025).

Today, the evolution of dog breeding from functionality to aesthetics raises significant ethical concerns regarding animal welfare. The emphasis on adhering to breed standards has contributed to an increase in inherited health issues, primarily driven by inbreeding and selective pressure for superficial traits, underscoring the need for a critical reevaluation of breeding practices. Prioritizing the health, longevity, and overall welfare of companion animals is essential to address these challenges.

See also: Breeds, Popularity Of; Cats; Dogs; Domestication

References and Further Reading

(2019). "Improving the Health of Pedigree Dogs." *Veterinary Record*, 184 (26): 782–82.

Åsbjer, E., Å. Hedhammar, K. Engdahl (2024). "Awareness, Experiences, and Opinions by Owners, Breeders, Show Judges, and Veterinarians on Canine Brachycephalic Obstructive Airway Syndrome (boas)." *Canine Medicine and Genetics*, 11 (1).

Collins, L., L. Asher, J. Summers, G. Diesel, P. McGreevy (2010). "Welfare Epidemiology as a Tool to Assess the Welfare Impact of Inherited Defects on the Pedigree Dog Population." *Animal Welfare*, 19 (S1): 67–75.

Felska-Błaszczyk, L., and P. ŁASKA (2024). "Health Problems in Selected Breeds of Large dogs." *Folia Pomeranae Universitatis Technologiae Stetinensis Agricultura Alimentaria Piscaria Et Zootechnica*, 373 (72): 1–11.

Leighton, E., E. Hare, S. Thomas, L. Waggoner, C. Otto (2018). "A Solution for the Shortage of Detection Dogs: A Detector Dog Center of Excellence and a Cooperative Breeding Program." *Frontiers in Veterinary Science*, 5.

Mäki, K., A. Liinamo, A. Groen, P. Bijma, and M. Ojala (2005). "The Effect of Breeding Schemes on the Genetic Response of Canine Hip Dysplasia, Elbow Dysplasia, Behaviour Traits and Appearance." *Animal Welfare*, 14 (2): 117–24.

Menor-Campos, D. (2024). "Ethical Concerns about Fashionable Dog Breeding." *Animals*, 14 (5): 756.

Packer, R., D. Murphy, and M. Farnworth (2017). "Purchasing Popular Purebreds: Investigating the Influence of Breed-Type on the Pre-Purchase Motivations and Behaviour of Dog Owners." *Animal Welfare*, 26 (2): 191–201.

Proschowsky, H., M. Arendt, B. Bonnett, C. Bruun, I. Czycholl, M. Fredholm, D. O'Neill, J. Serpell, and P. Sandøe (2025). "A New Future for Dog Breeding." *Animal Welfare*, 34.

Selba, M., G. Oechtering, H. Heng, and V. DeLeon (2018). "The Impact of Selection for Facial Reduction in Dogs." *The Faseb Journal*, 32 (S1).

Smith, C., and J. Laitman (2020). "Alterations to Vestibular Morphology in Highly Bred Domestic Dogs May Affect Balance." *The Anatomical Record*, 304 (1): 116–26.

Subramanian, S., and M. Kumar (2024). "The Association between the Abundance of Homozygous Deleterious Variants and the Morbidity of Dog Breeds." *Biology*, 13 (8): 574.

SHELTERS

Animal shelters, once known as pounds, are a nineteenth-century innovation that today seems like an inescapable part of the pet world. The establishment of the American Society for the Prevention of Cruelty to Animals (ASPCA) by Henry Bergh in 1866 marked a critical juncture in the treatment and welfare of animals in the United States, and while the ASPCA was not initially conceived of as a shelter, it eventually incorporated sheltering into its mission, and thus served as the model for every other animal shelter in the United States today.

Bergh's organization was founded to prevent cruelty toward animals, asserting that they deserved protection and humane treatment, similar to that afforded to humans. The ASPCA led campaigns against inhumane practices and pushed for legal frameworks to improve animal welfare (Guenther and Hassen, 2024). This paradigm shift contributed to the foundation of more structured animal shelters, which sought not only to house stray animals but also to promote their adoption and welfare.

Historically, what were once called pounds operated primarily as enforcers of local leash laws and ordinances, where lost or stray animals were held until claimed, via an impound fee, or euthanized. These pounds bore very little resemblance to today's animal shelters, as they were primarily designed to control stray populations rather than to promote animal welfare (Rowan and Kartal, 2018). By the late nineteenth and early twentieth centuries, animal shelters began to evolve alongside broader animal welfare sentiment, focusing more on the adoption and rehabilitation of animals rather than simply their containment and control (Guenther and Hassen, 2024). Consequently, the number of animals entering shelters and the rates of euthanasia have declined significantly, reflecting changes in public attitudes and policies concerning animal care (Spehar and Wolf, 2020).

Today, the landscape of animal sheltering in the United States has made substantial progress, with a focus on reducing euthanasia through innovative strategies such as Return-To-Field (RTF) and Trap-Neuter-Return (TNR) programs. These methods aim to manage feral cat populations effectively while enhancing the welfare of both domesticated and wild animals (Spehar and Wolf, 2019). Statistical reports indicate that shelter euthanasia rates in the United States have decreased dramatically, from about 90 percent of all incoming animals in the 1970s to approximately 20–30 percent in recent years (Spehar and Wolf, 2020). That has resulted in a drop from about twenty million dogs and cats euthanized each year to the current four to five million. Such programs and the expanding role of community involvement have helped curb the intake of animals into shelters, ultimately providing better outcomes for more animals.

Globally, animal sheltering practices vary widely. For example, in Taiwan, societal attitudes toward animal welfare are evolving, with increasing opposition to euthanasia and greater interest in animal protection laws (Weng et al., 2006). However, challenges remain, such as resource limitations and public education regarding responsible pet ownership and animal welfare (Hsu, Severinghaus, and Serpell, 2003). Similarly, various European countries have adopted a more progressive stance on sheltering, emphasizing the importance of adoption and fostering, thereby creating more humane environments for displaced animals (Montoya et al., 2017).

The evolution of animal shelters from mere containment units to organizations focused on welfare and rehabilitation exemplifies broader shifts in societal values regarding animal care. Looking forward, continued improvements in shelter management, community-based strategies, and legislative support will be essential to safeguard animal welfare and reduce euthanasia rates both in the United States and around the globe.

See also: Animal Protection Movement; Rescue Groups; Sanctuaries; United States, Pets In

References and Further Reading

Guenther, K., and K. Hassen (2024). "Coming to Terms with the Legacies of the Pound Model in Animal Sheltering in the United States." *Animals*, 14 (9): 1254.

Hsu, Y., L. Severinghaus, and J. Serpell (2003). "Dog Keeping in Taiwan: Its Contribution to the Problem of Free-Roaming Dogs." *Journal of Applied Animal Welfare Science*, 6 (1): 1–23.

Montoya, A., J. Rand, R. Greer, C. Alberthsen, and D. Vankan (2017). "Relationship between Sources of Pet Acquisition and Euthanasia of Cats and Dogs in an Animal Shelter: A Pilot Study." *Australian Veterinary Journal*, 95(6), 194–200.

Rowan, A., and T. Kartal (2018). "Dog Population & Dog Sheltering Trends in the United States of America." *Animals*, 8 (5): 68.

Spehar, D., and P. Wolf (2019). "Integrated Return-to-Field and Targeted Trap-Neuter-Vaccinate-Return Programs Result in Reductions of Feline Intake and Euthanasia at Six Municipal Animal Shelters." *Frontiers in Veterinary Science*, 6.

Spehar, D., and P. Wolf (2020). "The Impact of Return-to-Field and Targeted Trap-Neuter-Return on Feline Intake and Euthanasia at a Municipal Animal Shelter in Jefferson County, Kentucky." *Animals*, 10 (8): 1395. 5

Weng, H., P. Kass, L. Hart, and B. Chomel (2006). "Animal Protection Measures in Taiwan: Taiwanese Attitudes toward the Animal Protection Law and Animal Shelters." *Journal of Applied Animal Welfare Science*, 9 (4): 315–26.

SOCIAL CAPITAL, PETS AND

The relationship between pet ownership and social capital has garnered significant attention within the academic literature, highlighting the multifaceted benefits that pets provide as companions and facilitators of social connectivity. Numerous studies demonstrate that pet ownership correlates positively with social capital across various demographics and geographical contexts. For instance, Wood et al. demonstrate that pet ownership is associated with higher levels of social capital in diverse urban settings in the United States and Australia, suggesting a consistent pattern across differing cultural contexts (Wood et al., 2017). Similarly, McDonald et al. assert that pets serve as conversation starters and social connectors, allowing individuals to engage more readily with others, thereby enhancing social networks and relationships (McDonald et al., 2021). This notion is further supported by Arkow, who posits that pets can alleviate nature-deficit disorder within communities and enhance community-level social capital through their presence and the relationships they foster among owners (Arkow, 2015).

The developmental benefits of pet ownership also extend to younger populations, where Purewal et al. highlight that childhood pet ownership encourages social competencies, empathy, and communication skills, all of which are foundational for building social capital (Purewal et al., 2017). The effectiveness of pets in promoting social bonding is reinforced by research showing that pets act as facilitators of social interactions that enhance neighborhood cohesion, although specific studies focused on this mechanism may be limited (Crawford et al., 2023). Moreover, McConnell et al. observe that for individuals lacking human social support—often due to physical restrictions such as illness—pets can provide vital companionship, thus compensating for social deficits and enhancing psychological well-being (McConnell et al., 2011).

Furthermore, the psychological implications of pet ownership are significant; research indicates that the presence of pets correlates with positive mental health outcomes. For instance, Jiang et al. highlight that pet ownership may foster greater self-compassion and reduce feelings of isolation, particularly in professional environments such as nursing where high emotional demands exist (Jiang et al., 2023). Similarly, studies have shown that older adults who experience low social connectedness can benefit from pet ownership, which may mitigate feelings of loneliness and depression (Stanley et al., 2013). On a communal level, although the evidence is less than conclusive, it has been suggested that pet ownership could contribute to increased community engagement, linking mental health outcomes indirectly to enhanced social capital (Friedmann et al., 2020).

In conclusion, the literature consistently illustrates that pets are more than mere companions; they play a pivotal role in fostering social capital by enhancing community ties, facilitating social interactions, and improving mental health. The reciprocal relationship between pet ownership and social capital underscores the importance of considering pets as integral components of social frameworks within communities.

See also: Ethnicity, Pets And; Socioeconomic Class, Pets And; Status Pets

References and Further Reading

Arkow, P. (2015). "Animal Therapy on the Community Level: The Impact of Pets on Social Capital." In A. H. Fine (ed.), *Handbook on Animal-Assisted Therapy*, 43–51. Boston: Academic Press.

Crawford, C., J. Rand, V. Rohlf, R. Scotney, and P. Bennett (2023). "Solutions-Based Approach to Urban Cat Management—Case Studies of a One Welfare Approach to Urban Cat Management." *Animals*, 13 (21): 3423.

Friedmann, E., N. Gee, E. Simonsick, S. Studenski, B. Resnick, E. Barr, M. Kitner-Triolo, and A. Hackney (2020). "Pet Ownership Patterns and Successful Aging Outcomes in Community Dwelling Older Adults." *Frontiers in Veterinary Science*, 7.

Jiang, H., Y. Mei, X. Wang, W. Wang, B. Lin, Z. Zhao, and Z. Zhang (2023). "The Influence of Pet Ownership on Self-Compassion among Nurses: A Cross-Sectional Study." *Peerj*, 11: e15288.

McConnell, A., C. Brown, T. Shoda, L. Stayton, and C. Martin (2011). "Friends with Benefits: On The Positive Consequences of Pet Ownership." *Journal of Personality and Social Psychology*, 101 (6): 1239–52.

McDonald, S., A. Matijczak, N. Nicotera, J, Applebaum, L. Kremer, G. Natoli, R. Ryan, et al. (2021). ""he was like, my ride or die": Sexual and Gender Minority Emerging Adults' Perspectives on Living with Pets During the Transition to Adulthood." *Emerging Adulthood*, 10 (4): 1008–25.
Purewal, R., R. Christley, K. Kordas, C. Joinson, K. Meints, N. Gee, and C. Westgarth (2017). "Companion Animals and Child/Adolescent Development: A Systematic Review of the Evidence." *International Journal of Environmental Research and Public Health*, 14 (3): 234.
Stanley, I., Y. Conwell, C. Bowen, and K. Orden (2013). "Pet Ownership may Attenuate Loneliness among Older Adult Primary Care Patients who Live Alone." *Aging & Mental Health*, 18 (3): 394–9.
Wood, L., K. Martin, H. Christian, S. Houghton, I. Kawachi, S. Vallesi, and S. McCune (2017). "Social Capital and Pet Ownership – A Tale of Four Cities." *SSM - Population Health*, 3: 442–7.

SOCIALIZATION, PETS AND

The relationship between pets and childhood development has garnered considerable scholarly attention, particularly in understanding how pet companionship influences various aspects of children's socialization and emotional well-being. According to Purewal et al., ownership of companion animals presents children with opportunities to form social connections and enhance their social skills, leading to improved developmental outcomes in areas such as empathy and social communication. This research indicates that it is not merely ownership that matters, but the strength of the attachment between children and their pets, which emerges as a more significant determinant of positive developmental benefits (Purewal et al., 2017).

The role of pets in promoting responsibility and independence has also been emphasized across various age groups. Hara highlights that children acquire valuable life skills, including responsibility, companionship, and emotional regulation, from caring for pets. This preparedness facilitates more complex social interactions (Hara, 2007). Aydemir et al. further support this by stating that interaction with pets can foster cognitive and social skills, particularly during challenging periods, such as the Covid-19 pandemic, where interactions with pets were found to alleviate stress and enhance behavioral development (Aydemir et al., 2021).

Historically, the Victorian era emphasized "proper" child-rearing, which included instilling moral values, good manners, and emotional resilience. Children were often encouraged to engage with animals as part of their upbringing, fostering empathy and compassion (Yu, 2024). This historical perspective aligns with contemporary findings that demonstrate how pets not only serve as playmates but also play a crucial role in enhancing emotional literacy and social competence among children.

Modern studies indicate that children with pets frequently exhibit improved emotional expression and regulation. For instance, Sato et al. report that pet ownership correlates with better emotional outcomes, potentially offering protection against future psychological challenges (Sato et al., 2019). Moreover, evidence suggests that pets can act as buffers against negative socioeconomic factors affecting social relationships, thereby promoting resilience among children (Bryant and Donnellan, 2007).

Children with pets are also reported to possess a more positive attitude toward animals, indicating that early interactions can shape lifelong values concerning empathy and care for other living beings (Kerns et al., 2016). This nurturing bond is further evidenced by Kerns et al., who reveal that children often perceive their pets as friends, illustrating the meaningful connections forged and the support these relationships provide in navigating social contexts (Kerns et al., 2016).

In conclusion, the literature presents a comprehensive framework highlighting the multifaceted role of pets in childhood socialization, from improving social skills and emotional health to instilling a sense of responsibility and empathy. The synthesis of historical perspectives, particularly from the Victorian era, with modern psychological insights provides a robust understanding of why integrating pets into childhood development practices can be beneficial.

See also: Children and Pets; Family, Pets As; Link between Animal and Human Abuse

References and Further Reading

Aydemir, E., İ. Bilge, G. Chatzisouleiman, M. Musul, and T. Özel (2021). "Investigation of the Effects of the Interaction between Pets and Children on Development and Psychology During the COVID-19 Pandemic Period." *World Journal of Advanced Research and Reviews*, 10 (1): 203–9.

Bryant, B., and M. Donnellan (2007). "The Relation between Socio-Economic Status Concerns and Angry Peer Conflict Resolution is Moderated by Pet Provisions of Support." *Anthrozoös*, 20 (3): 213–23.

Hara, S. (2007). "Managing the Dyad between Independence and Dependence: Case Studies of the American Elderly and Their Lives with Pets." *International Journal of Japanese Sociology*, 16 (1): 100–14.

Kerns, K., A. Koehn, M. Dulmen, K. Stuart-Parrigon, and K. Coifman (2016). "Preadolescents' Relationships with Pet Dogs: Relationship Continuity and Associations with Adjustment." *Applied Developmental Science*, 21 (1): 67–80.

Purewal, R., R. Christley, K. Kordas, C. Joinson, K. Meints, N. Gee, and C. Westgarth (2017). "Companion Animals and Child/Adolescent Development: A Systematic Review of the Evidence." *International Journal of Environmental Research and Public Health*, 14 (3): 234.

Sato, R., T. Fujiwara, S. Kino, N. Nawa, and I. Kawachi (2019). "Pet Ownership and Children's Emotional Expression: Propensity Score-Matched Analysis of Longitudinal Data from Japan." *International Journal of Environmental Research and Public Health*, 16 (5): 758.

Yu, C. (2024). "A Systematic Review of the Positive Effects of Pets on Child Development." *Communications in Humanities Research*, 39 (1): 82–7.

SOCIOECONOMIC CLASS, PETS AND

The relationship between pet ownership and socioeconomic class is a relatively new area of study, illustrating how factors such as wealth and class have historically influenced the prevalence and perception of pet ownership. Initially, pets were considered a luxury for the affluent, often linked to displays of social status. The wealthy maintained a variety of pets, and specific breeds were markers of prestige (Westgarth et al., 2010; Westgarth et al., 2013).

As societies evolved and economic structures changed, especially during the late nineteenth and twentieth centuries, pets gradually transitioned from being exclusive to the wealthy to commonplace companions across various socioeconomic strata. This shift was facilitated in part by increased urbanization and the rise of disposable incomes, allowing broader access to pet ownership across classes (Mueller et al., 2021; Saunders et al., 2017). The resultant cultural acceptance of pets as family members rather than mere possessions has been reflected in changing attitudes toward animal welfare and care, transforming the way pets are regarded in society (Lopes-Conceição et al., 2023; Pikhartova et al., 2014). Moreover, studies have suggested that health and well-being outcomes associated with pet ownership might reflect broader socioeconomic privileges, wherein pet owners often enjoy better health care access and mental health stability than non-owners (Scoresby et al., 2021; Purewal et al., 2019).

Sociodemographic factors continue to influence pet ownership. For instance, research suggests that employment status, education, and family composition significantly affect pet ownership patterns (Purewal et al., 2019; Ikeuchi et al., 2022). Higher socioeconomic status correlates with an increased likelihood of pet ownership due to the resources required for pet care, including time, money, and housing accommodations (Chin et al., 2023; King et al., 2024). Furthermore, studies indicate that parents and those with children in

Equestrian Portrait of Cornelis (1639–80) and Michiel Pompe van Meerdervoort (1638–53) with Their Tutor and Coachman, by Dutch painter Aelbert Cuyp, c. 1652–53. Pets were historically considered a luxury for the affluent. (The Metropolitan Museum of Art, The Friedsam Collection, Bequest of Michael Friedsam, 1931)

the household are more likely to own pets, suggesting the role of family dynamics in the ownership landscape (Heyworth et al., 2006; King et al., 2024).

In conclusion, pets have transitioned from symbols of wealth to integral parts of family life, with ownership now influenced by a complex web of social and economic factors. Understanding these dynamics not only sheds light on pet ownership trends but also highlights the disparities in access and care that persist across different socioeconomic groups.

See also: Capitalism, Pets And; Ethnicity, Pets And; Status Pets

References and Further Reading

Chin, B., E. Casserly, S. Parsons, and B. Feeney (2023). "Pet Ownership and Mental Health in United States Adults During COVID-19." *Frontiers in Psychology*, 14.

Heyworth, J., H. Cutt, and G. Glonek (2006). "Does Dog or Cat Ownership Lead to Increased Gastroenteritis in Young Children in South Australia?" *Epidemiology and Infection*, 134 (5): 926–34.

Ikeuchi, T., Y. Taniguchi, T. Abe, Y. Yokoyama, S. Seino, M. Narita, M. Nishi, et al. (2022). "Pet Ownership and the Future Time Perspective of Older Adults." *Geropsych*, 35 (4): 226–33.

King, E., S. Dowling-Guyer, E. McCobb, and M. Mueller (2024). "Longitudinal Patterns of Companion Animals in Families with Children During the COVID-19 Pandemic: Findings from the Adolescent Brain Cognitive Development (abcd) Study®." *Frontiers in Veterinary Science*, 11.

Lopes-Conceição, L., B. Peleteiro, N. Araújo, T. Dias, F. Fontes, S. Pereira, S., and N. Lunet (2023). "Pet Ownership During the First 5 Years After Breast Cancer Diagnosis in the Neon-bc Cohort." *European Journal of Public Health*, 33 (3): 455–62.

Mueller, M., E. King, K. Callina, S. Dowling-Guyer, and E. McCobb (2021). "Demographic and Contextual Factors as Moderators of the Relationship between Pet Ownership and Health." *Health Psychology and Behavioral Medicine*, 9 (1): 701–23.

Pikhartova, J., A. Bowling, and C. Victor (2014). "Does Owning a Pet Protect Older People against Loneliness?." *BMC Geriatrics*, 14 (1).

Purewal, R., R. Christley, K. Kordas, C. Joinson, K. Meints, N. Gee, and C. Westgarth (2019). "Socio-Demographic Factors Associated with Pet Ownership Amongst Adolescents from a UK Birth Cohort." *BMC Veterinary Research*, 15 (1).

Saunders, J., L. Parast, S. Babey, and J. Miles (2017). "Exploring the Differences between Pet and Non-Pet Owners: Implications for Human-Animal Interaction Research and Policy." *Plos One*, 12 (6): e0179494.

Scoresby, K., E. Strand, Z. Ng, K. Brown, C. Stilz, K. Strobel, C. S. Barroso, and M. Souza (2021). "Pet Ownership and Quality of Life: A Systematic Review of the Literature." *Veterinary Sciences*, 8 (12): 332.

Westgarth, C., L. Boddy, G. Stratton, A. German, R. Gaskell, K. Coyne, P. Bundred, S. McCune, and S. Dawson (2013). "Pet Ownership, Dog Types and Attachment to Pets in 9–10 Year Old Children in Liverpool, U.K." *BMC Veterinary Research*, 9 (1): 102.

Westgarth, C., J. Heron, A. Ness, P. Bundred, R. Gaskell, K. Coyne, A. German, S. McCune, and S. Dawson (2010). "Family Pet Ownership During Childhood: Findings from a U.K. Birth Cohort and Implications for Public Health Research." *International Journal of Environmental Research and Public Health*, 7 (10): 3704–29.

STATUS PETS

Status pets are pets that are kept primarily or partially to provide status to their owner. The phenomenon of status pets, particularly in the context of pit bull-type dogs, is multifaceted, involving social dynamics, individual behaviors, and cultural representations.

Social status plays a significant role in how certain dog breeds are perceived and owned, particularly in youth cultures. Maher and Pierpoint highlight how pit bull-type dogs are frequently regarded as status symbols within urban youth gangs, where their ownership is aligned with social positioning and a means of assertive identification among peers (Maher and Pierpoint, 2011). This association with danger and aggression is reinforced by the perception that these breeds can enhance the social standing of their owners, irrespective of the actual behavior exhibited by individual dogs (Gunter, Barber, and Wynne, 2016).

The ownership of pit bull-type dogs among non-white youths suggests that these breeds may sometimes serve as proxies for identity in marginalized communities. This aligns with observations by Westgarth et al. regarding the correlation of dog ownership with socioeconomic factors, where lower social classes may gravitate toward the ownership of pit bulls as a means of socialization and protection in inner-city environments (Westgarth et al., 2013). Such relationships underline the intersectionality of race, class, and animal affiliation, challenging mainstream assumptions about pet ownership and breed-specific legislation (BSL) (Ly, Gordon, and Protopopova, 2021).

In the context of popular culture, figures like Paris Hilton have been instrumental in reshaping perceptions of certain dog breeds. Hilton's association with small dogs—especially chihuahuas—can be viewed as an attempt to project an image of luxury and social status, which starkly contrasts with the use of pit bulls as status symbols associated with toughness and resilience in certain social circles. This contrasting perception encapsulates broader societal narratives: while Hilton's choices mirror affluent trends, pit bulls are often emblematic of rebellion against normative pet ownership expectations (Gunter, Barber, and Wynne, 2016).

Moreover, the issue of breed-specific legislation (BSL) complicates the status of pit bull-type dogs in terms of social dynamics. Studies indicate that BSL is more closely linked to perceptions influenced by socioeconomic status and systemic racism than to actual safety concerns regarding these dogs (Ly, Gordon, and Protopopova, 2021; d'Angelo et al., 2022). The stigmatization of breed categories leads to disproportional rates of surrender and euthanasia among pit bulls compared to other breeds, reflecting societal biases that transcend breed and manifest in legislation (Ly, Gordon, and Protopopova, 2021).

Additionally, the misuse of pit bulls as weapons within organized dogfighting circles underscores a darker side of this status association, reinforcing negative stereotypes (McEwen, 2017; Nurse, 2021). The criminalization of these dogs for their perceived roles within gang cultures elevates their status but simultaneously subjects them to severe societal backlash and legislative aggression. This duality of status—while empowering for some, ultimately leads to the subjugation of the breed as a whole (McEwen, 2017; Nurse, 2021).

Today, pit bull-type dogs are experiencing a redefinition of sorts, as rescue groups, led by mostly female, mostly white, and mostly middle-class volunteers, have begun to reframe these dogs as loving family pets rather than symbols of violence. This reframing has allowed for more pit bulls to be adopted and has offered a counter-narrative to those that support BSL targeting these dogs. On the other hand, the "whitewashing" of pit bulls through their association with white, middle-class individuals, while benefiting the dogs themselves, has not done anything to reduce the stigma associated with urban minority youth (Tesler and McThomas, 2024).

See also: Capitalism, Pets And; Ethnicity, Pets And; Socioeconomic Class, Pets And; United Kingdom, Pets In; United States, Pets In

References and Further Reading:

d'Angelo, D., S. d'Ingeo, F. Napolitano, G. Perrotti, I. Maglione, V. Caputo, and A. Quaranta (2022). "Retrospective Analysis of Dog Bites in Southern Italy." Ciência Rural, 52: e20210421.

Gunter, L. M., R. T. Barber, and C. D. Wynne (2016). "What's In a Name? Effect of Breed Perceptions & Labeling on Attractiveness, Adoptions & Length of Stay for Pit-bull-type Dogs." *PloS one*, 11 (3): e0146857.

Ly, L., E. Gordon, and A. Protopopova (2021). "Exploring the Relationship between Human Social Deprivation and Animal Surrender to Shelters in British Columbia, Canada." *Frontiers in Veterinary Science*, 8: 597

Maher, J., and H. Pierpoint (2011). "Friends, Status Symbols and Weapons: The Use of Dogs by Youth Groups and Youth Gangs." *Crime Law and Social Change*, 55 (5): 405–20.

McEwen, B. (2017). "Eternally Vulnerable: The Pathology of Abuse in Domestic Animals." *Academic Forensic Pathology*, 7 (3): 353–69.

Nurse, A. (2021). "Green Criminological Perspectives on Dog-Fighting as Organised Masculinities-Based Animal Harm." *Trends in Organized Crime*, 24 (4): 447–66.

Tesler, M., and M. McThomas (2024). "The Racialization of Pit Bulls: What Dogs Can Teach Us about Racial Politics." *PLoS one*, 19 (6): e0305959.

Westgarth, C., L. Boddy, G. Stratton, A. German, R. Gaskell, K. Coyne, Peter Bundred, Sandra McCune, and S. Dawson (2013). "A Cross-Sectional Study of Frequency and Factors Associated with Dog Walking in 9–10 Year Old Children in Liverpool, UK." *BMC Public Health*, 13 (1).

T

TELEVISION, PETS ON

Animals have been on television since the very first days of TV. Research into animals on television focuses on a variety of concerns, from the animal actors themselves, including their training and treatment on set, the impact of television portrayals on real animals, and the ways that animals on television influence human viewers. Television has given us a window into the changing cultural significance and anthropomorphic representation of animals, including pets.

Historically, television emerged as an influential platform for depicting animal characters, evolving from early film and radio traditions where live animals were sometimes featured alongside human actors. Early televised representations frequently incorporated live-action animals into family oriented sitcoms and documentary-style adventure programs, establishing archetypal images of loyalty, playfulness, and intelligence that continue to influence audiences. More broadly, critical examinations of television performance by Donaldson & Walters (2018) underscore how the conventions of the medium enable a stylized reimagining of animal life that resonates with audiences while sometimes—often—diverging from naturalistic representations. For example, especially in children's programming, the anthropomorphization of animal characters has shaped not only how we see animals but has resulted in vast amounts of merchandise, public adoration, and changing pet ownership trends. Similarly, early television shows involving animal co-stars—where figures like Roy Rogers and Gene Autry (whose on-screen animal partners were critical to their cowboy personas) blurred the line between film and television—reflect an early recognition of the marketability and emotional impact of animal actors (Porst, 2019). These animal performances have contributed to a cultural lexicon where pet characters are idealized, thereby influencing the public's expectations of real animals.

Lassie (1954–1974) was a popular television show centered around the adventures of a dog named Lassie. In the 1969 episode "More Than Meets the Eye," Lassie assisted a blind student (played by Kathy Martinez) in overcoming her fears. (National Archives)

The impact of these representations on the lives of real animals is complicated. Television portrayals often imbue pets with exaggerated, anthropomorphic qualities that can lead to increased demand for certain breeds or animals perceived as "ideal" in terms of behavior and aesthetic traits. This phenomenon aligns with broader discussions in television studies that argue the mediated image of performance creates normative expectations among viewers. As audiences internalize the traits and narratives depicted on screen, there may be pressures on breeders and trainers to conform to these ideals, producing animals to meet new needs.

Furthermore, animal welfare concerns and ethical considerations in the training and use of animal actors have emerged as points of debate, with industry guidelines increasingly insisting on humane, reward-based training regimes that ensure the well-being of animal performers.

In terms of sourcing, television animal performers have historically been procured through a variety of channels. Early programs often relied on local (i.e., local to Los

Angeles, where most early programming was filmed) animals with ad hoc training regimens, whereas contemporary productions typically engage specialized animal acting agencies. These agencies supply animals that have been selectively bred and trained, ensuring that the animals can perform required behaviors under controlled circumstances while also hopefully receiving attention to their overall welfare. This evolution in sourcing and training reflects broader shifts in both the television industry and pet culture at large, emphasizing authenticity and the responsible depiction of animal life on screen.

Finally, the representation of pets on television involves a dynamic interplay between reality and fantasy. While on-screen animals often embody qualities that audiences admire and wish to see in their own pets, there remains an inherent tension between the scripted, idealized version of animal behavior and the complex, often unpredictable realities of living pets. This gap influences public perceptions and informs both the commercial aspects of pet industries and the ethical frameworks governing animal performance. As noted in studies of television performance and cultural representation (Donaldson & Walters, 2018), these iconic pet figures serve a dual function: they entertain and educate, subtly shaping the societal narrative surrounding animal companionship and responsible pet ownership.

Because popular culture like television, film, and other media both reflect as well as shape reality, animals on television are subject to these same forces. Writers and directors include animals in their productions, especially in family sitcoms, because that reflects the reality of most of our lives, which we share with pets. But at the same time, television productions then shape our understanding of reality, even when those productions are entirely fictional, which can and does lead to misunderstandings and misinterpretations of real animal lives.

See also: Advertising, Pets And; Film, Pets In; Literature, Pets In

References and Further Reading

Asher, L., E. Buckland, C. Phylactopoulos, M. Whiting, S. Abeyesinghe, L. Donaldson, and J. Walters (2018). "Inter(acting)." *Critical Studies in Television the International Journal of Television Studies*, 13 (3): 352–69.

Candea, M. (2010). "'I Fell in Love with Carlos the Meerkat': Engagement and Detachment in Human–Animal Relations." *American Ethnologist*, 37, no. (2): 241–58.

Milius, S. (2007). "Science Behind the Soap Opera: The Cute and the Shocking at Meerkat Manor." *Science News*, 171, no. (9): 138–40.

Mills, B. (2010). "Television Wildlife Documentaries and Animals' Right to Privacy." *Continuum*, 24, no. (2): 193–202.

Mills, B. (2017). *Animals on Television: The Cultural Making of the Non-Human*. SpringerLondon: Palgrave Macmillan.

Molloy, C. (2011). *Popular Media and Animals*. SpringerLondon: Palgrave Macmillan.

Porst, J. (2019). "Roy Rogers and Gene Autry Do Not Endorse This Project: Actors, Advertising, and Feature Films on Early Television." *Television & New Media*, 22 (4): 420–39.

THERAPY PET

One of the most recent forms of animal assistance is the field known as animal-assisted therapy (AAT), or, collectively, as animal-assisted interventions (AAI). AAT uses specially trained animals to help with the mental, physical, and emotional care of patients with a variety of physical and emotional complaints. The animals used in AAT are known as therapy animals and are chosen because of their gentle temperaments and their ability to help patients heal. Such animals include dogs, cats, rabbits, horses, birds, guinea pigs, and dolphins. Unlike service animals, who live with the person with the disability or condition, therapy animals typically live with handlers (and sometimes facilities) who train them and bring them to meet patients. These animals are companion animals as well as therapy animals.

AAT officially began during the Second World War when a Yorkshire Terrier named Smoky, who had already served in combat missions alongside his human partner, Corporal William Wynne, visited injured soldiers in a military hospital, and Smoky's presence was found to help the wounded soldiers feel better. It is most likely that this sort of activity has been occurring for as long as humans have lived with animals. For instance, dogs were used in some Greek temples to help heal the sick and wounded, and in ninth-century Belgium, animals were sometimes used to help the disabled cope with their conditions. But it wasn't until recently that animals have been formally recognized as playing a role in human healing.

The idea that animals could play a role in psychiatric treatment did not develop until the 1960s, when a New York psychotherapist named Boris Levinson brought his dog, Jingles, to work and found that patients would communicate with Jingles present, whereas they would not have done so in his absence (Serpell 2010).

Animal-assisted interventions have exploded in the past thirty years, with new organizations and college programs emerging to both study the role that animals play in human healing and to promote the use of therapy animals.

Today, therapy animals are used in hospitals, nursing homes, rehabilitation facilities, orphanages, and hospices. For instance, Gael, a golden retriever, recently won the 2018 Animal Therapy Award for her work while visiting patients at hospitals in Florida. Part of her job involves "hugging" patients with autism in order to make them feel better. Therapy animals like Gael are also used in a rehabilitation setting to improve physical health by improving motor skills and balance for the disabled or injured and those with neurological disorders.

Hippotherapy, or therapeutic horse riding, is the use of horse riding as a form of therapy for physically and emotionally disabled people, or for people recovering from conditions such as strokes. Learning to ride a horse in a safe setting teaches the disabled balance and flexibility to the disabled and brings a sense of accomplishment and companionship to those involved (Burgon 2003). Some programs are related to the care of horses as well, teaching responsibility and horsemanship skills.

Therapy animals are also used in areas where the healing is entirely emotional or psychological, rather than physical. For example, therapy animals are used in prisons and juvenile detention centers, in schools and libraries, and with people suffering from autism, speech disorders, Parkinson's, dementia, schizophrenia, and a whole host of other conditions. Because of the documentation demonstrating that the presence of animals can lower stress in people (see Hansen et al. 1999), animals are now starting to appear in courtrooms when children are asked to testify in stressful court cases. Cooper, a golden retriever from Hobbs, New Mexico, became New Mexico's first courthouse dog in 2010; he was available at the Lea County courthouse to visit with children who are either witnesses to or victims of a crime. Cooper's presence makes children feel safe, and when he accompanies a counselor or attorney to meet with a child, he also helps the child trust the adult as well.

Counseling and animal-assisted therapy professor Cynthia Chandler notes (2005) that one of the reasons for the success of therapeutic animals is that dogs act as a surrogate for therapeutic touch, as it is often inappropriate for human therapists to touch their patients during treatment.

Equine-facilitated psychotherapy (known as EFP) refers to the use of horses in psychotherapy for patients with psychiatric problems. For people with anger control problems, for example, learning to care for a horse without expressing anger is one benefit. Other patients with other conditions develop self-esteem, self-acceptance, trust, empathy, and communication skills (Rothe et al 2005; Meinersmann, Bradberry, and Roberts,. 2008; Bachi 2013).

Finally, the use of therapy animals has been shown to increase social health by facilitating verbal interaction, increasing attention skills, decreasing loneliness, and increasing self-esteem. Using animals in an educational setting can help children increase their vocabulary skills, improve reading, and assist with memorization (Hergovich et al., 2002; Gee, Crist, and Carr, 2010). Therapy animals provide comfort, someone safe to talk to, safety, and security, and can bring withdrawn children or adults suffering from PTSD out of their shells. It is not uncommon for patients who are uncommunicative and withdrawn to open up and begin talking after receiving a visit from a therapy animal. Some facilities, like nursing homes, are adopting their own cat, rabbit, or dog to live in the facility so that the patients can have access to an animal all the time. Even substance abuse treatment can benefit from the participation of animals in therapy sessions (Wesley, Minatrea, and Watson, 2009).

See also: Attachment; Bond, Human–Animal; Children, and Pets; Facility Pets; Human Health, Pets And; Mental Health, Pets And

References and Further Reading

Bachi, K. (2013). "Application of Attachment Theory to Equine-Facilitated Psychotherapy." *Journal of Contemporary Psychotherapy* 43, (3): 187–96.

Burgon, H. (2003). "Case Studies of Adults Receiving Horse-Riding Therapy." *Anthrozoös* 16 (3): 263–76.

Chandler, C. K. (2005). *Animal-Assisted Therapy in Counseling.* New York: Routledge.

Gee, N., E. Crist, and D. Carr (2010). "Preschool Children Require Fewer Instructional Prompts to Perform a Memory Task in the Presence of a Dog." *Anthrozoös* 23 (2): 173–84.

Hansen, K. M., C. J. Messinger, M. M. Baun, and Mary Erickson Megel (1999). "Companion Animals Alleviating Distress in Children." *Anthrozoös* 12 (3): 142–48.
Hergovich, A., B. Monshi, G. Semmler, and V. Zieglmayer (2002). "The Effects of the Presence of a Dog in the Classroom." *Anthrozoös* 15 (1): 37–50.
Meinersmann, K. M., J. Bradberry, and F. B. Roberts. (2008). "Equine-Facilitated Psychotherapy with Adult Female Survivors of Abuse." *Journal of Psychosocial Nursing and Mental Health Services* 46 (12): 36–42.
Rothe, E. Q., B. J. Vega, R. M. Torres, S. M. C. Soler, and R. M. Molina (2005). "From Kids and Horses: Equine Facilitated Psychotherapy for Children." *International Journal of Clinical and Health Psychology* 5 (2): 373–83.
Serpell, J. A. (2010). "Animal-Assisted Interventions in Historical Perspective." In A. Fine (ed.), *Handbook on Animal-Assisted Therapy: Theoretical Foundations and Guidelines for Practice*, 3rd ed., 17–32. San Diego, CA: Academic Press.
Wesley, M., N. Minatrea, and J. Watson (2009). "Animal-Assisted Therapy in the Treatment of Substance Dependence." *Anthrozoös* 22 (2): 137–48.

TRENDS IN PET OWNERSHIP

Trends in pet ownership reflect the increasing social, economic, and therapeutic importance of companion animals to humans. In the United States, multiple studies and periodic surveys by industry groups—most notably the American Pet Products Association (APPA) and the American Veterinary Medical Association (AVMA)—document that pet ownership remains highly prevalent, with estimates suggesting that approximately 60 percent to 67 percent of households include at least one pet (Gee, Mueller, and Curl, 2017). These groups not only track the numbers but also help gauge shifts in purchasing patterns, veterinary care usage, and lifestyle adaptations that accompany changes in pet demographics. Such coordinated efforts provide critical insights into the economic and cultural forces driving pet-related industries and public health initiatives (Gee, Mueller, and Curl, 2017).

Internationally, pet ownership trends reveal regional variability influenced by socioeconomic factors and cultural attitudes toward animals. For instance, in lower-middle-income countries such as rural Sri Lanka, research has indicated distinct household preferences for pet keeping that inform community-level planning for animal welfare and public health programs, thus highlighting similarities and differences in the global pet ownership landscape (Rathish et al., 2022). In developed countries, longitudinal research in the United Kingdom, as an example, has also evidenced stable pet ownership patterns over time, while noting that evolving demographic and generational factors continue to shape family dynamics and attitudes toward nontraditional pet species (Westgarth et al., 2010). These findings underscore the complex interplay of cultural, economic, and generational influences that drive both regional and global trends.

A notable recent development that has been widely investigated is the impact of the Covid-19 pandemic on pet acquisition behaviors. While early reports in popular media suggested a surge in pet adoptions during the lockdown periods, a national survey conducted in the United States in May 2021 indicated that the ownership rates of common pets such as dogs and cats had not increased as dramatically as initially expected (Hoffman et al., 2021). This counterintuitive trend, observed alongside a plethora of studies examining the human-animal bond during times of crisis, further illustrates the importance of reliable data sources and rigorous methodologies in tracking pet ownership dynamics over time (Bibbo et al., 2022; Hoffman, Thibault, and Hong, 2021).

In addition to tracking numerical trends, organizations such as the AVMA and APPA also highlight the broader social benefits of pet ownership. Their work emphasizes how pets contribute to improving family cohesion and even influencing public health outcomes through tempering loneliness and supporting emotional well-being—a perspective that is increasingly recognized both in the United States and worldwide (Gee, Mueller, and Curl, 2017). By combining industry data with academic research, these organizations help generate a comprehensive picture of the evolving role of pets in contemporary society, reinforcing the notion that pets are much more than mere animals; they are integral members of modern households with a measurable impact on economic trends, public health strategies, and social integration.

See also: Covid-19; Family, Pets As; United States, Pets In

References and Further Reading

Asher, L., E. Buckland, C. Phylactopoulos, M. Whiting, S. Abeyesinghe, and C. Wathes (2011). "Estimation of the Number and Demographics of Companion Dogs in the UK." *BMC Veterinary Research*, 7 (1): 74.

Bibbo, J., J. Johnson, J. Drost, M. Sanders, and S. Nicolay (2022). "Pet Ownership Issues Encountered by Geriatric Professionals: Preliminary Findings from an Interdisciplinary Sample." *Frontiers in Psychology*, 13.

Brown, H. (2007). *Netlines. BMJ: British Medical Journal*, 334 (7590): 156.

Gee, N., M. Mueller, A. Curl (2017). "Human–Animal Interaction and Older Adults: An Overview." *Frontiers in Psychology*, 8.

Hoffman, C., M. Thibault J. Hong (2021). "Characterizing Pet Acquisition and Retention During the COVID-19 Pandemic." *Frontiers in Veterinary Science*, 8.

Jung, C., and D. Pörtl (2019). "How Old Are (Pet) Dog Breeds?." *Pet Behaviour Science*, (7): 29–37.

Rathish, D., J. Rajapakse, and K. Weerakoon (2022). "Household Preferences for Pet Keeping: Findings from a Rural District of Sri Lanka." *Plos One*, 17 (11): e0277108.

Westgarth, C., J. Heron, A. R. Ness, P. Bundred, R. M. Gaskell, K. P. Coyne, ... and S. Dawson (2010). "Family Pet Ownership During Childhood: Findings from a UK Birth Cohort and Implications for Public Health Research." International Journal of Environmental Research and Public Health, 7 (10): 3704–29.

U

UNITED KINGDOM, PETS IN

Because so many of the economic, cultural, and social changes that allowed for the development of widespread pet keeping evolved in Europe, but especially the United Kingdom, understanding how pets developed there is critical. Although explicit documentation on medieval monks' roles is scarce in the current literature, scholars have noted that religious institutions in medieval Europe often maintained animals for companionship and utility, laying early foundations for a cultural tradition of pet keeping (Jung and Pörtl, 2019). This early engagement would later intertwine with elite practices, where the keeping of specially bred animals became a symbol of social prestige.

During the Medieval and early modern eras, elite pet keeping evolved into a sophisticated practice marked by a keen interest in canine breeds. Elite circles in the United Kingdom, starting in the eighteenth and nineteenth centuries, began to cultivate what would be recognized as the dog fancy. The emergence of specialized dog breeds and the establishment of breed standards were significantly influenced by both the stylish tastes of the British upper classes and the professionalization of care provided by specialized veterinarians in both Victorian and Edwardian London (Skipper, 2019). The formal codification of breed standards can also be seen as a response to the increasing demand from elite pet owners to maintain and display a refined aesthetic that reflected their social status.

The Victorian period, characterized by the cult of domesticity, brought a radical transformation in the role of pets within the family unit. During this time, a shift occurred as domestic ideologies promoted not only the ideal of the well-managed home but also the notion that pets could serve as companions and educators for children. In Victorian England, animal companionship was increasingly associated with moral values and domestic virtue, a view that echoed broader societal concerns for improving moral education through everyday interactions with animals. Such cultural shifts helped to normalize pet keeping as a central component of nurturing and domestic life.

In recent decades, pet keeping in the United Kingdom has created an industry marked by sophisticated population management and strong welfare considerations. Large-scale studies that estimate the size, density, and demographic distribution of the UK pet dog population provide crucial insights into contemporary ownership trends (McMillan et al., 2024; Asher et al., 2011). The modern pet owner's decision process is further informed by a diverse range of information sources, including online resources, published literature, and professional advice from breeders and veterinarians (Kuhl et al., 2021). This evolution reflects not only advances in animal welfare science and public health initiatives but also an ongoing societal emphasis on the human–animal bond.

Today, pet keeping in the United Kingdom is situated at the crossroads of tradition and modernity. Historic practices of monastic and elite pet keeping have given way to a more democratized model in which dogs—and pets more generally—are integrated into family life across the social spectrum. Contemporary challenges such as breed-specific legislation, public health concerns (e.g., dog bite prevention), and the regulation of commercial breeding further underscore the complexity of maintaining a humane and scientifically informed approach to pet keeping (Brown, 2007; McMillan et al., 2024). This enduring evolution highlights the continuous interplay between cultural heritage and modern societal values in shaping the lives of both humans and their animal companions.

See also: Adults and Pets; Children and Pets; Europe, Pets In; History of Pet Keeping: Victorian Era; Trends in Pet Ownership; United States, Pets In

References and Further Reading

Asher, L., E. Buckland, C. Phylactopoulos, M. Whiting, S. Abeyesinghe, and C. Wathes (2011). "Estimation of the Number and Demographics of Companion Dogs in the UK." *BMC Veterinary Research*, 7 (1): 74.

Brown, H. (2007). "Netlines." BMJ: British Medical Journal (Online), 334 (7585): 156.

Jung, C., and D. Pörtl (2019). "How Old Are (Pet) Dog Breeds?" Pet Behaviour Science, 7: 29–37.

Kuhl, C., R. Dean, C. Quarmby, and R. Lea (2021). "Information Sourcing by Dog Owners in the UK: Resource Selection and Perceptions of Knowledge." *Veterinary Record*, 190 (10).

McMillan, K., X. Harrison, M. Upjohn, R. Christley, and R. Casey (2024). "Estimation of the Size, Density, and Demographic Distribution of the UK Pet Dog Population." *Scientific Reports*, 14: 31746.

Skipper, A. (2019). "The 'Dog Doctors' of Edwardian London: Elite Canine Veterinary Care in the Early Twentieth Century." *Social History of Medicine*, 33 (4): 1233–58.

UNITED STATES, PETS IN

The history of pet keeping in the United States reflects an evolution from utilitarian relationships to deeply embedded cultural practices, affective relationships, and economic industries. Early pet keeping often had practical purposes—such as utilizing animals for labor or protection—but over time, animals increasingly became companions, symbols of status, and embodiments of emotional attachment. This transformation has been influenced by a complex interplay of biological, cultural, and economic forces. Herzog (2014) argues that pet keeping originated both as an adaptation and an evolutionary by-product, a view that helps explain the shifting preferences seen across historical fashion cycles. Such cycles have affected which species are favored—historically extending from dogs and cats to exotic species like nonhuman primates—and established a framework for the modern pet economy.

In the twentieth century and beyond, legal and regulatory frameworks have played a central role in mainstreaming and safeguarding the practice of pet keeping. State anti-cruelty laws, developed progressively since the early twentieth century, laid the groundwork for ensuring the humane treatment of companion animals, while federal statutes such as the Animal Welfare Act of 1966 have provided broader protections and standards for animal care (Mychka and Shkoliar, 2023) within a select set of contexts. These laws have contributed to a conceptual shift in the human–animal relationship, reinforcing the view of pets as family members rather than mere property. Sociological analyses underscore that this shift has stimulated changes in consumer behavior and in social norms around pet care, where pets are increasingly seen as sources of comfort, status, and even therapeutic benefits to their owners (Mychka and Shkoliar, 2023).

Today, pet keeping in the United States is characterized by high levels of economic investment and extensive integration into everyday life. Modern pet owners spend billions each year on veterinary care, food, accessories, and specialized services such as pet grooming and boarding. Economic studies reveal that the pet care industry has matured into a significant market sector reflective of broad urban and suburban demographic trends (Mychka and Shkoliar, 2023). Moreover, contemporary pet culture is increasingly facilitated by digital technologies and social media platforms. As illustrated by Hutchinson (Hutchinson, 2014), "petworking" has emerged as a culturally mediated process where pet owners create social networks dedicated to sharing images, advice, and experiences, further reinforcing the role of pets as integral members of the family and social identity.

While early practices were predominantly utilitarian, modern pet ownership is marked by emotional bonds, significant economic investment, and robust regulatory frameworks that ensure animal welfare. Together, these trends underscore a dynamic relationship where pets are not only cared for but are also central to the fabric of modern American life.

See also: Adults and Pets; Capitalism, Pets And; Children and Pets; Family, Pets As; Trends in Pet Ownership; United Kingdom, Pets In

References and Further Reading

Hutchinson, J. (2014). "I Can Haz Likes: Cultural Intermediation to Facilitate "Petworking"." *M/C Journal*, 17 (2).

Mychka, M., and M. Shkoliar (2023). "Sociological Discourse of the Domestic Animals Status Transformation in the Human-Animal Relations." In *Sociology – Social Work and Social Welfare: Regulation of Social Problems. Proceedings of The XIII International Scientific Conference*, 63–8. https://doi.org/10.23939/sosrsw2023.063

V

VETERINARY CARE

Today, veterinary care is critical to the survival, health, and happiness of our companion animals. But historically, veterinary medicine was primarily associated with large animal practices, focusing on food production and herd health. However, over recent decades, companion animal medicine has emerged as a distinct domain, characterized by its specialized training tracks, clinical guidelines, and infection control protocols (Volk et al. 2011), and has now overtaken large animal medicine in terms of numbers of veterinarians. Early developments in the field reflected a gradual recognition of the unique behavioral, physiological, and emotional needs of pets, prompting curriculum reforms and the creation of day one competencies, which include veterinary behavioral medicine elements (Golden and Hanlon 2018).

The history of veterinary medicine dates back to the history of medicine itself, with early scientists experimenting with dissection of both human and animal bodies in order to learn anatomy, physiology, and disease processes. The first actual veterinary practice was founded in France in 1761, but the field did not mature until the nineteenth century with the development of vaccines and specialized treatments, while the twentieth century brought further innovations like anesthesia, antibiotics, and modern surgical techniques.

As the scope of veterinary practice expanded, differences between small and large animal tracks became increasingly apparent. Companion animal practices, today largely located in urban settings, now include state-of-the-art diagnostic and therapeutic technologies, less emphasized in large animal care (Kustritz, Malone, and Rendahl, 2020). Meanwhile, while large animal tracks traditionally focused on herd management and production efficiency, small animal practices now address complex welfare issues, antimicrobial resistance (AMR), and client expectations that extend to routine, preventive, and emergency care (Hamilton et al., 2013; Pomba et al., 2017) as well as end of life or palliative care. This divergence has been reinforced by the development of separate clinical guidelines tailored specifically to the diseases and conditions prevalent in companion animals (Lappin et al., 2019).

The current practice of veterinary medicine in the companion animal sector is characterized by advancements in infection control and the incorporation of evidence-based standards. Observational studies have highlighted persistent issues with hand hygiene compliance in companion animal clinics, leading to educational campaigns and modifications in clinical protocols to minimize nosocomial infections (Anderson and Weese, 2016; Schmidt et al., 2021). Additionally, the integration of electronic health record systems has facilitated nationwide surveillance of emerging and zoonotic diseases in companion animals, thereby enhancing public health monitoring capabilities (Krone et al., 2014; Glickman et al., 2006). This shift toward digitalization and improved hygiene standards reflects an increased understanding of the risks associated with cross-species transmission of infectious agents.

In parallel, veterinary training programs have undergone significant reform to address the demands of modern companion animal care. Contemporary curricula now encompass modules on animal behavior, welfare, and antimicrobial stewardship, as evidenced by initiatives to define day one competencies in behavioral medicine (Golden & Hanlon 2018) and the increasing availability of continuing education resources (Endenburg et al., 2022; Endenburg & Lith, 2023). These educational enhancements are aimed at bridging the historical gap between theory and clinical practice, thereby ensuring that new graduates are better prepared to manage both routine care and emerging challenges (Fellman et al., 2024).

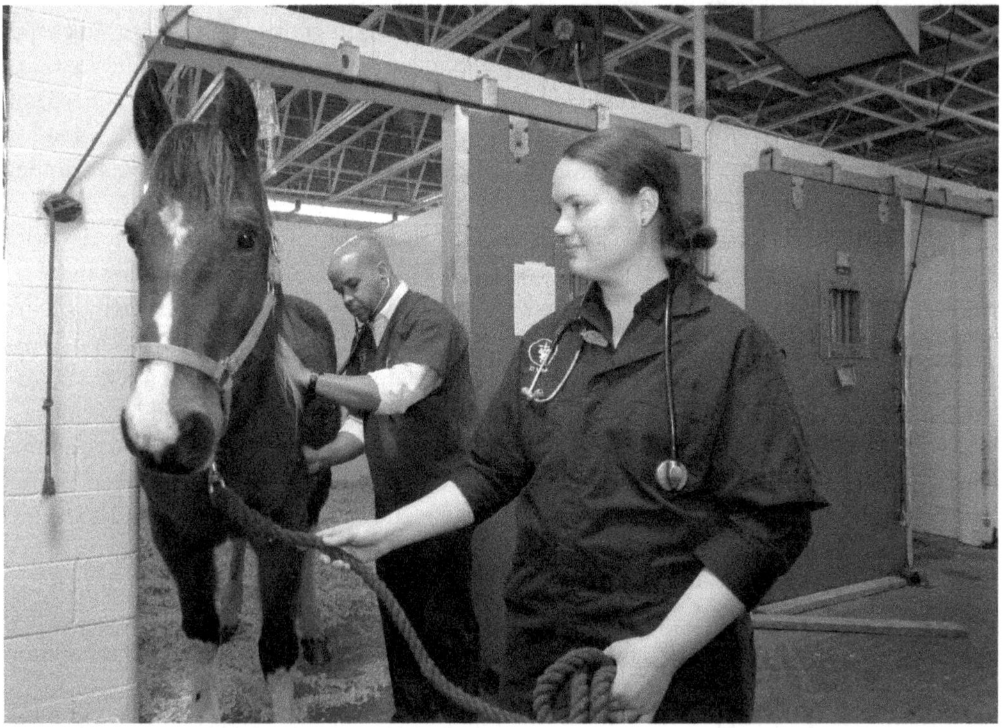

Students at the Tuskegee College of Veterinary Medicine, Nursing & Allied Health in Tuskegee, Alabama, receive hands-on experience with both small and large animals. (Library of Congress)

Overall, research indicates that while the veterinary care landscape for companion animals has seen remarkable progress through specialization, technology integration, and enhanced training, challenges remain. These challenges include ensuring compliance with hygiene protocols, addressing the public health implications of infectious diseases, and continuously adapting educational frameworks to meet the evolving needs of pet owners and their animals (Anderson and Weese, 2016; Schmidt et al., 2021; Fellman et al., 2024).

See also: Covid-19; Healthcare Spending on Pets; Trends in Pet Ownership; Zoonoses

References and Further Reading

Applebaum, J., C. Tomlinson, A. Matijczak, S. McDonald, and B. Zsembik (2020). "The Concerns, Difficulties, and Stressors of Caring for Pets During COVID-19: Results from a Large Survey of U.S. Pet Owners." *Animals*, 10 (10): 1882.

Anderson, M. E., and J. S. Weese (2016). "Self-Reported Hand Hygiene Perceptions and Barriers Among Companion Animal Veterinary Clinic Personnel in Ontario, Canada." The Canadian Veterinary Journal, 57 (3): 282.

Endenburg, N. and H. Lith (2023). "A Global Survey of the Views of Practicing Companion Animal Veterinarians on Their Undergraduate Curriculum and Their Access to Continuing Education Resources." *Journal of Veterinary Medical Education*, 50 (6): 713–31.

Endenburg, N., S. Ryan, and H. Lith (2022). "A Global Survey of Companion Animal Veterinary Practitioners on Animal Welfare Teaching – Focus on Undergraduate and Continuing Education, and Clients' Sources of Information." *Journal of Applied Animal Welfare Science*, 27 (1): 57–104.

Fellman, C., E. Janovyak, E. Bollig, I. DeStefano, D. Diaz-Campos, and J. Granick (2024). "Change Starts at Home: Summary of the 2023 Inaugural Small Animal Antimicrobial Stewardship Workshop for U.S. Veterinary Schools." *Journal of the American Veterinary Medical Association*, 262 (11): 1552–8.

Genosko, G. (1991). "The Modern Battlefront of Natural History and the Emergence of Animal Heroes." *Undercurrents Journal of Critical Environmental Studies*, 3: 28–32.

Glickman, L., G. Moore, N. Glickman, R. Caldanaro, D. AuCoin, and H. Lewis (2006). "Purdue University–Banfield National Companion Animal Surveillance Program for Emerging and Zoonotic Diseases." *Vector-Borne and Zoonotic Diseases*, 6 (1): 14–23.

Golden, O., and A. Hanlon (2018). "Towards the Development of Day One Competences in Veterinary Behaviour Medicine: Survey of Veterinary Professionals Experience in Companion Animal Practice in Ireland." *Irish Veterinary Journal*, 71 (1).

Hamilton, E., J. Kruger, W. Schall, M. Beal, S. Manning, and J. Kaneene (2013). "Acquisition and Persistence of Antimicrobial-Resistant Bacteria Isolated from Dogs and Cats Admitted to a Veterinary Teaching Hospital." *Journal of the American Veterinary Medical Association*, 243 (7): 990–1000.

Krone, L., C. Brown and J. Lindenmayer (2014). "Survey of Electronic Veterinary Medical Record Adoption and Use by Independent Small Animal Veterinary Medical Practices in Massachusetts." *Journal of the American Veterinary Medical Association*, 245 (3): 324–32.

Kustritz, M., E. Malone and A. Rendahl (2020). "Identification of Student Lifestyle Characteristics Associated with Training Choices to Drive Targeted Admissions in Veterinary Medicine." *Education in the Health Professions*, 3 (2): 70.

Lappin, M. R., T. Elston, L. Evans, C. Glaser, L. Jarboe, P. Karczmar, ... and M. Ray (2019). "2019 AAFP Feline Zoonoses Guidelines." Journal of Feline Medicine and Surgery, 21 (11): 1008–21.

Pomba, C., M. Rantala, C. Greko, K. E. Baptiste, B. Catry, E. Van Duijkeren, … and K. Törneke (2017). "Public Health Risk of Antimicrobial Resistance Transfer from Companion Animals." Journal of Antimicrobial Chemotherapy, 72 (4): 957–68.
Volk, J., K. Felsted, J. Thomas, and C. Siren (2011). "Executive Summary of Phase 2 of the Bayer Veterinary Care Usage Study." *Journal of the American Veterinary Medical Association*, 239 (10): 1311–16.
Schmidt, J. S., S. Hartnack, S. Schuller, S. P. Kuster, and B. Willi (2021). "Hand Hygiene Compliance in Companion Animal Clinics and Practices in Switzerland: An Observational Study." Veterinary Record, 189 (1).

VIRTUAL PETS

Virtual pets, also known as digital pets, are artificial pets that "live" on computers and mobile devices.

Early digital pet experiments, including Sony's Aibo (which is better defined as a robotic pet), Tamagotchi, and other interactive projects, were designed to mimic the companionship of real animals while addressing issues of social isolation and emotional support (Rault, 2015). Tamagotchi in particular emerged as one of the first widely recognized virtual pet phenomena, requiring users to care for an autonomous digital entity that demanded regular attention, evoking responsibilities similar to those associated with live pet ownership (Chen et al., 2011; Ruckenstein, 2013).

Scholars have documented that virtual pets and games not only serve an entertainment function but also promote learning, collaboration, and empathy among their users. Research in human-computer interaction points to the ability of these systems to foster emotional connections and nurture caring behaviors, as the continuous requirements of digital care routines encourage users—often children and adolescents—to engage in nurturing practices. Additionally, media analyses have shown that virtual pets like Tamagotchi have influenced cultural narratives about artificial intelligence and the nature of companionship, prompting broader discussions about the boundaries between life and simulation (Scott, 2023; Giddings, 2024).

Subsequent studies have further examined the long-term cultural impact and historical development of virtual pet communities. For instance, narrative analyses of online platforms dedicated to virtual pets, such as Tencent Pets, reveal that these digital companions have left lasting impressions on users' personal memories and social identities, manifesting in narratives of nostalgia, mourning, and even resistance (Wei, 2024). This enduring influence underscores how virtual pets have provided a framework within which traditional forms of pet ownership and care are reimagined in a digital context, blending interactive play with genuine emotional engagement.

Overall, the literature synthesizes that virtual pets like Tamagotchi have not only been technological innovations but also cultural artifacts that reflect evolving notions of care, responsibility, and the human–animal bond in the digital age. Their role continues to expand as new media, virtual reality, and smart toy platforms emerge, offering fresh avenues for exploring the intersections of technology, psychology, and social behavior.

See also: Japan, Pets In; Mental Health, Pets And; Robotic Pets

References and Further Reading

Chen, Z., C. Chou, G. Biswas, and T. Chan (2011). "Substitutive Competition: Virtual Pets as Competitive Buffers to Alleviate Possible Negative Influence on Pupils." *British Journal of Educational Technology*, 43 (2): 247–58.

Giddings, S. (2024). "Why Look at Toy Animals?." *Cultural Politics: An International Journal*, 20 (1): 60–74.

Rault, J. L. (2015). "Pets in the Digital Age: Live, Robot, or Virtual?" *Frontiers in Veterinary Science*, 2: 11.

Ruckenstein, M. (2013). "Playing Nintendogs: Desire, Distributed Agency and Potentials of Prosumption." *Journal of Consumer Culture*, 15 (3): 351–70.

Scott, D. (2023). "Retrospective Technological Mythmaking: Media Discourses of Furby and Artificial Intelligence." *TMG Journal for Media History*, 26 (2): 1–24.

Wei, T. (2024). ""my little penguin has come back to life": Chinese Nitizens' Digital Memory and Narrative Practice of Tencent Pets Game." *Galactica Media Journal of Media Studies*, 6 (4): 145–59.

W

WAR, PETS AND

The literature on animals in warfare, particularly focusing on pets, emphasizes that domesticated animals have played both utilitarian and emotional roles during conflicts. During the First World War and the Second World War, many dogs and horses used in military operations were once household pets that had been "donated" to or requisitioned for the war effort. Historical accounts indicate that numerous pet horses were indeed donated by civilians for military use, serving essential functions in transport and logistics (Genosko, 1991; Bruneau, 2005). These animals were valued not only for their operational utility but also for their symbolic role as "animal heroes," adding an emotional dimension to wartime narratives (Genosko, 1991; Bruneau, 2005). Military historiography notes that the mobilization of companion animals was part of a broader mobilization of national resources, with the donation of beloved pets reflecting the societal sacrifices expected during times of national crisis (Genosko, 1991).

Subsequent research shows that the role of pet animals in warfare has evolved considerably since the two world wars. With advancements in military mechanization and technology, the direct use of pet animals in combat and logistical roles has decreased significantly. Contemporary military contexts now emphasize the specialized use of trained working animals, such as military working dogs, which are selected for specific operational tasks rather than their status as typical household pets. This evolution reflects broader changes in military strategy and acknowledges that the historical attachment between humans and animals has long-lasting social and psychological implications. The role of pets in boosting morale and providing comfort during conflicts has been well documented, particularly during urban bombings such as the Blitz during the Second World War, where the companionship of animals helped mitigate the psychological toll of warfare on civilians (Kean, 2017).

Conflicts have mobilized animals for military purposes while imposing significant hardships on civilian pet populations. Wars have often led to displacement and neglect of pets, as well as instances of mass culling during extreme crises (Colerick, 2024). Research indicates that these tragic outcomes have long-term ramifications for animal welfare, prompting humanitarian initiatives to address the protection and care of animals in conflict zones (Colerick, 2024). Furthermore, the psychological stress experienced by pet owners is exacerbated by witnessing the suffering of animals to whom they are deeply attached, highlighting the importance of integrating animal welfare into post-conflict recovery programs (Colerick, 2024).

Overall, historical and contemporary research underscores that while the practical military roles of pets have shifted from the direct use of donated animals in the world wars to the specialized functions of trained working animals today, the impact of war on companion animals remains substantial. The legacy of using pet animals in early twentieth-century conflicts informs current approaches to military strategy—where animal contributions are now more technologically integrated—and the development of policies aimed at mitigating the adverse effects of war on animal welfare (Genosko, 1991; Colerick, 2024). This synthesis illustrates the dual role of pets as both functional military assets and central figures in the socioemotional landscape of wartime, necessitating continued scholarly attention and policy measures reflecting their enduring significance.

See also: Family, Pets As; Mental Health, Pets And; Immigration, Pets and; Natural Disasters and Pets

References and Further Reading

Bruneau, T. C. (2005). "Civil-Military Relations in Latin America: The Hedgehog and the Fox Revisited." *Revista Fuerzas Armadas y Sociedad*, 19(1–2), 111–31.

Colerick, E. (2024). "The Child and the Animal in Times of War: Halahmy's the Emergency Zoo and Contemporary News Reports on the War in Ukraine." *International Research in Children's Literature*, 17 (3): 264–77.

Genosko, G. (1991). "The Modern Battlefront of Natural History and the Emergence of Animal Heroes." *UnderCurrents: Journal of Critical Environmental Studies*, 3: 28–32.

Kean, H. (2017). *The Great Cat and Dog Massacre.* Chicago: University of Chicago Press.

Z

ZOONOSES

Zoonoses are diseases that can transmit from humans to nonhuman animals, and vice versa. In recent years, infectious diseases originating in animals represent a substantial threat to global public health, with companion animals playing a role within these dynamics. Zoonotic diseases, which constitute approximately 75 percent of emerging infectious diseases (Chakraborty, Fama, and Sander, 2024), arise through complex interactions among ecological, socioeconomic, and environmental factors.

Within the context of the Covid-19 pandemic, considerable attention has been paid to the role of companion animals as potential hosts for zoonotic pathogens. Covid-19 has been a catalyst for intensified research into natural infections among pets. Systematic reviews have documented that although dogs and cats may contract Covid through close contact with infected humans, the overall detection rates remain low, with cats sometimes exhibiting relatively higher positivity than dogs (Heydarifard et al., 2024; Morais et al., 2020). Surveillance studies further elucidate that the risk of pets acting as significant reservoirs or vectors remains minimal, even though isolated cases, including infections with variants of concern, have been reported (Murphy and Ly, 2021; Hamer et al., 2021). Concurrently, public communication campaigns have been essential to address pet owners' concerns about zoonotic transmission, thereby preventing unnecessary anxiety and potential negative repercussions such as pet abandonment (Laconi et al., 2023; Applebaum et al., 2020).

The One Health paradigm advocates for interdisciplinary collaboration to address the human–animal-environment interface, which is critical for mitigating both traditional and emerging zoonotic risks (Rahman et al., 2020; Overgaauw et al., 2020). We now know that integrating veterinary and human public health initiatives—ranging from routine parasitic and infectious disease surveillance in pets Pereira et al. (2016) to strategic policy engagement with stakeholders Reaser et al. (2008)—can substantially enhance early detection and control of zoonoses. Furthermore, several studies highlight the necessity of addressing reverse zoonoses, where pathogens may spill back from humans to animals, reinforcing that a bidirectional surveillance framework is essential for global health security (Jia et al., 2021; Vijayaraghavan et al., 2020).

Overall, the literature synthesizes the evidence that while companion animals may occasionally be infected with zoonotic pathogens, the risk of substantial transmission remains low under current epidemiological conditions (Heydarifard et al., 2024; Morais et al., 2020; Murphy and Ly, 2021). Nonetheless, the broader threat posed by zoonotic diseases demands continued investment in One Health strategies and public education to manage both established and emerging risks effectively (Rahman et al., 2020; Overgaauw et al., 2020). This intersection of zoonoses, Covid-19, and companion animal health emphasizes the importance of collaborative, evidence-based policies that safeguard public health without undermining the benefits of the human–animal bond.

See also: Covid-19; Health Risks of Pet Ownership; Human Health, Pets And

References and Further Reading

Applebaum, J., C. Tomlinson, A. Matijczak, S. McDonald, and B. Zsembik (2020). "The Concerns, Difficulties, and Stressors of Caring for Pets During COVID-19: Results from a Large Survey of U.S. Pet Owners." *Animals*, 10 (10): 1882.

Chakraborty, S., A. Fama, and W. Sander (2024). "Zoonoses-Specific Resources, Collaborative Networks, and Enhanced Communication Can Help Us Veterinarians Tackle Zoonotic Diseases: Results from a National Survey." *Journal of the American Veterinary Medical Association*, 262 (7): 877–86.

Hamer, S., R. Ghai, I. Zecca, L. Auckland, C. Roundy, E. Davila, R. Busselman, et al. (2021). "Sars-cov-2 b.1.1.7 Variant of Concern Detected in a Pet Dog and Cat after Exposure to a Person With COVID--19, USA." *Transboundary and Emerging Diseases*, 69 (3): 1656–58.

Heydarifard, Z., A. Chegeni, F. Heydarifard, B. Nikmanesh, and V. Salimi (2024). "An Overview of Sars-cov2 Natural Infections in Companion Animals: A Systematic Review of the Current Evidence." *Reviews in Medical Virology*, 34 (1).

Jia, P., S. Dai, T. Wu, and S. Yang (2021). "New Approaches to Anticipate the Risk of Reverse Zoonosis." *Trends in Ecology & Evolution*, 36 (7): 580–90.

Laconi, A., B. Saracino, E. Fattorini, G. Pellegrini, M. Bucchi, L. Bailoni and A. Piccirillo (2023). "Sars-cov-2 and Companion Animals: Sources of Information and Communication Campaign During the COVID-19 Pandemic in Italy." *Veterinary Sciences*, 10 (7): 426.

Morais, H., A. Santos, N. Nascimento, L. Kmetiuk, D. Barbosa, P. Brandão, A. M. S. Guimarães, C. Pettan-Brewer and A. Biondo (2020). "Natural Infection by Sars-cov-2 in Companion Animals: A Review of Case Reports and Current Evidence of Their Role in the Epidemiology of COVID-19." *Frontiers in Veterinary Science*, 7.

Murphy, H., and H. Ly (2021). "Understanding the Prevalence of Sars-cov-2 (COVID-19) Exposure in Companion, Captive, Wild, and Farmed Animals." *Virulence*, 12 (1): 2777–86.

Overgaauw, P., C. Vinke, M. Hagen, and L. Lipman (2020). "A One Health Perspective on the Human–Companion Animal Relationship with Emphasis on Zoonotic Aspects." *International Journal of Environmental Research and Public Health*, 17 (11): 3789.

Pereira, A., Â. Martins, H. Brancal, H. Vilhena, P. Silva, P. Pimenta, D. Diz-Lopes, et al. (2016). "Parasitic Zoonoses Associated with Dogs and Cats: A Survey of Portuguese Pet Owners' Awareness and Deworming Practices." *Parasites & Vectors*, 9 (1).

Rahman, M., M. Sobur, M. Islam, S. Ievy, M. Hossain, M. Zowalaty, A. T. Rahman, and H. Ashour. (2020). "Zoonotic Diseases: Etiology, Impact, and Control." *Microorganisms*, 8 (9): 1405.

Reaser, J., E. Clark, and N. Meyers. (2008). "All Creatures Great and Minute: A Public Policy Primer for Companion Animal Zoonoses." *Zoonoses and Public Health*, 55 (8–10): 385–401.

Vijayaraghavan, R., V. Naveenkumar, B. Nag, and K. Porteen. (2020). "The Possible Risk of Reverse Zoonosis in COVID-19: An Epidemiological Driving Approach for the One Health Future Challenges: A Review." *Asian Journal of Dairy and Food Research*, (OF).

ABOUT THE AUTHOR

Margo DeMello received her PhD in cultural anthropology from UC Davis in 1995 and is an assistant professor in the Anthrozoology program at Carroll College. She served as the Program Director for Human-Animal Studies at Animals & Society Institute from 2004 to 2019 and is the past president of House Rabbit Society.

Her books include *Bodies of Inscription: A Cultural History of the Modern Tattoo Community* (2000), *Stories Rabbits Tell: A Natural and Cultural History of a Misunderstood Creature* (with Susan Davis, 2003), *Low-Carb Vegetarian* (2004), *Why Animals Matter: The Case for Animal Protection* (with Erin Williams, 2007), *The Encyclopedia of Body Adornment* (2007), *Feet and Footwear* (2009), *Teaching the Animal: Human-Animal Studies across the Disciplines* (2010), *Faces around the World* (2012), *Animals and Society: An Introduction to Human-Animal Studies* (2012), *Speaking for Animals: Animal Autobiographical Writing* (2012), *Inked* (2014), *Body Studies: An Introduction* (2014), *Mourning Animals: Rituals and Practices Surrounding Animal Death* (2016), *On the Job: An Encyclopedia of Unique Occupations around the World* (2021), a second edition of *Animals and Society: An Introduction to Human-Animal Studies* (2021), and her latest, *Bigfoot to Mothman* (2024). She lives in Montana with her husband, four Chihuahuas, five cats, and nine house rabbits.

INDEX